Introduction to Network Simulator NS2

Teerawat Issariyakul • Ekram Hossain

Introduction to Network Simulator NS2

Springer

Teerawat Issariyakul
TOT Public Company Limited
89/2 Moo 3 Chaengwattana Rd.
Thungsonghong, Laksi
Bangkok, Thailand 10210
teerawas@tot.co.th
iteerawat@hotmail.com

Ekram Hossain
Department of Electrical &
    Computer Engineering
University of Manitoba
75A Chancellor's Circle
Winnipeg MB R3T 5V6
Canada
ekram@ee.umanitoba.ca

ISBN: 978-1-4419-4412-2   e-ISBN: 978-0-387-71760-9
DOI: 10.1007/978-0-387-71760-9

Printed on acid-free paper

springer.com

To our families

Preface

NS2 is an open-source event-driven simulator designed specifically for research in computer communication networks. Since its inception in 1989, NS2 has continuously gained tremendous interest from industry, academia, and government. Having been under constant investigation and enhancement for years, NS2 now contains modules for numerous network components such as routing, transport layer protocol, application, etc. To investigate network performance, researchers can simply use an easy-to-use scripting language to configure a network, and observe results generated by NS2. Undoubtedly, NS2 has become the most widely used open source network simulator, and one of the most widely used network simulators.

Unfortunately, most research needs simulation modules which are beyond the scope of the built-in NS2 modules. Incorporating these modules into NS2 requires profound understanding of NS2 architecture. Currently, most NS2 beginners rely on online tutorials. Most of the available information mainly explains how to configure a network and collect results, but does not include sufficient information for building additional modules in NS2. Despite its details about NS2 modules, the formal documentation of NS2 is mainly written as a reference book, and does not provide much information for beginners. The lack of guidelines for extending NS2 is perhaps the greatest obstacle, which discourages numerous researchers from using NS2. At this moment, there is no guide book which can help the beginners understand the architecture of NS2 in depth.

The objective of this textbook is to act as a primer for NS2 beginners. The book provides information required to install NS2, run simple examples, modify the existing NS2 modules, and create as well as incorporate new modules into NS2. To this end, the details of several built-in NS2 modules are explained in a comprehensive manner.

NS2 by itself contains numerous modules. As time elapses, researchers keep developing new NS2 modules. This book does not include the details of all NS2 modules, but does so for selected modules necessary to understand the basics of NS2. For example, it leaves out the widely used modules such as wireless node or web caching. We believe that once the basics of NS2 are grasped, the readers can go through other documentations, and readily understand the details of other NS2 components. The details of Network AniMator (NAM) and Xgraph are also omitted here. We understand that these two tools are nice to have and could greatly facilitate simulation and analysis of computer networks. However, we believe that they are not essential to the understanding of the NS2 concept, and their information are widely available through most of the online tutorials.

This textbook can be used by researchers who need to use NS2 for communication network performance evaluation based on simulation. Also, it can be used as a reference textbook for laboratory works for a senior undergraduate level course or a graduate level course on telecommunication networks offered in Electrical and Computer Engineering and Computer Science Programs. Potential courses include "Network Simulation and Modeling", "Computer Networks", "Data Communications", "Wireless Communications and Networking", "Special Topics on Telecommunications". In a fifteen-class course, we suggest the first class for an introduction to programming (Appendix A), and other 14 classes for each of the 14 chapters. Alternately, the instructor may allocate 10 classes for teaching and 5 classes for term projects. In this case, we suggest that the materials presented in this book are taught in the following order: Chapters 1–2, 3, 12, 4–5, 6, 7–8, 9–11, 13 and 14. When the schedule is really tight, we suggest the readers to go through Chapters 2, 4–7, and 9–10. The readers may start by getting to know NS2 in Chapter 2, and learn the main concepts of NS2 in Chapters 4–5. Chapters 6–7 and 9–10 present the details of most widely used NS2 modules. From time to time, the readers may need to visit Chapter 3, 8, and 12 for further information. If tracing is required, the readers may also have to go through Chapter 13. Finally, Chapter 14 would be useful for those who need to extend NS2 beyond it scopes.

We recommend the readers who intend to go through the entire book to proceed chapter by chapter. A summary of all the chapters in this book is provided below.

As the opening chapter, **Chapter 1** gives an introduction to computer networks and network simulation. The emphasis is on event-driven simulation from which NS2 is developed.

An overview of Network Simulator 2 (NS2) is discussed in **Chapter 2**. Here, we briefly show the two-language NS2 architecture, NS2 directory and the conventions used in this book, and NS2 installation guidelines for UNIX and Windows systems. We also present a three-step simulation formulation as well as a simple example of NS2 simulation. Finally, we demonstrate how to use the `make` utility to incorporate new modules into NS2.

Chapter 3 explains the details of the NS2 two language structure, which consists of the following six main C++ classes: `Tcl`, `Instvar`, `TclObject`,

`TclClass`, `TclCommand`, and `EmbeddedTcl`. **Chapters 4–5** present the very main simulation concept of NS2. While Chapter 4 explains implementation of event-driven simulation in NS2, Chapter 5 focuses on network objects as well as packet forwarding mechanism.

Chapters 6–11 present the following six most widely used NS2 modules. First, nodes (Chapter 6) act as routers and computer hosts. Secondly, links, particularly `SimpleLink` objects (Chapter 7), deliver packets from one network object to another. They model packet transmission time as well as packet buffering. Thirdly, packets (Chapter 8) contain necessary information in its header. Fourthly, agents (Chapters 9–10) are responsible for generating packets. NS2 has two main transport-layer agents: TCP and UDP agents. Finally, applications (Chapter 11) model the user demand for data transmission.

Chapter 12 presents three helper modules: timers, random number generators, and error models. It also discusses the concepts of two bit-wise operations, namely, bit masking and bit shifting, which are used throughout NS2.

Chapter 13 summarizes the post-simulation process, which consists of three main parts: debugging, variable and packet tracing, and result compilation.

After discussing all the NS components, **Chapter 14** demonstrates how a new module is developed and integrated into NS2 through two following examples: Automatic Repeat reQuest (ARQ) and packet schedulers.

Appendices A and B provide programming details which could be useful for the beginners. These details include an introduction to Tcl, OTcl, and AWK programming languages as well as a review of the polymorphism OOP concept.

As the final words, we would like to express sincere gratitude to our colleagues, especially, Surachai Chieochan, at the University of Manitoba, and the colleagues at TOT Public Company Limited, Bangkok, Thailand, for their continuous support. Last but not the least, we would like to acknowledge our families as well as our partners – Wannasorn and Rumana – for their incessant moral support and patient understanding throughout this endeavor.

TOT Public Company Limited   *Teerawat Issariyakul*
University of Manitoba   *Ekram Hossain*
July 2008

Contents

1

Simulation of Computer Networks

People communicate. One way or another, they exchange some information among themselves all the times. In the past several decades, many electronic technologies have been invented to aid this process of exchanging information in an efficient and creative way. Among these are the creation of fixed telephone networks, the broadcasting of television and radio, the advent of computers, and the emergence of wireless sensation. Originally, these technologies existed and operated independently, serving their very own purposes. Not until recently that these technological wonders seem to converge, and it is a well-known fact that a computer communication network is a result of this convergence.

This chapter presents an overview of computer communication networks, and the basics of simulation of such a network. Section 1.1 introduces a computer network along with the reference model which is used for describing the architecture of a computer communication network. A brief discussion on designing and modeling a complex system such as a computer network is then given in Section 1.2. In Section 1.3, the basics of computer network simulation are discussed. Section 1.4 presents one of the most common type of network simulation-time-dependent simulation. An example simulation is given in Section 1.5. Finally, Section 1.6 summarizes the chapter.

1.1 Computer Networks and the Layering Concept

A computer network is usually defined as a collection of computers interconnected for gathering, processing, and distributing information. *Computer* is used as a broad term here to include devices such as workstations, servers, routers, modems, base stations, wireless extension points, etc. These computers are connected by communications links such as copper cables, fiber optic cables, and microwave/satellite/radio links. A computer network can be built as a nesting and/or interconnection of several networks. The Internet is a good example of computer networks. In fact, it is a network of networks,

T. Issariyakul, E. Hossain, *Introduction to Network Simulator NS2*,
DOI: 10.1007/978-0-387-71760-9_1, © Springer Science+Business Media, LLC 2009

within which, tens of thousands of networks interconnect millions of computers worldwide.

1.1.1 Layering Concept

A computer network is a complex system. To facilitate design and flexible implementation of such a system, the concept of *layering* is introduced. Using a layered structure, the functionalities of a computer network can be organized as a stack of layers. There is a peer-to-peer relationship (or virtual link) between the corresponding layers in two communicating nodes. However, actual data flow occurs in a vertical fashion – from the highest layer to the lowest layer in a node, and then through the physical link to reach the lowest layer at the other node, and then following upwards to reach the highest layer in the stack. Each layer represents a well-defined and specific part of the system and provides certain **services** to the above layer. Accessible (by the upper layers) through so-called **interfaces**, these services usually define *what* should be done in terms of network operations or primitives, but does not specifically define *how* such things are implemented. The details of how a service is implemented is defined in a so-called **protocol**. For example, the transmitter at a source computer can use a specific protocol (e.g., a data encoding scheme) at the physical layer to transmit information bits to the receiving computer, which should be able to decode the received information based on the protocol rules. The beauty of this layering concept is the layer independency. That is, a change in a protocol of a certain layer does not affect the rest of the system as long as the interfaces remain unchanged. Here, we highlight the words services, protocol, and interface to emphasize that it is the interaction among these components that makes up the layering concept.

Figure 1.1 graphically shows an overall view of the layering concept used for communication between two computer hosts: a source host and a

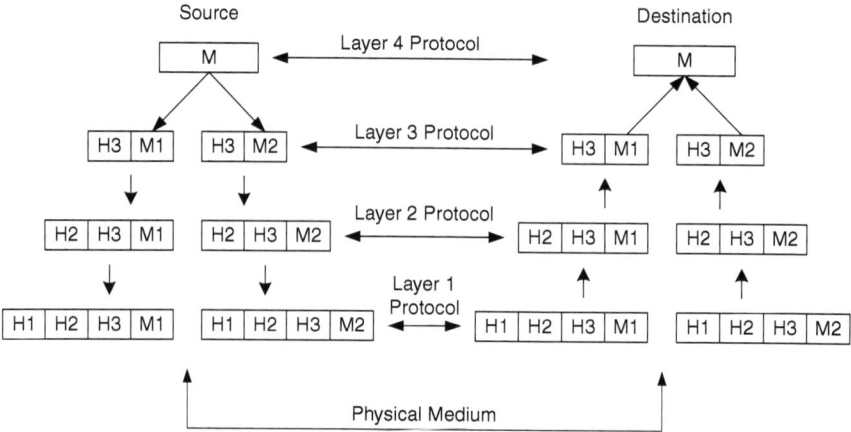

Fig. 1.1. Data flow in a layered network architecture.

destination host. In this figure, the functionality of each computer host is divided into four layers.[1] When virtually linked with the same layer on another host, these layers are called *peers*.[2] Although not directly connected to each other, these peers virtually communicate with one another using a protocol represented by an arrow. As has already been mentioned, the actual communication needs to propagate down the stack and use the above layering concept.

Suppose an application process running on Layer 4 of the source generates data or messages destined for the destination. The communication starts by passing a generated message M down to Layer 3, where the data are segmented into two chunks (M1 and M2), and control information called *header* (H3) specific to Layer 3 is appended to M1 and M2. The control information are, for example, sequence numbers, packet sizes, and error checking information. These information are understandable and used only by the peering layer on the destination to recover the data (M). The resulting data (e.g., H3+M1) is handed to the next lower layer, where some protocol-specific control information are again added to the message. This process continues until the message reaches the lowest layer, where transmission of information is actually performed over a physical medium. Note that, along the line of these processes, it might be necessary to further segment the data from upper layers into smaller segments for various purposes. When the message reaches the destination, the reverse process takes place. That is, as the message is moving up the stack, its headers are ripped off layer by layer. If necessary, several messages are put together before being passed to the upper layer. The process continues until the original message (M) is recovered at Layer 4.

1.1.2 OSI and TCP/IP Reference Models

The OSI (Open Systems Interconnection) model was the first reference model developed by ISO (International Standards Organization) to provide a standard framework in order to describe the protocol stacks in a computer network. Its consists of seven layers where each layer is intended to perform a well-defined function [1]. These are physical layer, data link layer, network layer, transport layer, session layer, presentation layer, and application layer. The OSI model only specifies what each layer should do; it does not specify the exact services and protocols to be used in each layer. Although not implemented in current systems, the OSI model philosophy (i.e., the layering concept) lays a strong foundation for further developement in computer networking.

---

[1] For the sake of illustration only four layers are shown. In the real world systems, the number of layers may vary, depending on the functionality and objectives of the networks.

[2] A peering host of a source and a destination are the destination and the source, respectively.

The TCP (Transmission Control Protocol)/IP (Internet Protocol) reference model, which is based on the two primary protocols, namely, TCP and IP, is used in the current Internet. These protocols have proven very powerful, and as a result, have experienced widespread use and implementation in the existing computer networks. It was developed for ARPANET, a research network sponsored by the U.S. Department of Defense, which is considered as the grandparent of all computer networks. In the TCP/IP model, the protocol stack consists of five layers – physical, data link, network, transport, and application – each of which is responsible for certain services as will be discussed shortly. Note that the application layer in the TCP/IP model can be considered as the combination of session, presentation, and application layers of the OSI model.

Application Layer

The application layer sits on top of the stack, and uses services from the transport layer (discussed below). This layer supports several higher-level protocols such as HTTP (Hypertext Transfer Protocol) for World Wide Web applications, SMTP (Simple Mail Transfer Protocol) for electronic mail, TELNET for remote virtual terminal, DNS (Domain Name Service) for mapping comprehensible host names to their network addresses, and FTP (File Transfer Protocol) for file transfer.

Transport Layer

The objective of a transport layer is to transport the messages from the application layer of the source host to that of the destination host. To accomplish this goal, two well-known protocols, namely, TCP and UDP (User Datagram Protocol), are defined in this layer. While TCP is responsible for a reliable and connection-oriented communication between the two hosts, UDP supports an unreliable connectionless transport. TCP is ideal for applications that prefer accuracy over prompt delivery and the reverse is true for UDP.

Generally, control information related to flow control and error control need to be embedded into the messages. Also, before adding any header, fragmentation is usually performed to break a long message into segments. For this reason, the protocol data units in this layer are normally called *segments*.

Network Layer

This layer provides routing services to the transport layer. Network layer is designed to deliver the data units, usually called *packets*, along the paths they are meant to traverse from a source host to a destination host. Again, to facilitate routing, headers containing information such as source and destination network addresses are added to the transport protocol data units to formulates network-layer data unit.

Link Layer

The packets are generally routed through several communication links and nodes before they actually reach the destination node. To successfully route these packets all the way to the destination, a mechanism is required for node-to-node delivery across each of the communication links. A link layer protocol is responsible for data delivery across a communication link.

A link layer protocol has three main responsibilities. First, flow control regulates the transmission speed in a communication link. Secondly, error control ensures the integrity of data transmission. Thirdly, flow multiplexing/demultiplexing combines multiple data flows into and extracts data flows from a communication link. Choices of link layer protocols may vary from host to host and network to network. Examples of widely-used link layer protocols/technologies include Ethernet, Point-to-Point Protocol (PPP), IEEE 802.11 (i.e., WiFi), and Asynchronous Transfer Mode (ATM).

Physical Layer

The physical layer deals with the transmission of data bits across a communication link. Its primary goal is to ensure that the transmission parameters (e.g., transmission power, modulation scheme) are set appropriately to achieve the required transmission performance (e.g., to achieve the target bit error rate performance).

Finally, we point out that the five layers discussed above are common to the OSI layer. As has already been mentioned, the OSI model contains two other layers sitting on top of the transport layer, namely, session and presentation layers. The session layer simply allows users on different computers to create communication sessions among themselves. The presentation layer basically takes care of different data presentations existing across the network. For example, a unified network management system gathers data with different format from different computers and converts their format into a uniform format.

1.2 System Modeling

System modeling refers to an act of representing an actual system in a simply way. System modeling is extremely important in system design and development, since it gives an idea of how the system would perform if actually implemented. With modeling, the parameters of the system can be changed, tested, and analyzed. More importantly, modeling, if properly handled, can save costs in system development. To model a system, some simplifying assumptions are often required. It is important to note that too many assumptions would simplify the modeling but may lead to an inaccurate representation of the system.

Traditionally, there are two modeling approaches: analytical approach and simulation approach.

1.2.1 Analytical Approach

The general concept of analytical modeling approach is to first come up with a way to describe a system mathematically with the help of applied mathematical tools such as queuing and probability theories, and then apply numerical methods to gain insight from the developed mathematical model. When the system is simple and relatively small, analytical modeling would be preferable (over simulation). In this case, the model tends to be mathematically tractable. The numerical solutions to this model in effect require lightweight computational efforts.

If properly employed, analytical modeling can be cost-effective and can provide an abstract view of the components interacting with one another in the system. However, if many simplifying assumptions on the system are made during the modeling process, analytical models may not give an accurate representation of the real system.

1.2.2 Simulation Approach

Simulation is widely-used in system modeling for applications ranging from engineering research, business analysis, manufacturing planning, and biological science experimentation, just to name a few. Compared to analytical modeling, simulation usually requires less abstraction in the model (i.e., fewer simplifying assumptions) since almost every possible detail of the specifications of the system can be put into the simulation model to best describe the actual system. When the system is rather large and complex, a straightforward mathematical formulation may not be feasible. In this case, the simulation approach is usually preferred to the analytical approach.

In common with analytical modeling, simulation modeling may leave out some details, since too much details may result in an unmanageable simulation and substantial computation effort. It is important to carefully consider a measure under consideration and not to include irrelevant detail into the simulation.

In the next section, we describe the basic concepts of simulation in more detail with particular emphasis on simulation of a computer network.

1.3 Basics of Computer Network Simulation

A simulation is, more or less, a combination of art and science. That is, while the expertise in computer programming and the applied mathematical tools account for the science part, the very skill in analysis and conceptual model formulation usually represents the art portion. A long list of steps in executing a simulation process, as given in [2], seems to reflect this popular claim. Basically, all these steps can be put into three main tasks each of which carries different degrees of importance.

According to Shannon [2], it is recommended that 40 percent of time and effort be spent on defining a problem, designing a corresponding model, and devising a set of experiments to be performed on the simulation model. Further, it was pointed out that a portion of 20 percent should be used to program the conceptual elements obtained during the first step. Finally, the remaining 40 percent should be utilized in verifying/validating the simulation model, experimenting with designed inputs (and possibly fine-tuning the experiments themeselves), and analyzing the results. We note that this formula is in no way a strict one. Any actual simulation may require more or less time and effort, depending on the context of interest and, definitely, on the modeler himself/herself.

A simulation can be thought of as a flow process of network entities (e.g., nodes, packets). As these entities move through the system, they interact with other entities, join certain activities, trigger events, cause some changes to the state of the system, and leave the process. From time to time, they contend or wait for some type of resources. This implies that there must be a logical execution sequence to cause all these actions to happen in a comprehensible and manageable way. An execution sequence plays an important role in supervising a simulation and is sometimes used to characterize the types of simulation (see Section 1.4).

1.3.1 Simulation: The Formal Definition

According to Shannon [2], simulation is "the process of designing a model of a real system and conducting experiments with this model for the purpose of understanding the behavior of the system and/or evaluating various strategies for the operation of the system." With the dynamic nature of computer networks, we thus actually deal with a *dynamic* model of a real *dynamic* system.

1.3.2 Elements of Simulation

According to Ingalls [3], the structural components of a simulation consist of the following:

Entities

Entities are objects which interact with one another in a simulation program to cause some changes to the state of the system. In the context of a computer network, entities may include computer nodes, packets, flows of packets, or non-physical objects such as simulation clocks. To distinguish the different entities, unique attributes are assigned to each of them. For instance, a packet entity may have attributes such as packet length, sequence number, priority, and the header.

Resources

Resources are a part of complex systems. In general, a limited supply of resources has to be shared among a certain set of entities. This is usually the case for computer networks, where bandwidth, air time, the number of servers, for instance, represent network resources which have to be shared among the network entities.

Activities and Events

From time to time, entities engage in some activities. This engaging creates events and triggers changes in the system states. Common examples of activities include delay and queuing. When a computer needs to send a packet but find the medium busy, it waits until the medium is free. In this case, the packet is to be sent over the air but the medium is busy, the packet is said to be engaged in a waiting activity.

Scheduler

A scheduler maintains the list of events and their execution time. During a simulation, it runs a simulation clock creates events, and executes them.

Global Variables

In simulation, a global variable is accessible by any function or entity in the system, and basically keeps track of some common values of the simulation. In the context of computer networks, such variables might represent, for example, the length of the packet queue in a single-server network, the total busy air time of the wireless network, or the total number of packets transmitted.

Random Number Generator

A Random number generator (RNG) is required to introduce randomness in a simulation model. Random numbers are generated by sequentially picking numbers from a deterministic sequence of psudo-random number [4], yet the numbers picked from this sequence appear to be random. In most case, a psudo-random sequence is predefined and is used by every RNG.

In many situations, several statistically results are required. An RNG needs to start picking numbers from different location (i.e., seed) in the (same) predefined psudo-random sequence. Otherwise, the results for every run would be the same. In an actual implementation, an RNG is initialized with a seed. A seed identifies the starting location in a psudo-random sequence, where an RNG starts picking numbers. Different simulation initialized with different seeds therefore generates different results (but statistically identical).

In a computer network simulation, for example, a packet arrival process, waiting process, and service process are usually modeled as random processes.

A random process is expressed by sequences of random variables. These random process are usually implemented with the aids of an RNG. For a comprehensive treatment on random process implementation (e.g., those having the uniform, exponential, Gaussian, Poisson, Binomial distribution functions), the readers are referred to [5, 6].

Statistics Gatherer

The main responsibility of a statistics gatherer is to collect data generated by the simulation so that meaningful inferences can be drawn from such data.

1.4 Time-Dependent Simulation

A main type of simulation is time dependent simulation which proceeds chronologically. This type of simulation maintains a simulation clock which keeps track of the current simulation time. In most cases, the simulation is run until the clock reaches a predefined threshold.

Time-dependent simulation can be further divided into time-driven simulation and event-driven simulation. A time-driven simulation induces and executes events for every fixed time interval. In other words, the simulation advances from one time interval to another, and executes events (if any) until it reaches a certain limit. An event-driven simulation, on the other hand, induces events at arbitrary time. The simulation moves from one event to another, and again executes the event (if any) until the simulation terminates.

There is an important note for time-dependent simulation: The simulation *must* progress in a chronological order. While this note is fairly straightforward for a time-driven simulation, [7] specifies two important points for an implementation of event-driven simulation. First, every new event scheduled into the event list must be tagged with a timestamp equal to or greater than that of the current event. In other words, no outdated events can be scheduled. Secondly, the next event the simulation always executes is that event with the smallest timestamp in the event list. It will never jump over chronologically ordered events or Jump back to the past event.

1.4.1 Time-Driven Simulation

In time-driven simulations, the simulation clock is advanced exactly by a fixed interval of Δ time units. After each advancement of the clock, the simulation looks for events that may have occurred during this fixed interval. If so, such events are treated as if they occurred at the end of this interval.

Figure 1.2 shows the basic idea behind time advancement in a time-driven simulation. The curved arrows here represent such advances, and a, b, and c mark the occurrences of particular events. During the first interval, no event occurs, whereas the second interval contains event a, which is not handled until the end of the interval. One disadvantage of time-driven simulation is illustrated in the fifth interval, where events b and c are considered to occur exactly at the end of the interval (at time 5Δ). This calls for a procedure that determines which event should be handled first. One solution to get around this situation is to narrow down a simulation time interval such that every interval contains only one event. This, however, puts substantial computational burden on the simulator. Time-driven simulation is therefore not recommended for system models whose events tend to occur over a random period of time.

Fig. 1.2. Clock advancement in a time-driven simulation.

Example 1.1. Program 1.1 shows time-driven simulation pseudo codes. Lines 1 and 2 initializes the system state variables and the simulation clock, respectively. Line 3 specifies the stopping criterion. Here, Lines 4-7 are run as long as the simulation clock (i.e., simClock) is less than a predefined threshold (i.e., stopTime). These lines collect statistics, executes events, and advance the simulation to the current event time.

Program 1.1 Skeleton of the event-processing loop in a time-driven simulation.

```
1 initialize {system states}
2 SimClock := startTime;
3 while {SimClock < stopTime}
4     collect statistics from current state;
5     execute all events that occurred during
6     [SimClock, SimClock + step];
7     SimClock := SimClock + step;
8 end while
```

1.4.2 Event-Driven Simulation

As the name suggests, an event-driven simulation is initiated and run by a set of events. A list of all scheduled events are usually maintained and updated throughout the simulation process. Technically speaking, the main loop in the simulation program actually has to sequence through this list, and handle one event after another until either the list is empty or the stopping criterion is

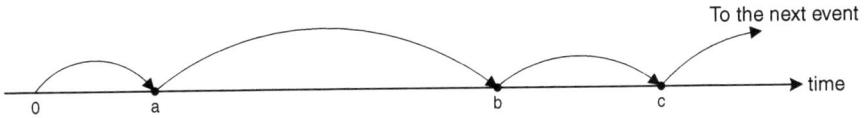

Fig. 1.3. Clock advancement in an event-driven simulation.

satisfied. The mechanism of handling events is shown graphically in Fig. 1.3, where events a, b, and c are executed in order. The time gap between two events is not fixed. The simulation advance from one event to another, as opposed to one interval to another in a time-driven simulation. Except for the time advancing mechanism, the event-driven simulation is quite similar to the time-driven mechanism.

In an event-driven simulation, all the events in an entire simulation may not be created at the initialization. As the simulation advances, one event may induce one or more events. The new event is usually inserted into the *chain* (i.e., list) of events arranged chronologically. An event-driven simulation ignores the intervals of inactivity by advancing the simulation clock from one event time to another. This process goes on and on until all the events are executed, or until the system reaches a specific state (e.g., the simulation time reaches a predefined value). Along the way, we certainly need a way to gather some statistics or states of the system for analysis purposes. This process of gathering information can take place right after every event execution. Alternatively, it can be done using a specialized entity which gathers statistics during the simulation.

Example 1.2. Program 1.2 shows the skeleton of a typical event-driven simulation program. Lines 1 and 2 initializes the system state variables and the list of events, respectively. Line 3 specifies the stopping criterion. Lines 4–6 are executed as along as Line 3 returns **true**. Here, the previously executed event is removed from the list, the simulation clock is set to the scheduled time of the current event, and the current event is executed. Within such a loop, the system state variables may be modified to capture those changes that occur in the system according to the executed event.

Program 1.2 Skeleton of the event-processing loop in an event-driven simulation.

```
1 initialize {system states}
2 initialize {list of events}
3 while {state != finalState}   % or while {this.event != Null}
4     expunge the previous event from list of events;
5     set SimClock := time of current event;
6     execute this.event
7 end while
```

1.5 A Simulation Example: A Single-Channel Queuing System

This section demonstrates a simulation of a single-channel queuing system, as an example. Consider a point-to-point wired communication link as shown in Fig. 1.4. For simplicity, we consider only a one-way communication from node A to node B. In particular, we are interested in an intra-station packet queuing system at node A, where a packet is retrieved from the queue and transmitted (or served) one at a time – the transmission time depends on the bandwidth or capacity of the link.

Futhermore, we assume that packets, whose inter-arrival time follows some probability distribution, are unlimited and randomly generated from a set of applications. Since a packet can be of any random length and the conditions of the channel may vary, the service time of each packet is also random and follows some probability distribution. In our case, it is defined as the elapsed time from the moment a packet is transmitted to the moment it is successfully received by node B.

Next, the queuing discipline employed at node A is First-In-First-Out (FIFO), i.e., packets are enqueued and transmitted (served) in the order of their arrival. For simplicity, the queuing mechanism at node B is ignored. Additionally, for the system to be stable, we assume that the arrival rate is less than the service rate. Otherwise, the queue will build up with no bound.

Entities

The primary entities in this simulation include the following:

- Server (medium availability) with *idle* and *busy* attributes,

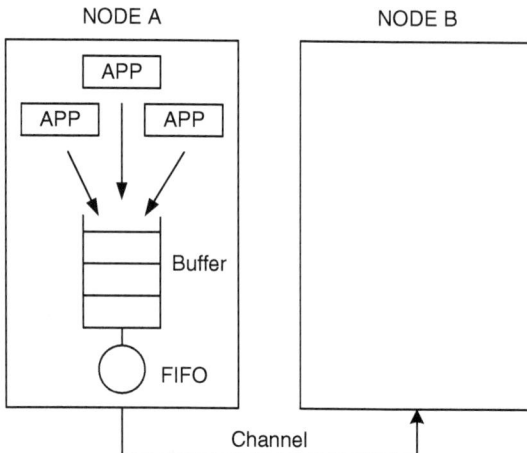

Fig. 1.4. Illustration of a single-channel queuing system.

- Packets with *arrival time* and *service time* attributes, and
- Queue with *empty* and *non-empty* attributes.

Resource

Obviously, the only resource in this example is the transmission time in the channel.

System State Variables and Events

- Two system state variables:
  - (i) `num_system` is the number of packets in the system, i.e., the one being served and those waiting in the queue.
  - (ii) `channel_free` is the status of the channel (server) which is either idle or busy.
- Two events:
  - (i) `pkt_arrival` corresponds to a *packet arrival* event. This event occurs when a packet arrives at the queue. As shown in Fig. 1.5, once entered, the packet may either go directly to service or wait in the queue, depending on whether the channel is busy or idle.
  - (ii) `pkt_complete` corresponds to a *successful packet transmission* event. This event indicates that a packet has been received successfully by node B. At the completion, node A begins to transmit (serve) another packet waiting in the queue. If there is no more packet to be sent, the channel becomes idle. The flow diagram of such a process is shown in Fig. 1.6.

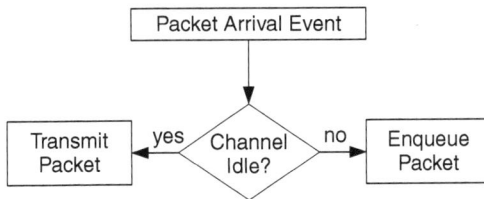

Fig. 1.5. Packet arrival event.

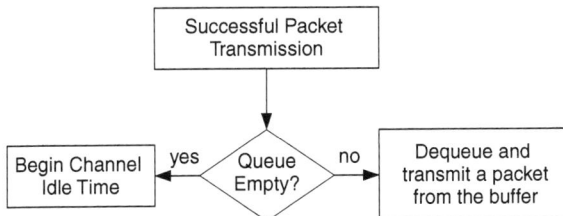

Fig. 1.6. Successful packet transmission (service completion) event.

Two other important elements in an event-driven simulation are a simulation clock and an event list. A simulation clock maintains the current simulation time, as the simulation advances. An event list is a chain of scheduled events (e.g., *packet arrival* and *successful packet transmission*) connecting in a chronological order. Again, the simulation executes an event after another down the event list, and updates the simulation clock based on the time specified in the executed event.

Simulation Performance Measures

Here, we consider three following performance measures are:

- *Mean waiting time* is the average time that a packet spends in the queue. In the simulation, we define a global variable which keeps track of the total time all the transmitted packets spent in the queue. At the end of the simulation, we divide this value by the total number of packets transmitted to obtain the mean waiting time.
- *Mean packet transmission latency* is the average time that a packet spends (from its arrival to its departure) in the system. It is the total time of all the packets spend the system divided with the total number of transmitted packets.
- *Mean server utilization* is the percentage time where the server is busy. During the simulation we measure the time where the server is busy. At the end of the simulation, we divide this busy time by the total simulation time, and obtain the mean server utilization.

It is important to note that all the above measures are the average values taken over time, implying that the longer the simulation, the more accurate the statistics.

Program 1.3 shows a skeleton of the simulation program that can be used to implement the single-channel queuing system described above.

The program starts with the initialization of system state variables as defined above. Additionally, we define `num_queue` (Line 3) and `num_system` (Line 4) to store the number of waiting packets and the number of all packets currently in the system (i.e., both the queue and the channel), respectively. The variable `SimClock` is also initialized to zero at the beginning of the simulation. Next, Line 7 creates an event list by invoking the procedure `create_list()`. We assume that this function automatically generates packets and associates each packet with the random inter-arrival and service times. Further we assume that the `event_list` here is implemented using some appropriate data structure that usually indicates the event type (arrival or completion) and the associated timestamp (i.e., either inter-arrival time and service time). Initially, only the arrival events are put into the `event_list`.

Now we define a main loop which continuously checks whether the simulation should be terminated. The stopping criteria in Line 9 are (1) the event list is exhausted and (2) the simulation clock has reached a predefined threshold.

Program 1.3 Simulation skeleton of a single-channel queuing system.

```
1   % Initialize system states
2   channel_free = true; %Channel is idle
3   num_queue = 0;       %Number of packets in queue
4   num_system = 0;      %Number of packets in system
5   SimClock = 0;        %Current time of simulation

6   %Generate packets and schedule their arrivals
7   event_list = create_list();

8   % Main loop
9   while {event_list != empty} & {SimClock < stopTime}
10      expunge the previous event from event list;
11      set SimClock := time of current event;
12      call current event;
13  end while

14  %Define events
15  pkt_arrival(){
16      if(channel_free)
17          channel_free = false;
18          num_system = num_system + 1;
19          % Update "event_list": Put "successful packet tx event"
20          % into "event_list," T is random service time.
21          schedule event "pkt_complete" at SimClock + T;
22      else
23          num_queue = num_queue + 1; %Place packet in queue
24      num_system = num_queue + 1;
25  }

26  pkt_complete(){
27      num_system = num_system - 1;
28      num_queue = num_queue - 1;
29      if(num_queue > 0)
30          schedule event "pkt_complete" at SimClock + T;
31      else
32          channel_free = true;
33          num_system = 0;
34          num queue = 0;
35  }
```

If not, Lines 10–12 keep on executing the next event by invoking either the procedure pkt_arrival() in Lines 15–25 or the procedure pkt_complete() in Lines 26–35.

The procedure pkt_arrival() (Lines 15–25) checks whether the channel (server) is idle when a packet arrives. If it is idle, the channel is set to *busy*, and a *successful packet transmission* event is inserted into the `event_list` for future execution. The timestamp associated with the event is equal to the current clock time (`SimClock`) plus the packet's randomly generated service time (`T`). If the channel is busy, on the other hand, the packet is simply put in the queue whose counter (`num_queue`) is incremented by one unit. The number of packets in the system is also updated accordingly.

When `SimClock` advances to a *successful packet transmission* event, the procedure `pkt_complete()` is executed (Lines 26–35). Here, the number of packets in the system (`num_system`) is updated. The queue counter `num_queue` is decremented by one unit. Upon any successful packet transmission, it is also necessary to check whether the queue is empty. If not, the head-of-the-line packet will be served. This is done by feeding the packet to the channel and scheduling it for transmission completion at time `SimClock + T`. However, if the queue is empty, the channel is set to idle and the numbers of packets in the queue and system are set to zero.

Suppose that the inter-arrival time and the service time comply with the probability mass functions specified in Table 1.1. Table 1.2 shows the simulation results for 10 packets. The inter-arrival time and service time of each packet are shown in the first and second columns, respectively. The third and fourth columns specify the time where the packet arrives and starts to be served. The fifth column represents the packet waiting time, the time that a packet spends in the queue. It is computed as the time difference between when the service starts and when the packet arrives. Finally, the sixth column represents the packet transmission latency, the time that a packet spends in both the queue and the channel. It is computed as the summation of the waiting time and the service time.

Based on the result in Table 1.2, we compute the average waiting time and the average packet transmission latency by averaging the sixth and seventh columns (i.e., adding all the values and dividing the result by 10). The average waiting time and the average packet transmission latency are therefore 1.0 and 3.5 time units, respectively.

Table 1.1. Probability mass functions of inter-arrival time and service time.

Time unit	Inter-arrival (probability mass)	Service (probability mass)
1	0.2	0.5
2	0.2	0.3
3	0.2	0.1
4	0.2	0.05
5	0.1	0.05
6	0.05	
7	0.05	

Table 1.2. Simulation of a single-channel queuing system.

Packet	Interarr. time	Service time	Arrival time	Service starts	Time spent in-queue	Packet trans- mission latency
1	-	5	0	0	0	5
2	2	4	2	5	3	7
3	4	1	6	9	3	4
4	1	1	7	10	3	4
5	6	3	13	13	0	3
6	7	1	20	20	0	1
7	2	1	22	22	0	1
8	1	4	23	23	0	4
9	3	3	26	27	1	4
10	5	2	31	31	$\underline{0}$	$\underline{2}$
					$\underline{\underline{10}}$	$\underline{\underline{3.5}}$

Table 1.3. Evolution of number of packets in the queue over time.

Event	Packet No.	Simulation clock
Arrival	1	0
Arrival	2	2
Completion	1	5
Arrival	3	6
Arrival	4	7
Completion	2	9
Completion	3	10
Completion	4	11
Arrival	5	13
Completion	5	16

Based on the information in Table 1.2, we also show in Table 1.3 how each event for the first five packets occurs in a chronological order with respect to the *Simulation Clock* (SimClock).

Figure 1.7 depicts the evolution of the number of packets in the queue over time, which is also shown in Table 1.3. As shown in Fig. 1.7, at various instances, the number of packets in the system differs. When the first packet is being transmitted, another packet arrives in the queue at time 2. The number of packets in the system becomes 2. That is, it includes the one that has been served plus the one just arrived. In Fig. 1.7, this event causes a jump in the graph at time 2. At time 5, when the first packet is successfully received at

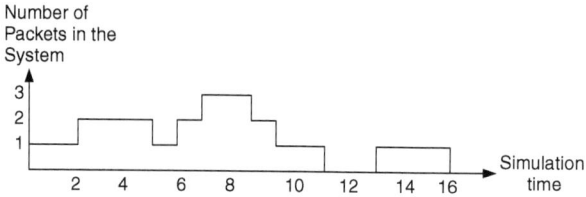

Fig. 1.7. Number of packets in the system at various instances.

node B, the next packet in queue is transmitted. Therefore, the graph drops to level 1, indicating that there is only one packet in the system. This dynamics continues until all the packets are transmitted. Based on Fig. 1.7, the mean server utilization can be computed from the ratio of the time where the server is in use and the simulation time, which is $14/16 = 0.875$ in this case.

1.6 Chapter Summary

A computer network is a complex system that requires a careful treatment in design and implementation. Simulation, regarded as one of the most powerful performance analysis tools, is usually used in carrying out such a treatment to complement the analytical tools.

This chapter focuses mainly on time-dependent simulation, which advances in a time domain. The time-dependent simulation can be divided into two categories. Time-driven simulation advances the simulation by fixed time intervals, while event-driven simulation proceeds from one event to another. NS2 is an event-driven simulation tool. Designing event-driven simulation models using NS2 is the theme of the rest of the book.

2

Introduction to Network Simulator 2 (NS2)

2.1 Introduction

Network Simulator (Version 2), widely known as NS2, is simply an event-driven simulation tool that has proved useful in studying the dynamic nature of communication networks. Simulation of wired as well as wireless network functions and protocols (e.g., routing algorithms, TCP, UDP) can be done using NS2. In general, NS2 provides users with a way of specifying such network protocols and simulating their corresponding behaviors.

Due to its flexibility and modular nature, NS2 has gained constant popularity in the networking research community since its birth in 1989. Ever since, several revolutions and revisions have marked the growing maturity of the tool, thanks to substantial contributions from the players in the field. Among these are the University of California and Cornell University who developed the REAL network simulator,[1] the foundation which NS is based on. Since 1995 the Defense Advanced Research Projects Agency (DARPA) supported development of NS through the Virtual InterNetwork Testbed (VINT) project [9].[2] Currently the National Science Foundation (NSF) has joined the ride in development. Last but not the least, the group of researchers and developers in the community are constantly working to keep NS2 strong and versatile.

Again, the main objective of this book is to provide the readers with insights into the NS2 architecture. This chapter gives a brief introduction to NS2. NS2 Beginners are recommended to go thorough the detailed introductory online resources. For example, NS2 official website [10] provides NS2 source code as well as detailed installation instruction. The web pages in [11] and [12] are among highly recommended ones which provide tutorial and

---

[1] REAL was originally implemented as a tool for studying the dynamic behavior of flow and congestion control schemes in packet-switched data networks.

[2] Funded by DARPA, the VINT project aimed at creating a network simulator that will initiate the study of different protocols for communication networking.

T. Issariyakul, E. Hossain, *Introduction to Network Simulator NS2*,
DOI: 10.1007/978-0-387-71760-9_2, © Springer Science+Business Media, LLC 2009

examples for setting up basic NS2 simulation. A comprehensive list of NS2 codes contributed by researchers can be found in [13]. These introductory online resources would be helpful in understanding the material presented in this book.

In this chapter an introduction to NS2 is provided. In particular, Section 2.2 presents the basic architecture of NS2. The information on NS2 installation is given in Section 2.3. Section 2.4 shows NS2 directories and conventions. Section 2.5 shows the main steps in NS2 simulation. A simple simulation example is given in Section 2.6. Section 2.7 describes how to include C++ modules in NS2. Finally, Section 2.8 concludes the chapter.

2.2 Basic Architecture

Figure 2.1 shows the basic architecture of NS2. NS2 provides users with an executable command **ns** which takes on input argument, the name of a Tcl simulation scripting file. Users are feeding the name of a Tcl simulation script (which sets up a simulation) as an input argument of an NS2 executable command **ns**. In most cases, a simulation trace file is created, and is used to plot graph and/or to create animation.

NS2 consists of two key languages: C++ and Object-oriented Tool Command Language (OTcl). While the C++ defines the internal mechanism (i.e., a backend) of the simulation objects, the OTcl sets up simulation by assembling and configuring the objects as well as scheduling discrete events (i.e., a frontend). The C++ and the OTcl are linked together using TclCL. Mapped to a C++ object, variables in the OTcl domains are sometimes referred to as *handles*. Conceptually, a handle (e.g., n as a **Node** handle) is just a string (e.g., _o10) in the OTcl domain, and does not contain any functionality. Instead, the functionality (e.g., receiving a packet) is defined in the mapped C++ object (e.g., of class **Connector**). In the OTcl domain, a handle acts as a frontend which interacts with users and other OTcl objects. It may defines its own procedures and variables to facilitate the interaction. Note that the member procedures and variables in the OTcl domain are called instance procedures

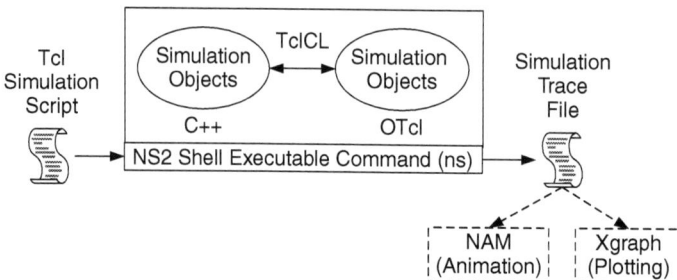

Fig. 2.1. Basic architecture of NS.

(instprocs) and instance variables (instvars), respectively. Before proceeding further, the readers are encouraged to learn C++ and OTcl languages. We refer the readers to [14] for the detail of C++, while a brief tutorial of Tcl and OTcl tutorial are given in Appendices A.1 and A.2, respectively.

NS2 provides a large number of built-in C++ objects. It is advisable to use these C++ objects to set up a simulation using a Tcl simulation script. However, advance users may find these objects insufficient. They need to develop their own C++ objects, and use a OTcl configuration interface to put together these objects.

After simulation, NS2 outputs either text-based or animation-based simulation results. To interpret these results graphically and interactively, tools such as NAM (Network AniMator) and XGraph are used. To analyze a particular behavior of the network, users can extract a relevant subset of text-based data and transform it to a more conceivable presentation.

2.3 Installation

NS2 is a free simulation tool, which can be obtained from [9]. It runs on various platforms including UNIX (or Linux), Windows, and Mac systems. Being developed in the Unix environment, with no surprise, NS2 has the smoothest ride there, and so does its installation. Unless otherwise specified, the discussion in this book is based on a Cygwin (UNIX emulator) activated Windows system.

NS2 source codes are distributed in two forms: the all-in-one suite and the component-wise. With the all-in-one package, users get all the required components along with some optional components. This is basically a recommended choice for the beginners. This package provides an "install" script which configures the NS2 environment and creates NS2 executable file using the "make" utility.

The current all-in-one suite consists of the following main components:

- NS release 2.30,
- Tcl/Tk release 8.4.13,
- OTcl release 1.12, and
- TclCL release 1.18.

and the following are the optional components:

- NAM release 1.12: NAM is an animation tool for viewing network simulation traces and packet traces.
- Zlib version 1.2.3: This is the required library for NAM.
- Xgraph version 12.1: This is a data plotter with interactive buttons for panning, zooming, printing, and selecting display options.

The idea of the component-wise approach is to obtain the above pieces and install them individually. This option save considerable amount of downloading time and memory space. However, it could be troublesome for the beginners, and is therefore recommended only for experienced users.

2.3.1 Installing an All-In-One NS2 Suite on Unix-Based Systems

The all-in-one suite can be installed in the Unix-based machines by simply running the `install` script and following the instructions therein. The only requirement is a computer with a C++ compiler installed. The following commands show how the all-in-one NS2 suite can be installed and validated, respectively:

```
shell>./install
shell>./validate
```

Validating NS2 involves simply running a number of working scripts that verify the essential functionalities of the installed components.

2.3.2 Installing an All-In-One NS2 Suite on Windows-Based Systems

To run NS2 on Windows-based operating systems, a bit of tweaking is required. Basically, the idea is to make Windows-based machines emulate the functionality of the Unix-like environment. A popular program that performs this job is Cygwin.[3] After getting Cygwin to work, the same procedure as that of Unix-based installation can be followed. For ease of installation, it is recommended that the all-in-one package be used. The detailed description of Windows-based installation can be found online at NS2's Wiki site [9], where the information on post-installation troubles can also be found.

Note that by default Cygwin does not install all packages neccessary to run NS2. A user needs to manually install the addition packages shown in Table 2.1[4].

Table 2.1. Additional Cygwin packages required to run NS2.

Category	Packages
Development	`gcc, gcc-objc, gcc-g++, make`
Utils	`patch`
X11	`xorg-x11-base, xorg-x11-devel`

---

[3] Cygwin is available online and comes free. Information such as how to obtain and install Cygwin is available online at the Cygwin website (www.cygwin.com).

[4] Different versions may install different default packages. Users may need to install more or less packages depending on the version of Cygwin.

2.4 Directories and Convention

2.4.1 Directories

Suppose that NS2 is installed in directory `nsallinone-2.30`. Figure 2.2 shows the directory structure under directory `nsallinone-2.30`. Here, directory `nsallinone-2.30` is on the Level 1. On the Level 2, directory `tclcl-1.18` contains classes in TclCL (e.g., `Tcl`, `TclObject`, `TclClass`). All NS2 simulation modules are in directory `ns-2.30` on the Level 2. Hereafter, we will refer to directories `ns-2.30` and `tclcl-1.18` as ~*ns*/ and ~*tclcl*/, respectively.

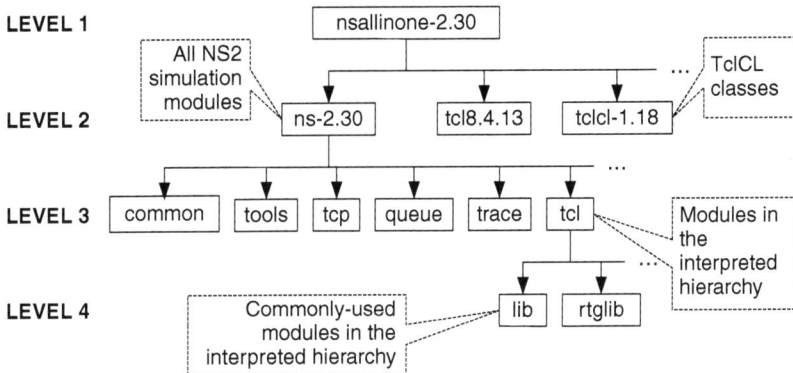

Fig. 2.2. Directory structure of NS2 [12].

On Level 3, the modules in the interpreted hierarchy are under directory `tcl`. Among these modules, the frequently-used ones (e.g., `ns-lib.tcl`, `ns-node.tcl`, `ns-link.tcl`) are stored under directory `lib` on Level 4. Simulation modules in the compiled hierarchy are classified in directories on Level 2. For example, directory `tools` contains various helper classes such as random variable generators. Directory `common` contains basic modules related to packet forwarding such as the simulator, the scheduler, connector, packet. Directories `queue`, `tcp`, and `trace` contain modules for queue, TCP (Transmission Control Protocol), and tracing, respectively.

2.4.2 Convention

The terminologies and formats which are used in NS2 and in this book hereafter are shown below:

Terminology

- An NS2 simulation script (e.g., `myfirst_ns.tcl`) is referred to as a *Tcl simulation script* .
- C++ and OTcl class hierarchies, which have one-to-one correspondence, are referred to as *the compiled hierarchy* and *the interpreted hierarchy*, respectively. Class (or member) variables and class (or member) functions are the variables and functions which belong to a class. In the compiled hierarchy, they are referred to simply as variables and functions, respectively. Those in the interpreted hierarchy are referred to as *instance variables* (*instvars*) and *instance procedures* (*instprocs*), respectively. As we will see in Section 3.4.4, *command*, is a special instance procedure, whose implementation is in the compiled hierarchy (i.e., written in C++). An OTcl object is, therefore, associated with instance variables, instance procedures, and commands, while a C++ object is associated with variables and functions.
- Despite their minor differences, the terms "OTcl" and "interpreted" are used interchangeably throughout the book. Likewise, "C++" and "compiled" are used interchangeably. These terms can be used as adjectives to indicate the domain under consideration. For example, both OTcl variables and interpreted variables refer to variables in the interpreted hierarchy. Similarly, both C++ functions and compiled functions refer to functions in the compiled hierarchy. Also, we will refer to the C++ compiler and the OTcl interpreter simply as the compiler and the interpreter, respectively.
- A "MyClass" object is a shorthand for an object of class `MyClass`. A "MyClass" pointer is a shorthand for a pointer which points to an object of class `MyClass`. For example, based on the statements "Queue q" and "Packet* p", "q" and "p" are said to be a "Queue" object and a "Packet pointer", respectively. Also, suppose further that class `DerivedClass` and `AnotherClass` derive from class `MyClass`. Then, the term a `MyClass` object refers to any object which is instantiated from class `MyClass` or its derived classes (i.e., `DerivedClass` or `AnotherClass`).
- *Objects* and *instances* are instantiated from a C++ class and an OTcl class, respectively. However, the book uses these two terms interchangeably.
- NS2 consists of two languages. Suppose that objects "A" and "B" are written in each language and correspond to one another. Then, "A" is said to be *the shadow object* of "B". Similarly "B" is said to be *the shadow object* of "A" .
- Consider two consecutive nodes in Fig. 3.2. In this configuration, an object (i.e., node) on the left always sends packets to the object on the right. The object on the right is referred to as a *downstream object* or a *target*, while the object on the right is referred to as an *upstream object*. In a general case, an object can have more than one target. However, a packet must be forwarded to one of these targets. From the perspective of an upstream object, a downstream object which receive the packet is also referred to as a *forwarding object*.

Notations

- As in C++, we use "`::`" to indicate the scope of functions and instprocs (e.g., `TcpAgent::send(...)`).
- Most of the texts in this book are written in regular letters. NS2 codes are written in "`this font type`". The quotation marks are omitted if it is clear from the context. For example, the Simulator is a general term for the simulating module in NS2, while a `Simulator` object is an object of class `Simulator`.
- A value contained in a variable is embraced with `<>`. For example, if a variable `var` stores an integer 7, `<var>` will be 7.
- A command prompt or an NS2 prompt is denoted by "`>>`" at the beginning of a line.
- In this book, codes shown in figures are *partially* excerpted from NS2 file. The file name from which the codes is excerpted is shown in the first line of the figure. For example, the codes in Program 2.1 are from file "`myfirst_ns.tcl`".
- A class name may consist of several words. All the words in a class name are capitalized. In the interpreted hierarchy, a derived class is named by having the name of its parent class and a slash character ("`/`") as a prefix, while that in the compiled hierarchy is named by having the name of its base class as a suffix. Examples of NS2 naming convention are given in Table 2.2.
- In the interpreted hierarchy, an instproc name is written in lower-case. If the instproc name consists of more than one word, each word except for the first one will be capitalized. In the compiled hierarchy, all the words are written in lower case and separated by an underscore "`_`" (see Table 2.2).
- The naming convention for variables is similar to that for functions and instprocs. However, the last character of the names of class variables in both the hierarchies is always an underscore ("`_`"; see Table 2.2). Note that this convention is only a guideline that a programmer should (but does not *have to*) follow.

Table 2.2. Examples of NS2 naming convention

	The interpreted hierarchy	The compiled hierarchy
Base class	`Agent`	`Agent`
Derived class	`Agent/TCP`	`TcpAgent`
Derived class (2nd level)	`Agent/Tcp/Reno`	`RenoTcpAgent`
Class functions	`installNext`	`install_next`
Class variables	`windowOption_`	`wnd_option_`

Exercise 2.1. Design C++ and OTcl classes (e.g., Class My TCP). Derive this class from the TCP Reno classes shown in Table 2.2. Use the convention defined above to name the class names, variables/instvars, and functions/instprocs in both the domain.

2.5 Running NS2 Simulation

2.5.1 NS2 Program Invocation

After the installation and/or recompilation (see Section 2.7), an executable file **ns** is created in the NS2 home directory. NS2 can be invoked by executing the following statement from the shell environment:

```
>>ns [<file>] [<args>]
```

where `<file>` and `<args>` are optional input argument. If no argument is given, the command will bring up an NS2 environment, where NS2 waits to interpret commands from the standard input (i.e., keyboard) line-by-line. If the first input argument `<file>` is given, NS2 will interpreted the input scripting `<file>` (i.e., a so-called Tcl simulation script) according to the Tcl syntax. The detail for writing a Tcl scripting file is given in Appendix A.1. Finally, the input arguments `<args>`, each separated by a white space, are fed to the Tcl file `<file>`. From within the file `<file>`, the input argument is stored in the built-in variable `argv` (see Appendix A.1.1).

2.5.2 Main NS2 Simulation Steps

The followings show the three key step guideline in defining a simulation scenario in a NS2:

Step 1: Simulation Design

The first step in simulating a network is to design the simulation. In this step, the users should determine the simulation purposes, network configuration and assumptions, the performance measures, and the type of expected results.

Step 2: Configuring and Running Simulation

This step implements the design in the first step. It consists of two phases:

- *Network configuration phase*: In this phase network components (e.g., node, TCP and UDP) are created and configured according to the simulation design. Also, the events such as data transfer are scheduled to start at a certain time.

• *Simulation Phase*: This phase starts the simulation which was configured in the Network Configuration Phase. It maintains the simulation clock and executes events chronologically. This phase usually runs until the simulation clock reached a threshold value specified in the Network Configuration Phase.

In most cases, it is convenient to define a simulation scenario in a Tcl scripting file (e.g., `<file>`) and feed the file as an input argument of an NS2 invocation (e.g., executing "`ns <file>`").

Step 3: Post Simulation Processing

The main tasks in this steps include verifying the integrity of the program and evaluating the performance of the simulated network. While the first task is referred to as *debugging*, the second one is achieved by properly collecting and compiling simulation results (see Chapter 13).

2.6 A Simulation Example

We demonstrate a network simulation through a simple example. Again, a simulation process consists of three steps.

Step 1: Simulation Design

Figure 2.3 shows the configuration of a network under consideration. The network consists of five nodes n0 to n4. In this scenario, node n0 sends constant-bit-rate (CBR) traffic to node n3, and node n1 transfers data to node n4 using

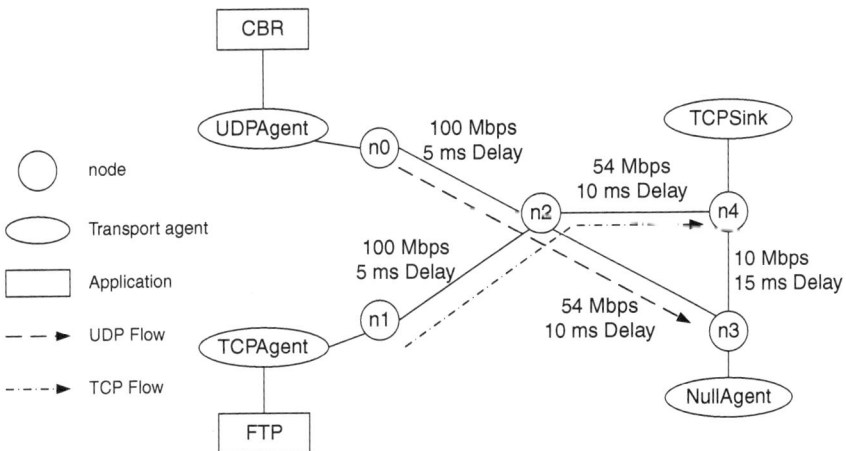

Fig. 2.3. A sample network topology.

a file transfer protocol (FTP). These two carried traffic sources are carried by transport layer protocols User Datagram Protocol (UDP) and Transmission Control Protocol (TCP), respectively. In NS2, the transmitting object of these two protocols are a UDP agent and a TCP agent, while the receivers are a Null agent and a TCP sink agent, respectively.

Step 2: Configuring and Running Simulation

Programs 2.1–2.2 show two portions of a Tcl simulation script which implements the scenario in Fig. 2.3.

Consider Program 2.1. This program creates a simulator instance in Line 1. It creates a trace file and a NAM trace file in Lines 2–3 and 4–5, respectively. It defines procedure finish{} in Lines 6–13. Finally, it creates nodes and links them together in Lines 14–18 and 19–24, respectively.

The Simulator is created in Line 1 by executing "new Simulator". The returned Simulator handle is stored in a variable ns. Lines 2 and 4 open files out.tr and out.nam, respectively, for writing. The variables myTrace and myNAM are the file handles for these two files, respectively. Lines 3 and 5 inform NS2 to collect all trace information for a regular trace and a NAM trace, respectively.

The procedure finish{} is invoked immediately before the simulation terminates. The keyword global informs the Tcl interpreter that the variables ns, myTrace, myNAM are those defined in the global scope (i.e., defined outside the procedure). Line 8 flushes the buffer of the packet tracing variables. Lines 9–10 close the file associated with handles myTrace and myNAM. Line 11 executes the statement "nam out.nam &" from the shell environment. Finally, Line 12 tells NS2 to exit with code 0.

Lines 14–18 creates Nodes using the instproc node of the Simulator whose handle is ns. Lines 19–23 connects each pair of nodes with a bi-directional link using an instproc duplex-link {src dst bw delay qtype} of class Simulator, where src is a beginning node, dst is an terminating node, bw is the link bandwidth, delay is the link propagation delay, and qtype is the type of the queues between the node src and the node dst. Similar to the instproc duplex-link{...}, Line 23 create a uni-directional link using an instproc simplex-link{...} of class Simulator. Finally, Line 24 sets the queue size of the queue between node n2 and node n3 to be 40 packets.

Next, consider the second portion of the Tcl simulation script in Program 2.2. A UDP connection, a CBR traffic source, a TCP connection, and an FTP session are created and configured in Lines 25–30, 31–34, 35–40, and 41–42, respectively. Lines 43–47 schedules discrete events. Finally, the simulator is started in Line 48 using the instproc run{} associated with the simulator handle ns.

To create a UDP connection, a sender udp and a receiver null are created in Lines 25 and 27, respectively. Taking a node and an agent as input

Program 2.1 First NS2 Program

```
# myfirst_ns.tcl
# Create a Simulator
1  set ns [new Simulator]

# Create a trace file
2  set mytrace [open out.tr w]
3  $ns trace-all $mytrace

# Create a NAM trace file
4  set myNAM [open out.nam w]
5  $ns namtrace-all $myNAM

# Define a procedure finish
6  proc finish { } {
7       global ns mytrace myNAM
8       $ns flush-trace
9       close $mytrace
10      close $myNAM
11      exec nam out.nam &
12      exit 0
13 }

# Create Nodes
14 set n0 [$ns node]
15 set n1 [$ns node]
16 set n2 [$ns node]
17 set n3 [$ns node]
18 set n4 [$ns node]

# Connect Nodes with Links
19 $ns duplex-link  $n0 $n2 100Mb 5ms DropTail
20 $ns duplex-link  $n1 $n2 100Mb 5ms DropTail
21 $ns duplex-link  $n2 $n4 54Mb 10ms DropTail
22 $ns duplex-link  $n2 $n3 54Mb 10ms DropTail
23 $ns simplex-link $n3 $n4 10Mb 15ms DropTail
24 $ns queue-limit $n2 $n3 40
```

argument, an instproc attach-agent{...} of class Simulator in Line 26 attaches a UDP agent udp and a node n0 together. Similarly, Line 28 attaches a Null agent null to a node n3. The instproc connect{from_agt to_agt} in Line 29 informs an agent from_agt to send the generated traffic to an agent to_agt. Finally, Line 30 sets the UDP flow ID to be 1. The construction of a TCP connection in Lines 35–40 is similar to that of a UDP connection in Lines 25–30.

Program 2.2 First NS2 Program (Continued)

```
#   Create a UDP agent
25 set udp [new Agent/UDP]
26 $ns attach-agent $n0 $udp
27 set null [new Agent/Null]
28 $ns attach-agent $n3 $null
29 $ns connect $udp $null
30 $udp set fid_ 1

#   Create a CBR traffic source
31 set cbr [new Application/Traffic/CBR]
32 $cbr attach-agent $udp
33 $cbr set packetSize_ 1000
34 $cbr set rate_ 2Mb

#   Create a TCP agent
35 set tcp [new Agent/TCP]
36 $ns attach-agent $n1 $tcp
37 set sink [new Agent/TCPSink]
38 $ns attach-agent $n4 $sink
39 $ns connect $tcp $sink
40 $tcp set fid_ 2

#   Create an FTP session
41 set ftp [new Application/FTP]
42 $ftp attach-agent $tcp

#   Schedule events
43 $ns at 0.05 "$ftp start"
44 $ns at 0.1  "$cbr start"
45 $ns at 60.0 "$ftp stop"
46 $ns at 60.5 "$cbr stop"
47 $ns at 61 "finish"

#   Start the simulation
48 $ns run
```

A CBR traffic source is created in Line 31. It is attached to a UDP agent udp in Line 32. The packet size and generation rate of the CBR connection are set to 1000 bytes and 2 Mbps, respectively. Similarly, an FTP session handle is created in Line 41 and is attached to a TCP agent tcp in Line 42.

In NS2, discrete events can be scheduled using an instproc at of class Simulator, which takes two input arguments: time and str. This instproc schedules an execution of str when the simulation time is time. Lines 43 and 44 start the FTP and CBR traffic at $0.05th$ second and $1st$ second, respectively. Lines 45 and 46 stop the FTP and CBR traffic at 60.0th second and 60.5th second, respectively. Line 47 terminates the simulation by invoking the procedure finish{} at 61*st* second. Note that the FTP and CBR traffic source can be started and stopped by invoking its commands start{} and stop{}, respectively.

We run the above simulation script by executing

```
>>ns myfirst_ns.tcl
```

from the shell environment. At the end of simulation, the trace files should be created and NAM should be running (since it is invoked from within the procedure finish{}).

Step 3: Post Simulation Processing–Packet Tracing

Packet tracing records the detail of packet flow during a simulation. It can be classified into a text-based packet tracing and a NAM packet tracing.

Text-Based Packet Tracing

Text-based packet tracing records the detail of packets passing through network checkpoints (e.g., nodes and queues). A part of the text-based trace obtained by running the above simulation (myfirst_ns.tcl) is shown below.

```
. . .
+ 0.110419 1 2 tcp 1040 ------- 2 1.0 4.0 5 12
+ 0.110419 1 2 tcp 1040 ------- 2 1.0 4.0 6 13
- 0.110431 1 2 tcp 1040 ------- 2 1.0 4.0 5 12
- 0.110514 1 2 tcp 1040 ------- 2 1.0 4.0 6 13
r 0.11308 0 2 cbr 1000 ------- 1 0.0 3.0 2 8
+ 0.11308 2 3 cbr 1000 ------- 1 0.0 3.0 2 8
- 0.11308 2 3 cbr 1000 ------- 1 0.0 3.0 2 8
r 0.11316 0 2 cbr 1000 ------- 1 0.0 3.0 3 9
+ 0.11316 2 3 cbr 1000 ------- 1 0.0 3.0 3 9
- 0.113228 2 3 cbr 1000 ------- 1 0.0 3.0 3 9
r 0.115228 2 3 cbr 1000 ------- 1 0.0 3.0 0 6
r 0.115348 1 2 tcp 1040 ------- 2 1.0 4.0 3 10
+ 0.115348 2 4 tcp 1040 ------- 2 1.0 4.0 3 10
- 0.115348 2 4 tcp 1040 ------- 2 1.0 4.0 3 10
r 0.115376 2 3 cbr 1000 ------- 1 0.0 3.0 1 7
r 0.115431 1 2 tcp 1040 ------- 2 1.0 4.0 4 11
. . .
```

Figure 2.4 shows the format of each trace line, which consists of 12 columns.

The general format of each trace line is shown in Fig. 2.4, where 12 columns make up a complete trace line. The *type identifier* field corresponds to four possible event types that a packet has experienced: r (received), + (enqueued), – (dequeued), and d (dropped). The *time* field denotes the time at which

Type Identifier	Time	Source Node	Destination Node	Packet Name	Packet Size	Flags	Flow ID	Source Address	Destination Address	Sequence Number	Packet Unique ID

Fig. 2.4. Format of each line in a normal trace file.

such event occurs. Fields 3 and 4 are the starting and the terminating nodes, respectively, of the link at which a certain event takes place. Fields 5 and 6 are packet type and packet size, respectively. The next field is a series of flags, indicating any abnormal behavior. Note the output "-------" denotes no flag. Following the flags is a packet flow ID. Fields 9 and 10 mark the source and the destination addresses, respectively, in the form of node.port. For correct packet assembly at the destination node, NS also specifies a packet sequence number in the second last field. Finally, to keep track of all packets, a packet unique ID is recorded in the last field.

Now, having this trace at hand would not be useful unless meaningful analysis is performed on the data. In post-simulation analysis, one usually extracts a subset of the data of interest and further analyzes it. For example, the average throughput associated with a specific link can be computed by extracting only the columns and fields associated to that link from the trace file. Two of the most popular languages that facilitate this process are AWK and Perl. The basic structures and usage of these languages are described in Appendix A.

Text-based packet tracing is activated by executing "$ns trace-all $file", where ns is the Simulator handle and file is a handle associated with the file which stores the tracing text. This statement simply informs NS2 of the need to trace packets. When an object is created, a tracing object is also created to collect the detail of traversing packets. Hence, the "trace-all" statement must be executed prior to object creation. We shall discuss the detail of text-based packet tracing later in Chapter 13.

Network AniMation (NAM) Trace

NAM trace is records simulation detail in a text file, and uses the text file the play back the simulation using animation. NAM trace is activated by the command "$ns namtrace-all $file", where ns is the Simulator handle and file is a handle associated with the file (e.g., out.nam in the above example) which stores the NAM trace information. After obtaining a NAM trace file, the animation can be initiated directly at the command prompt through the following command (See Line 11 in Program 2.2):

    >>nam filename.nam

Many visualization features are available in NAM. These features are for example animating colored packet flows, dragging and dropping nodes (positioning), labeling nodes at a specified instant, shaping the nodes, coloring a specific link, and monitoring a queue.

2.7 Including C++ Modules into NS2 and the *make* Utility

In developing an NS2 simulation, very often it is necessary to create the customized C++ modules to complement the existing libraries. As such, the developer is faced with the task of keeping track of all the created files as a part of NS2. When a change is made to one file, usually it requires recompilation of some other files that depend on it. Manual recompilation of each of such files may not be practical. In Unix, a utility tool called `make` is available to overcome such difficulties. In this section we introduce this tool and discuss how to use it in the context of NS2 simulation development.

As a Unix utility tool `make` is very useful for managing the development of software written in any compilable programming language including C++. Generally, the `make` program automatically keeps track of all the files created throughout the development process. By *keeping track*, we mean recompiling or relinking wherever interdependencies exist among these files, which may have been modified as a part of the development process.

2.7.1 An Invocation of a Make Utility

A "`make`" utility can be invoked form a UNIX shell with the following command:

```
>>make [-f mydescriptor]
```

where "`make`" is mandatory, while the text inside the bracket is optional. By default (i.e., without optional input arguments), the `make` utility recompiles and `relinks` the source codes according to what specified in the default descriptor file `Makefile`. If the descriptor file `mydescriptor` is specified, the utility is use this file in place of the default file `Makefile`.

2.7.2 A Make Descriptor File

A descriptor file contains an instructor of how the codes should be recompiled and relinked. Again, the default descriptor file is the file named "`Makefile`". A descriptor file contains the names of the files that make up the executable, their interdependencies, and how each file should be rebuilt or recompiled. Such descriptions are specified through a series of so-called "dependency rules". Each rule takes three components, i.e., targets, dependencies, and commands. The following is the format of the dependency rule:

```
<target1> [<target2> ...] : [<dependency1> ...]
    [<command>]
```

A target with a colon sign is mandatory. Everything else inside the brackets are optional. A target is usually the name of the file which needs to be *remade* if any modification is done to dependency files specified after the mandatory colon (`:`). If any change is noticed, the second line `executes` to regenerate the target file.

Example 2.2 (Example of a Descriptor File). Assume that we have a main executable file **channel** consisting of three separate source files named **main.c**, **fade.c**, and **model.c**. Also assume that **model.c** depends on **model.h**. The Makefile corresponding to this example is shown below.

```
# makefile of channel
channel : main.o fade.o model.o
    cc -o channel main.o fade.o model.o

main.o : main.c
    cc -c main.c

fade.o : fade.c
    cc -c fade.c

model.o : model.c model.h
    cc -c model.c

clean :
    rm main.o fade.o model.o
```

The first line is a comment beginning with a pound ("**#**") sign. When **make** is invoked, it starts checking the targets one by one. The target **channel** is examined first, and **make** finds that **channel** depends on the object files **main.o**, **fade.o**, and **model.o**. The **make** utility next checks to see if any of these object files is designated as a target file. If this is the case, **make** further checks the **main.o** object file's dependency, and finds that it depends on **main.c**. Again, **make** proceeds to check whether **main.c** is listed as a target. If not, the command under the **main.o** target is executed if any change is made to **main.c**. In the command line "cc -c main.c",[5] main.c is simply compiled to obtain the **main.o** object. Next, **make** proceeds in a similar manner with the **fade.o** and **model.o** targets. Once any of these object files is updated, **make** returns to the **channel** target and executes its command, which merely compiles all of its dependent objects. Finally, we note a special target known as *phony target* which is not really the name of any file in the dependency hierarchy. This target is "**clean**", and usually performs a housekeeping function such as cleaning up all the object files no longer needed after the compilation and linking.

In Example 2.2 we notice several occurrences of certain sequences such as **main.o fade.o model.o**. To avoid a repetitive typing, which may introduce typos or omissions, a macro can be defined to represent such a long sequence.

---

[5] The UNIX command "cc -c file.c" compiles the file **file.c** and creates an object file **file.o**, while the command "cc -o file.o" links the object file **file.o** and create an executable file **file**.

For example, we may define a macro to represent `main.o fade.o model.o` as follows:

```
OBJS = main.o fade.o model.o
```

After defining the macro, we refer to "`main.o fade.o model.o`" by either parentheses or curly brackets and precede that with a dollar sign (e.g., `$(OBJS)` or `${OBJS}`). With this macro, Example 2.2 becomes a bit more handy as shown in Example 2.3.

Example 2.3 (Example of makefile/Makefile with Macros.).

```
# makefile of channel
OBJS = main.o fade.o model.o
COM = cc
channel : ${OBJS}
    ${COM} -o channel ${OBJS}

main.o : main.c
    ${COM} -c main.c

fade.o : fade.c
    ${COM} -c fade.c

model.o : model.c model.h
    ${COM} -c model.c

clean :
    rm ${OBJS}
```

2.7.3 NS2 Descriptor File

The NS2 descriptor file is defined in a file `Makefile` located in the home directory of NS2. It contains the details needed to recompile and relink NS2. The key relevant details are those beginning with the following keywords.

- `INCLUDES = :` The items behind this keyword are the directory which should be *included* into the NS2 environment.
- `OBJ_CC =` and `OBJ_STL = :` The items behind these two keywords constitute the entire NS2 object files. When a new C++ module is developed, its corresponding object file name should be added here.
- `NS_TCL_LIB = :` The items bind this keywords are the Tcl file of NS2. Again, when a new OTcl module is developed, its corresponding Tcl file name should be added here.

Suppose a module consisting of C++ files `myc.cc` and `myc.h` and a Tcl file `mytcl.tcl`. Suppose further that these files are created in a directory `myfiles` under the NS2 home directory. Then this module can be incorporated into NS2 using the following steps:

(i) Include a string "`-I./myfiles`" into the Line beginning with `INCLUDES = ` in the `Makefile`.
(ii) Include a string "`myfile/myc.o`" into the Line beginning with `OBJ_CC = ` or `OBJ_STL = ` in the `Makefile`.
(iii) Include a string "`myfile/mytcl.tcl`" into the Line beginning with `NS_TCL_LIB = ` in the `Makefile`.
(iv) Run `make` from the shell.

After running "`make`", an executable file **ns** is created. We can now use this file **ns** to run simuation.

2.8 Chapter Summary

This chapter introduces Network Simulator (Version 2), NS2. In particular, information on the installation of NS2 in both Unix and Windows-based systems is provided. The basic architecture of NS2 is described. These materials are essential for understanding NS2 as a whole and would help to get one started working with NS2.

NS2 consists of OTcl and C++. The C++ objects are mapped to OTcl handles using TclCl. To run a simulation, a user needs to define a network scenario in a Tcl Simulation script, and feeds this script as an input to an executable file **ns**. During the simulation, the packet flow information can be collected through text-based tracing or NAM tracing. After the simulation, an AWK program or a perl program can be used to analyze a text-based trace file. The NAM program, on the other hand, utilizes a NAM trace file to replay the network simulation using animation.

Simulation using NS2 consists of three main steps. First, the simulation design is probably the most important step. Here, we need to clearly specify the objectives and assumptions of the simulation. Secondly, configuring and running simulation implements the concept designed in the first step. This step also includes configuring the simulation scenario and running simulation. The final step in a simulation is to collect the simulation result and trace the simulation if necessary.

Written mainly in C++, NS2 employs a `make` utility to compile the source code, to link the created object files, and create an executable file **ns**. It follows the instruction specified in the default descriptor file `Makefile`. The `make` utility provides a simple way to incorporate a newly developed modules into NS2. After developing a C++ source code, we simply add an object file name into the dependency, and re-run `make`.

3

Linkage Between OTcl and C++ in NS2

NS2 is an object oriented simulator written in OTcl and C++ languages. While OTcl acts as the frontend (i.e., user interface), C++ acts as the backend running the actual simulation (Fig. 3.1). As can be seen from Fig. 3.1, class hierarchies of both languages can be either standalone or linked together using an OTcl/C++ interface called TclCL [15]. There are two types of classes in each domain. The first type includes classes which are linked between the C++ and OTcl domains. In the literature, these OTcl and C++ class hierarchies are referred to as *the interpreted hierarchy* and *the compiled hierarchy*, respectively. The second type includes OTcl and C++ classes which are not linked together. These classes are neither a part of the interpreted hierarchy nor a part of compiled hierarchy. This chapter discusses how OTcl and C++ languages constitute NS2.

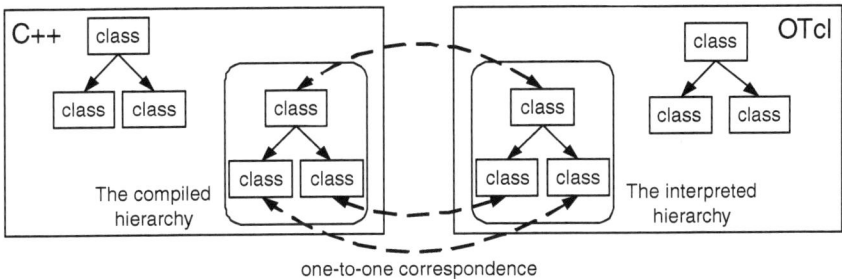

Fig. 3.1. Two language structure of NS2 [12]. Class hierarchies in both the languages may be standalone or linked together. OTcl and C++ class hierarchies which are linked together are called *the interpreted hierarchy* and *the compiled hierarchy*, respectively.

Written in C++, TclCL consists of the following six main classes:

- Class Tcl provides methods to access the interpreted hierarchy (from the compiled hierarchy; Defined in files ˜*tclcl*/tclcl.h and ˜*tclcl*/Tcl.cc).

T. Issariyakul, E. Hossain, *Introduction to Network Simulator NS2*,
DOI: 10.1007/978-0-387-71760-9_3, © Springer Science+Business Media, LLC 2009

- Class InstVar binds member variables in both the hierarchies together (Defined in file ˜*tclcl*/Tcl.cc).
- Class TclObject is the base class for all C++ simulation objects in the compiled hierarchy (defined in file ˜*tclcl*/Tcl.cc).
- Class TclClass maps class names in the interpreted hierarchy to class names in the compiled hierarchy (Defined in files ˜*tclcl*/tclcl.h and ˜*tclcl*/Tcl.cc).
- Class TclCommand provides a global access to the compiled hierarchy from the interpreted hierarchy (Defined in files ˜*tclcl*/tclcl.h and ˜*tclcl*/Tcl.cc).
- Class EmbeddedTcl translates OTcl scripts into C++ codes (Defined in files ˜*tclcl*/tclcl.h, ˜*tclcl*/Tcl.cc, and ˜*tclcl*/tclAppInit.cc).

The organization of this chapter is as follows. Section 3.1 describes the concept behind the two language structure of NS2. Sections 3.2 through 3.7 discuss the six main components of TclCL, namely, class Tcl, class InstVar, class TclObject, class Tclclass, class TclCommand, and class EmbeddedTcl. Finally, the chapter summary is given in Section 3.8.

3.1 The Two-Language Concept in NS2

Why two languages? Loosely speaking, NS2 uses OTcl to create and configure a network, and uses C++ to run simulation. All C++ codes need to be compiled and linked to create an executable file. Since the body of NS2 is fairly large, the compilation time is not negligible. A typical Pentium 4 computer requires few seconds (long enough to annoy most programmers) to compile and link the codes with a small change such as including "int i=0;" into the codes.

OTcl, on the other hand, is an interpreter, not a compiler. Any change in a OTcl file does not need compilation. Nevertheless, since OTcl does not convert all the codes into machine language, each line needs more execution time. In summary, C++ is fast to run but slow to change. It is suitable for running a large simulation. OTcl, on the other hand, is slow to run but fast to change. It is therefore suitable to run a small simulation over several repetitions (each may have different parameters). NS2 is constructed by combining the advantages of these two languages.

NS2 manual provides the following guidelines to choose a coding language:

- Use OTcl
  - for configuration, setup, or one time simulation, or
  - to run simulation with existing NS2 modules.

  This option is preferable for most beginners, since it does not involve complicated internal mechanism of NS2. Unfortunately, existing NS2 modules are fairly limited. This option is perhaps not sufficient for most researchers.
- Use C++
  - when you are dealing with a *packet*, or

– when you need to modify existing NS2 modules.
This option perhaps discourages most of the beginners from using NS2. This book particularly aims at helping the readers understand the structure of NS2 and feel more comfortable in modifying NS2 modules.

In principle, one can develop a C++ program in three styles. The first style–namely "Basic C++"–is the simplest form and involves basic C++ instructions only. This style has a flexibility problem, since any change in system parameters requires a compilation (which takes non-negligible time) of the entire program. Addressing the flexibility problem, the second coding style–namely "C++ coding with input arguments"–takes the system parameters as input arguments. As the system parameters change, we can simply change the input arguments, and do not need to recompile the entire program. The main problem of the second style is that the invocation could be quite lengthy for a large number of input arguments. The last coding style–"C++ coding with configuration files"–puts all system parameters in a configuration file, and has the C++ code read the system parameters from the configuration file. This style does not have the flexibility problem, and it facilitates program invocation. To change system parameters, we can simply change the content of the configuration file. In fact, this style acts as a foundation from which NS2 develops.

Recall from Section 2.5 that we write a *Tcl simulation script* and feed it as an input argument to NS2 when running a simulation (e.g., executing "ns myfirst_ns.tcl"). Here, "ns" is a C++ executable file obtained from the compilation, while myfirst_ns.tcl is an input configuration file specifying system parameters and configuration such as nodes, link, and how they are connected. Analogous to reading a script file through C++, NS2 reads the system configuration from the Tcl simulation script. Again, when we change the parameters, we do not need to re-compile the entire NS2 code. All we have to do is to modify the Tcl simulation script and re-run the simulation.

Example 3.1. Consider the network topology in Fig. 3.2. Define overall packet delivery delay as the time needed to carry a packet from the leftmost node to the rightmost node, where delay in link i is d_i and total number of nodes is num_nodes. We would like to measure the overall packet delivery delay and show the result on the screen.

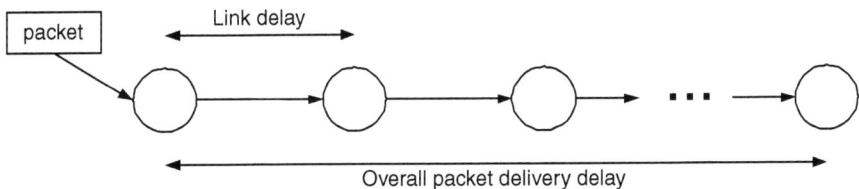

Fig. 3.2. A chain topology for network simulation.

Basic C++ Coding

Program 3.1 Basic C++ codes which simulates Example 3.1, where the delay for each of the links is 1 unit and the number of nodes is 11.

```
    //sim.cc
1   main(){
2       float delay = 0, d_i = 1;
3       int i, num_nodes = 11;
4       for(i = 1; i < num_nodes; i++)
5           delay += d_i;
6       printf("Overall Packet Delay is %2.1f seconds.\n",delay);
7   }
```

Suppose that every link has the same delay of 1 second (i.e., `d_i = 1` second for all `i`), and the number of nodes is 11 (`num_nodes = 11`). Program 3.1 shows the C++ codes written in this style (the filename is "`sim.cc`"). Since the link delay is fixed, we simply increment `delay` for `num_nodes-1` times (Lines 4-5). After compiling and linking file `sim.cc`, we obtain an executable file `sim`. By executing "`./sim`" at the command prompt, we will see the following statement on the screen:

```
>>./sim
Overall Packet Delay is 10.0 seconds.
```

Despite its simplicity, this coding style has a flexibility problem. Suppose link delay is changed to 2 seconds. Then, we need to modify, compile, and link the file `sim.cc` to create a new executable file `sim`. After that, we can run "`./sim`" to generate another result (for `d_i = 2` seconds).

C++ Coding with Input Arguments

We can avoid the above need for re-compilation and re-linking by feeding system parameters as input arguments of the program. Program 3.2 shows the codes which feed link delay and the number of nodes as the first and the second arguments, respectively. Line 1 specifies that the codes take input arguments. Variable `argc` is the number of input arguments. Variable `argv` is an argument vector which contains all input arguments provided by the caller (See the details on C++ coding with input arguments in [14]).

With this style, we only need to compile and link the program once. After obtaining an executable file `sim`, we can obtain results by simply changing the input arguments. For example,

```
>> ./sim 1 11
```

```
Overall Packet Delay is 10.0 seconds.
>> ./sim 2 11
Overall Packet Delay is 20.0 seconds.
```

Program 3.2 C++ coding with input arguments: C++ codes which simulate Example 3.1. The first and second arguments are link delay (d_i) and the number of nodes (num_nodes), respectively.

```
    //sim.cc
1   int main(int argc, char* argv[]) {
2       float delay = 0, d_i = atof(argv[0]);
3       int i, num_nodes = atoi(argv[1]);
4       for(i = 1; i < num_nodes; i++)
5           delay += d_i;
6       printf("Overall Packet Delay is %2.1f seconds\n",delay);
7   }
```

Though this coding style solves the flexibility problem, it suffers from a large number of input arguments. For example, if delays in all the links in Example 3.1 are different, we will have to type in all values of link delay every time we run the program.

C++ Coding with Configuration Files

Program 3.3 C++ coding style with configuration files: C++ code which simulates Example 3.1. A sample configuration file (config.txt) is given in Lines 10–11.

```
    //sim.cc
1   int main(int argc, char* argv[]) {
2       float delay = 0, d[10];
3       FILE* fp = fopen(argv[1],"w");
4       int i, num_nodes = readArgFromFile(fp,d);
5       for(i = 1; i < num_nodes; i++)
6           delay += d[i-1];
7       printf("Overall Packet Delay is %2.1f seconds\n", delay);
8       fclose(fp);
9   }

    //config.txt
10  Number of node = 11
11  Link delay = 1 2 3 4 5 6 7 8 9 10\vspace*{-3pt}
```

Program 3.3 shows C++ simulation codes for Example 3.1. The program takes only one input argument which is the configuration file name (See C++ file input/output in [14]). Function `readArgFromFile(fp,d)` reads the configuration file associated with a file pointer `fp`, and sets variables `num_node` and `d` (the details are not shown here). In this case, the configuration file (`config.txt`) is shown in Lines 10–11. When invoking "`./sim config.txt`", the screen will show the following result.

```
>>./sim config.txt
Overall Packet Delay is 55.0 seconds.
```

To change the system parameters, we can simply modify the file "`config.txt`". Clearly, this coding style removes the necessity for compiling the entire code and the lengthy invocation process.

3.2 Class Tcl

Class `Tcl` is a C++ class which acts as an interface to the OTcl domain. Declared in file `~tclcl/Tcl.cc`, it provides methods for the following operations:

(i) Obtain the Tcl instance (using function `instance`),
(ii) Invoke OTcl (instance) procedures from within the C++ domain (using functions `eval(...)`, `evalc(...)`, and `evalf(...)`),
(iii) Pass or receive results to/from the interpreter (using functions `result(...)` and `resultf(...)`),
(iv) Report error and quit the program in a uniform manner (using function `error(...)`), and
(v) Retrieve the reference to TclObjects (using functions `enter(...)`, `delete(...)`, and `lookup(...)`).

3.2.1 Obtain a Reference to the Tcl Instance

In C++, class functions are invoked through a class object (e.g., function "fn" can be invoked by "`object.fn`"). To invoke the above functions (e.g., `eval(...)` and `result(...)`) of class `Tcl`, we need to have an object of class `Tcl`. Class `Tcl` provides function "`instance()`" to obtain a static `Tcl` variable:

```
Tcl& tcl = Tcl::instance();
```

Here, function `instance()` of class `Tcl` returns the static variable `instance_` of class `Tcl`. Since it is static, in a simulation, there is only one Tcl object, `instance_`. Therefore, any attempt to retrieve a Tcl object using the above statement returns the same Tcl object. After obtaining the Tcl object, we can invoke class functions through the Tcl instance (e.g., `eval(...)` and `result(...)`).

3.2.2 Invoking a Tcl Procedure

We may need to invoke an OTcl instance procedure (instproc) when programming in C++. For example, we may obtain the current simulation time (see the definition in Chapter 4) by invoking instproc now{} of class Simulator in the interpreted hierarchy. Class Tcl provides four following functions to invoke OTcl procedures. For example, the following C++ codes tell OTcl to print out "Overall Packet Delay is 10.0 seconds" on the screen[1].

- Tcl::eval(char* str): executes the command string stored in a variable "str" through the interpreter. For example,

```
Tcl& tcl = Tcl::instance();
char s[128];
strcpy(s,"puts [Overall Packet Delay is 10.0 seconds]");
tcl.evalc(s);
```

- Tcl::evalc(const char* str): executes the command string "str". For example,

```
Tcl& tcl = Tcl::instance();
tcl.eval("puts [Overall Packet Delay is 10.0 seconds]");
```

This function is different from the former one in that the former one takes a "string variable" as an input variable (char*), while this one take a "string" as an input variable (const char*).

- Tcl::eval(): executes the command which has already been stored in the internal variable bp_. For example,

```
Tcl& tcl = Tcl::instance();
char s[128];
sprintf(tcl.buffer(),"puts [Overall
                  Packet Delay is 10.0 seconds]");
tcl.eval();
```

where tcl::buffer() returns the internal variable bp_. The third line above prints the string stored in the variable bp_.

- Instproc Tcl::evalf(const char* fmt,...): uses the format fmt of printf(...) in C++ to formulate a command string, and passes the formulated string to the interpreter. For example,

```
Tcl& tcl=Tcl::instance();
float delay = 10.0;
tcl.evalf("puts [Overall
                  Packet Delay is %2.1f seconds]",delay);
```

---

[1] You can save the sample codes in *any* C++ file and compile NS2 to create an executable ns file. When NS2 is invoked, the message "Overall Packet Delay is 10.0 seconds" should appear on the screen.

3.2.3 Pass or Receive Results to/from the Interpreter

After executing few statements, we may need to pass or receive values to/from the interpreter. For example, in Example 3.1, we may want to pass the value of the overall packet delivery delay to the interpreter instead of printing it to the screen. Class `Tcl` provides three functions to pass values back and forth between the two hierarchies.

- `Tcl::result(const char* fmt)`: passes the string `result` as the result to the interpreter. For example, the following statement returns 10 to the interpreter.

  ```
  Tcl& tcl=Tcl::instance();
  tcl.result("10");
  return TCL_OK;
  ```

- `Tcl::resultf(const char* result,...)`: uses the format of `printf` (...) in C++ to formulate a result string, and passes the formulated string to the interpreter.

  Example 3.2. Let command `returnDelay` of class `Chain` returns the value in C++ variable `delay` with one decimal digit to the interpreter. The implementation of the command `returnDelay` is given below:

  ```
  Tcl& tcl=Tcl::instance();
  tcl.resultf("%1.1f",delay);
  return TCL_OK;
  ```

  From OTcl, the following statement stores the value of the variable "delay" of the C++ `Chain` object in the variable "d".

  ```
  set chain [new Chain]
  set d [$chain returnDelay]
  ```

  sets the variable d to be the same as the variable `delay` in C++.

- `Tcl::result(void)`: retrieves the result from the interpreter as a string. For example, the following statements stores the value of the OTcl variable "d" in the C++ variable "delay".

  ```
  Tcl& tcl=Tcl::instance();
  tcl.evalc("$d");
  char* delay = tcl.result();
  ```

Class `Tcl` uses a private member variable `tcl_->result(...)` to pass results between the two hierarchies. Here, `tcl_` is a member of class `Tcl`, and is a pointer to a `Tcl_Interp` object. NS2 protects `tcl_->result(...)` from being accessed externally, and provides three functions to access this variable. Functions `Tcl::result(const char* result)` and `Tcl::resultf(const char* result)` set the value of `tcl->result(...)`. After setting the value of tcl->result(...) in C++, NS2 may return to OTcl with a certain return value (e.g., TCL_OK, TCL_ERROR) We will discuss the details of this return mechanism in Section 3.4.4. After returning to OTcl, the interpreter reads the value of tcl->result(...) for a certain purpose (e.g., setting the delay value or reporting error).

Similarly, after executing an OTcl statement (e.g., tcl.evalc("$delay")), the execution result is stored in the variable tcl->result(...). Function Tcl::result(void) in the compiled hierarchy returns the value stored in tcl->result(...) by the interpreter.

3.2.4 Reporting Error and Quitting the Program

Class Tcl provides function "error(...)" to exit the program in a uniform way. This function simply prints a string stored in "str" and tcl->result(...) to the screen, and exits with code 1.

    Tcl::error(const char* str)

The difference between Tcl::error(str) and return TCL_ERROR is as follows. Function Tcl::error(str) simply prints out the error message and exits. When returning TCL_ERROR, NS2 traps the error, which may occur in more than one point. In the end, the user may use the trapped errors to recover from the error, to locate the error, or to print all error messages in the error stack.

3.2.5 Retrieve the Reference to TclObjects

Recall that an interpreted object always has a shadow compiled object. In some cases, we may need to obtain a shadow compiled object which corresponds to an interpreted object.

NS2 creates the association between objects in two hierarchies by means of a hash table. Class Tcl provides the following functions to enter, delete, and retrieve an entry to/from the hash table.

- Tcl::enter(TclObject* o): inserts the object "*o" to the hash table, and associates "*o" to the OTcl name string stored in a protected variable name_. This function is invoked by function TclClass:create_shadow(...), when an object is created.
- Tcl::delete(TclObject* o): deletes the entry associated with TclObject "*o" from the hash table. This function is invoked by function TclClass:delete_shadow(...), when an object is destroyed.
- Tcl::lookup(char* str): returns the TclObject whose name is "str".

Example 3.3. Consider the C++ codes in Program 3.4. Here, argv[2] is an input argument passed from OTcl (in this case argv[2] is an interpreted object). Line 8 uses function TclObject::lookup(argv[2]) to retrieve the

---

Program 3.4 Function `Connector::command`.

```
//~ns/common/connector.cc
1  int Connector::command(int argc, const char*const* argv)
2  {
3      Tcl& tcl = Tcl::instance();
4      ...
5      if (argc == 3) {
6          if (strcmp(argv[1], "target") == 0) {
7              ...
8              target_ = (NsObject*)TclObject::lookup(argv[2]);
9              ...
10         }
11     ...
12     }
13     return (NsObject::command(argc, argv));
14 }
```

---

shadow compiled object corresponding to the interpreted object `argv[2]`.
The retrieved object is converted to an object of type `NsObject` and stored in
variable* `target_`. Note that the details of function `command` will be discussed
later in Section 3.4.4.

3.3 Class InstVar

Class `InstVar` acts as a glue which binds a member variable of a C++ class
to an instproc of an OTcl class. When a C++ variable is bound to an OTcl
instvar, any change in the C++ variable or the OTcl instvar will result in
an automatic update the OTcl instvar or the C++ variable, respectively.
NS2 supports variable binding for 5 following NS2 data types: real, integer, bandwidth, time, and boolean. These 5 data types are neither a C++
data type nor an OTcl data type.[2] They are defined here to facilitate NS2
value assignment. As shown in Table 3.1, these data types are defined in the
C++ classes `InstVarReal`, `InstVarInt`, `InstVarBandwidth`, `InstVarTime`,
and `InstVarBool`, respectively, which derive from class `InstVar`. Among these
five data types, real, bandwidth, and time data types make use of a double
C++ data type, while integer and boolean employ `int` and `bool` C++ data
types, respectively.

---

[2] As indicated in Appendix A.1.3, Tcl stores everything in strings. Therefore, OTcl
variables have no data type.

Table 3.1. OTcl bindable data types and C++ binding classes.

OTcl data type	C++ binding class
Real	InstVarReal
Integer	InstVarInt
Bandwidth	InstVarBandwidth
Time	InstVarTime
Boolean	InstVarBool

3.3.1 Real and Integer Variables

These two NS2 data types are specified as real-valued and integer-valued, respectively. Optionally, we can also use "e<x>" as "$\times 10^{<x>}$", where <x> denotes the value stored in the variable x.

Example 3.4. Let `realvar` and `intvar` be instvars of an OTcl object "obj" and be of real and integer NS2 data types, respectively. Different ways of setting[3] `realvar` and `intvar` to 1200 are shown below.

```
$obj set realvar 1.2e3
$obj set realvar 1200
$obj set intvar 1200
```

3.3.2 Bandwidth

Bandwidth is specified as real-valued. By default, the unit of bandwidth is bits per second (bps). Optionally, we can add the following suffixes to bandwidth setting.

- "k" or "K" means kilo or $\times 10^3$,
- "m" or "M" means mega or $\times 10^6$, and
- "B" changes the unit from bits per second to bytes per second.

NS2 only considers leading character of valid suffixes. Therefore, the suffixes "M" and "Mbps" are the same to NS2.

Example 3.5. In Example 3.4, let `bwvar` be an instvar of "obj" whose NS2 data type is bandwidth. The followings show different ways of setting `bwvar` to be 8 Mbps (megabits per second).

```
$obj set bwvar 8000000
$obj set bwvar 8m
$obj set bwvar 8Mbps
$obj set bwvar 800k
$obj set bwvar 1MB
```

---

[3] See the OTcl value assignment in Appendix A.2.

3.3.3 Time

Time is specified as real-valued. By default, the unit of time is second. Optionally, we can add the following suffixes to change the unit.

- "m" means milli or $\times 10^{-3}$,
- "n" means nano or $\times 10^{-9}$, and
- "p" means pico or $\times 10^{-12}$.

Again, NS2 only reads the leading character of valid suffixes. Therefore, the suffixes "p" and "ps" are the same to NS2.

Example 3.6. From Example 3.4, let `timevar` also be a time variable of "obj". The following shows different ways of setting `timevar` to be 2 micro seconds.

```
$obj set timevar 2m
$obj set timevar 2e-3
$obj set timevar 2e6n
$obj set timevar 2e9ps
```

3.3.4 Boolean

Boolean is specified as either `true` (or a positive number) or `false` (or a zero). A boolean variable will be `true` if the first letter of the value is greater than 0, is "t", or is "T". Otherwise, the variable will be `false`.

Example 3.7. In Example 3.4, let `boolvar` be a boolean variable of "obj". The following show different ways of setting `boolvar` to be `true` and `false`.

```
# set boolvar to be TRUE
$obj set boolvar 1
$obj set boolvar T
$obj set boolvar true
$obj set boolvar tasty
$obj set boolvar 20
$obj set boolvar 3.37
$obj set boolvar 4xxx

# set boolvar to be FALSE
$obj set boolvar 0
$obj set boolvar f
$obj set boolvar false
$obj set boolvar something
$obj set boolvar 0.9
$obj set boolvar -5.29
```

NS2 ignores all letters except for the first one. As can be seen from Example 3.7, there are several strange ways for setting a boolean variable (e.g., `tasty`, `something`, `-5.29`). For better understanding, the readers are encouraged to experiment with boolean variable `debug_` and real variable `rate_` in the following codes[4]:

```
# Create a Simulator instance
set ns [new Simulator]

# Create an error model object
set err [new ErrorModel]

# Set values for class variables
$err set debug_ something
$err set rate_ 12e3

# Show the results
puts "debug_(bool) is [$err set debug_]"
puts "rate_(double) is [$err set rate_]"
```

The results of execution of the above codes are as follows:

```
>>debug_(bool) is 0
>>rate_(double) is 12000
```

After assigning a value to an OTcl variable, NS2 converts the string value to the corresponding type in C++. Except for boolean, NS2 converts the string to either `double` or `int`. During the conversion, valid suffixes are also converted (e.g., "M" is converted by multiplying 10^6 to the value). For boolean data type, NS2 retrieves the first character in the string and throws away all other characters. If the retrieved character is an integer, NS2 will do nothing. If the retrieved character is a non-integer, NS2 will convert the character to one if it is "t" or "T" and to zero otherwise. After converting the string to a one-digit integer, NS2 casts the converted integer to boolean and updates the bound compiled variable.

3.4 Class TclObject

Class `TclObject` provides an instruction to create a compiled shadow object, when an interpreted object is created. The C++ class `TclObject` is mapped to the OTcl class `SplitObject`. These two classes are the base classes from which all classes (excluding the standalone classes) in their hierarchies develop. When an object is instantiated from the OTcl domain, the constructor of class `SplitObject` is invoked to initialize the object. One of the initialization is

---

[4] Save the codes to a file (e.g., `test.tcl`) and run it (e.g., `ns test.tcl`).

the shadow object construction whose instruction, which will be discuss in this section, Section 3.4.1 shows how a TclObject is referred to in both the hierarchies. Section 3.4.2 explains the shadow object creation and deletion procedure. The variable binding process performed during object construction is discussed in Section 3.4.3. Finally, Section 3.4.4 discusses a special function command(...), which provides an access to the compiled class from the OTcl domain.

3.4.1 Reference to a TclObject

OTcl and C++ employ different method to access their objects. As a compiler, C++ directly accesses the memory space allocated for a certain object (e.g., 0xd6f9c0). As an interpreter, OTcl does not directly access the memory. Rather, it uses a string (e.g., _o10) as a reference (or a handle)[5] to the object. By convention, the name string of a SplitObject is of format _<NNN>, where <NNN> is a number uniquely generated for each SplitObject.

Example 3.8. Let the variables c_obj and otcl_obj be C++ and OTcl objects, respectively. Table 3.2 shows examples of the reference value of C++ and OTcl objects.

Table 3.2. Examples of reference to (or handle of) TclObjects.

Domain	Variable name	Example
C++	c_object	0xd6f9c0
OTcl	otcl_object	_o10

We can see the format of a value stored in an OTcl object by running the following codes[6]:

```
set ns [new Simulator]
set tcp [new Agent/TCP]
puts "The value of tcp is $tcp"
```

which show the following line on the screen.

```
The value of tcp is _o10
```

3.4.2 Creating and Destroying a Shadow TclObject

In most cases, objects are created and destroyed in the OTcl domain, or more precisely in a Tcl simulation script. Again, the OTcl commands create and

---

[5] In NS2, the term "handle" means the object itself.
[6] Put the sample codes in a file (e.g, test.tcl) and run the file (e.g., ns test.tcl).

`destroy` can be used to create and destroy, respectively, a standalone OTcl object. However, these commands are rarely used in NS2, since they do not create the shadow compiled object. In NS2, the global procedures `new{...}` and `delete{...}` are used to create and delete, respectively, an OTcl object as well as a shadow compiled object.

Creating a TclObject

A TclObject is created by using the global procedure `new{...}`, whose syntax is

> `new<classname> [<args>]`

The details of procedure "`new{...}`" are shown in Program 3.5. The procedure "`new{className args}`" takes two input arguments. The first argument `<className>` (mandatory) is the OTcl class name. The subsequent arguments `<args>` (optional) is fed as input arguments to the OTcl constructor. The procedure "`new{className args}`" creates an object whose OTcl class is `<className>` as well as its corresponding shadow compiled object. It will return the reference string (Line 11) if the construction process is successful. Otherwise, it will show an error message on the screen (Line 9).

Program 3.5 Global instance procedure `new`.

```
    //~tclcl/tcl-object.tcl
1   proc new { className args } {
2       set o [SplitObject getid]
3       if [catch "$className create $o $args" msg] {
4           if [string match "__FAILED_SHADOW_OBJECT_" $msg] {
5               delete $o
6               return ""
7           }
8           global errorInfo
9           error "class $className: constructor failed:
                                        $msg" $errorInfo
10      }
11      return $o
12  }
```

The internal mechanism of the procedure "`new{className args}`" proceeds as follows. First, Line 2 retrieves a reference string for an object, and stores the string in variable "o". Instproc `getid{}` of class `SplitObject` creates a reference string according to the naming format defined in Section 3.4.2. Next, Line 3 creates an object whose OTcl class is `className` and associates the created object with the string stored in "o". Finally, if the object is successfully created, Line 11 returns the reference string "o" to the caller[7]. Otherwise, an error message (Line 9) will be shown on the screen.

The OTcl command **create** in Line 3 invokes instproc alloc{...} to allocate a memory space for an object of class className, and instproc init{...} to initialize the object. In most cases, instproc init{...} is referred to an OTcl constructor. Each class overrides function init{...} and defines its own initialization in this function.

Program 3.6 Samples of object constructor: Classes Agent/TCP and SimpleLink.

```
   //~ns/tcl/lib/ns-agent.tcl
1  Agent/TCP instproc init {} {
2      eval $self next
3      set ns [Simulator instance]
4      $ns create-eventtrace Event $self
5  }

   //~tclcl/tcl-object.tcl
6  SplitObject instproc init args {
7      $self next
8      if [catch "$self create-shadow $args"] {
9          error "__FAILED_SHADOW_OBJECT_" ""
10     }
11 }
```

Program 3.6 shows an example of the OTcl constructor. Instproc next{...} in Line 2 invokes the instproc with the same name (i.e., init{...} in this case) of the parent class. This is a common concept in an Object Oriented Programming, where the constructor of the parent class needs to be called earlier. The construction therefore moves up the hierarchy until it reaches class SplitObject (see Lines 6–11 in Program 3.6). Here, Line 8 creates a shadow compiled object by invoking the command **create-shadow**, which will be discussed later in Section 3.5.

We now conclude this section with an example of a creation of a Agent/TCP OTcl object as well as its shadow compiled object.

Example 3.9. To create an OTcl Agent/TCP object, we can execute "new Agent/TCP" from a Tcl Simulation Script. In the interpreted hierarchy, class Agent/TCP derives from class Agent, which derives from class SplitObject. In the compiled hierarchy, these three classes correspond to class TcpAgent, Agent, and TclObject, respectively.

---

[7] Note that Line 11 returns a reference string stored in o, not the variable o. Hence the procedure **new** returns a reference string stored in the variable o.

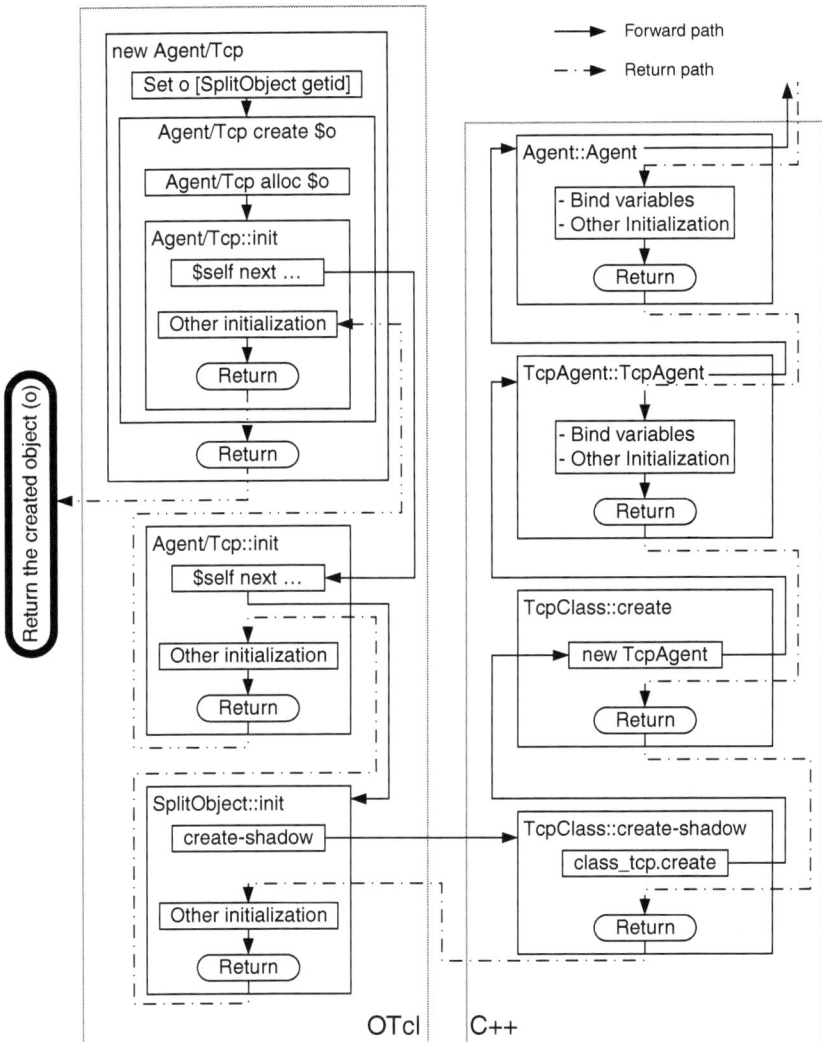

Fig. 3.3. Object creation diagram: Class **Agent/TCP** derives from class **Agent**, which derives from class **SplitObject**.

Figure 3.3 shows the creation process of an object (o) of class **Agent/TCP**. Again, the first step is to retrieve a reference string by invoking instproc **getid{}** of class **SplitObject**. The next step is to invoke instproc **init{...}** up the hierarchy. On the top level, class **SplitObject** invokes command **create-shadow** to create a shadow compiled object (on the right hand side of Fig. 3.3 which will be discussed in Section 3.5).

After returning from instproc **create-shadow**, the process performs the rest of initialization and moves (or returns) down the interpreted hierarchy

until it reaches class `Agent/TCP`. Then, it returns to procedures `create{...}`
and `new{...}`, respectively, where the reference string corresponding to the
created object "o" is returned to the caller (of procedure `new{...}`).

Note that the above procedures are used to create an interpreted TclObject
which is linked to the compiled hierarchy. Standalone C++ or OTcl objects
do not need any shadow object, and do not have to go through the above
procedures. They can be constructed in a normal way.

Destroying a TclObject

OTcl uses instproc `delete{...}` to destroy an interpreted object as well as its
shadow compiled object (by invoking instproc `delete-shadow`). Program 3.7
shows a sample usage of instproc `delete{...}`. Instproc `Simulator::use-sch`
`eduler{...}` removes the existing scheduler (if any; Line 3) by using instproc `delete{...}`, and creates an object of class `Scheduler/$type` using the global procedure `new{...}`. We will discuss the details of instproc
`Simulator::use-scheduler{...}` in Chapter 4.

Program 3.7 An example usage of global procedures `new` and `delete`.

```
      //~ns/tcl/lib/ns-lib.tcl
1   Simulator instproc use-scheduler type {
2       $self instvar scheduler_
3       delete $scheduler_
4       set scheduler_ [new Scheduler/$type]
5   }
```

3.4.3 Binding Variables in the Compiled and Interpreted Hierarchies

In general, both interpreted and compiled objects have their own class variables, and they are not allowed to directly access one another's class variables.
NS2, therefore, provides a mechanism which binds class variables in both hierarchies together. After the binding, a change in a class variable in one hierarchy
will result in an automatic change in the bound variable in another hierarchy.

Binding Variables in Both Hierarchies

NS2 binds an interpreted class variable to a compiled variable during shadow
object construction. More specifically, class `TclObject` invokes the following
functions in the constructor to bind variables in both hierarchies (see file
~*tclcl*/tclcl.h).

```
bind("iname",&cname)
bind_bw("iname",&cname)
bind_time("iname",&cname)
bind_bool("iname",&cname)
```

where iname and cname are the names of the class variables in the interpreted and compiled hierarchies, respectively. Essentially, the first and second arguments of the above functions are the name string of the interpreted variable and the address of the compiled variable, respectively.

Example 3.10. Let class Test in both hierarchies be bound together. Let icount_, idelay_, ispeed_, ivirtual_time_, iis_running_ be OTcl class variables whose types are integer, real, bandwidth, time, and boolean, respectively. The following codes show declaration and the constructor of C++ class Test.

```
class Test { /* Declaration */
    public:
    int count_;
    double delay_,virtual_time_,speed_;
    bool is_running_;
    Test() { /* Constructor */
        bind("icount_",&count_);
        bind("idelay_",&delay_);
        bind_bw("ispeed_",&speed_);
        bind_time("ivirtual_time_",&virtual_time_);
        bind_bool("iis_running_",&is_running_);
    };
};
```

All class variables are bound in the compiled constructor (i.e., Test()). By convention, we use the same variable name for both hierarchies. Here, however, we would like to show that bound variables do not need to have the same names.

Setting the Default Values

NS2 sets the value of bound variables as specified in file ~*ns*/tcl/lib/ns-default.tcl. The syntax for setting a default value is similar to the value assignment syntax in OTcl. That is,

```
<className> set <instvar> <def_value>
```

which sets the instvar <instvar> of class <className> to be <def_value>. As an example, a part of file ~*ns*/tcl/lib/ns-default.tcl is shown in Program 3.8.

To set the default values for the variables, NS2 invokes instproc init-instvar{...} of class SplitObject (see file ~*tclcl*/tcl-object.tcl). Instproc init-instvar{...} takes variables' default values from file ~*ns*/tcl/lib/

Program 3.8 An example for specifying default values in NS2: A part of file ˜*ns*/tcl/lib/ns-default.tcl.

```
    //~ns/tcl/lib/ns-default.tcl
 1  ErrorModel set enable_ 1
 2  ErrorModel set markecn_ false
 3  ErrorModel set delay_pkt_ false
 4  ErrorModel set delay_ 0
 5  ErrorModel set rate_ 0
 6  ErrorModel set bandwidth_ 2Mb
 7  ErrorModel set debug_ false

 8  Classifier set offset_ 0
 9  Classifier set shift_ 0
10  Classifier set mask_ 0xffffffff
11  Classifier set debug_ false
```

ns-default.tcl, and assigns them to the bound variables. If we bind a variable but do not specify the default value, instproc `SplitObject::warn-instvar` {. . .} invoked from within `SplitObject::init-instvar`{. . .} will show a warning message on the screen. A warning message will *not* be shown, if a default value is assigned to an invalid variable (e.g., not-bound or does not exist).

3.4.4 OTcl Commands

Section 3.2.2 showed an approach to access the interpreted hierarchy from the compiled hierarchy. This section discusses the reverse: a method to access the compiled hierarchy from the interpreted hierarchy called "*command*".

Review of Instance Procedure Invocation Mechanism

Before we proceed further, let us review the OTcl instproc invocation mechanism. An instproc is invoked according to the following syntax:

    $obj <instproc> [<args>]

where the instproc name `<instproc>` and the input argument `<args>` are mandatory and optional, respectively, for such an invocation. The internal mechanism of the above instproc invocation proceeds as follows:

(i) Look for a matching instproc in the object class. If found, execute the matched instproc and return. If not, proceed to the next step.
(ii) Look for instproc "`unknown`{. . .}". If found, execute "`unknown`{. . .}" and return. If not, proceed to the next step. The instproc "`unknown`{. . .}" is the default instproc which will be invoked if no matching instproc is found.

(iii) Repeat steps (i) and (ii) for the base class of the object.

(iv) If the top class is reached but neither the input instproc nor the instproc unknown is found, report an error and exit the program.

OTcl Command Invocation

The syntax of a command is the same as that of an instproc, i.e.,

    $obj <cmd_name> [<args>]

The main difference is that <instproc> is replaced with <cmd_name>. Since the syntax for invoking a command is the same as that for invoking an instproc, OTcl executes the command as if it is an instproc. In the following, we will explain the command invocation mechanism of an OTcl Agent/TCP object (see Program 3.9). Figure 3.4 shows the internal mechanism of the command invocation process, which proceeds as follows:

  (i) Execute the statement "$tcp <cmd_name> <args>".

 (ii) Look for an instproc <cmd_name> in the OTcl class Agent/TCP. If found, invoke the instproc and complete the process. Otherwise, proceed to the next step.

(iii) Look for an instproc unknown{...} in the OTcl class Agent/TCP. If found, invoke the instproc unknown{...} and complete the process. Otherwise, proceed to the next step.

(iv) Repeat steps (ii) and (iii) until reaching class SplitObject. If the instprocs unknown{...} is not found in any class in the inheritance tree, the following statement will be executed.

    SplitObject unknown

The instproc unknown{...} of class SplitObject is defined in file ~tclcl/tcl-object.tcl. Here, the statement "$self cmd $args" is executed, where args are the input arguments of instproc unknown{...}. Based on the above invocation, this statement interpolates to

    SplitObject cmd <cmd_args>

where <cmd_args> is "<cmd_name> <args>".

 (v) Instproc cmd passes the entire statement (i.e., "cmd <cmd_args>") as an input argument vector (argv) to function "command(argc,argv)" of the shadow object (TcpAgent in this case).

As shown in Program 3.9, function "command(argc,argv)" always takes two input arguments. The second input argument (argv) is an argument vector, which is an array of strings containing arguments passed from the instproc "cmd". The first input argument (argc) is the total number of input arguments (i.e., the number of non-empty elements of argv). The first and second elements of argv are "cmd" and the command name (<cmd_name>), respectively. The subsequent elements contain the input arguments (<args>) of the original invocation, each of which is separated by one or more white spaces (see Table 3.3).

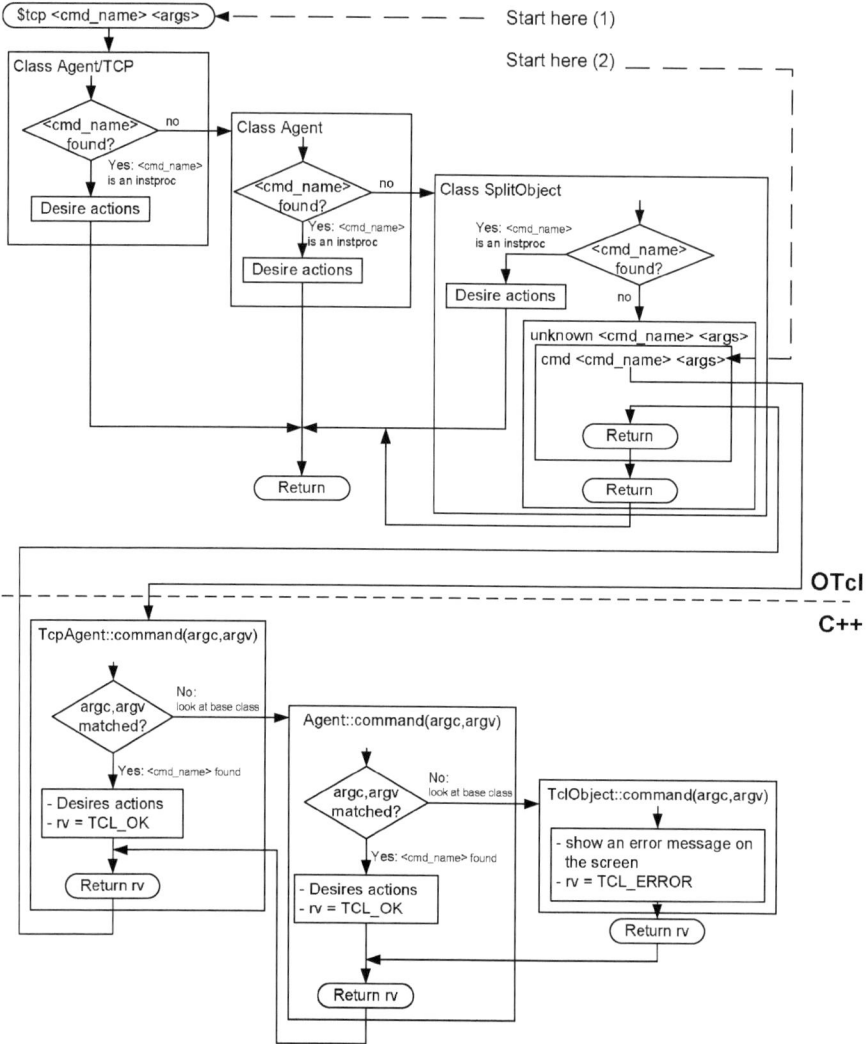

Fig. 3.4. Command invocation process.

(vi) Function command(argc,argv) checks for the matching number of arguments (stored in argc) and command name (stored in argv[1]). If found, it takes the desired actions (e.g., Lines 6–7 in Program 3.9), and returns TCL_OK. If no criterion matches with (argc,argv), it will skip to the last line (Line 12).

(vii) Line 12 in Program 3.9 invokes function command(argc,argv) of the base class (i.e., class Agent::command(argc,argv)), feeding (argc,argv) as input arguments.

Program 3.9 Function `TcpAgent::command`.

```
     //~ns/tcp/tcp.cc
1    int TcpAgent::command(int argc, const char*const* argv)
2    {
3        ...
4        if (argc == 3) {
5            if (strcmp(argv[1], "eventtrace") == 0) {
6                et_ = (EventTrace *)TclObject::lookup(argv[2]);
7                return (TCL_OK);
8            }
9            ...
10       }
11       ...
12       return (Agent::command(argc, argv));
13   }
```

Table 3.3. Description of elements of array `argv` of function `command`.

index (i)	Element (`argv[i]`)
1	cmd
2	The command name (`<cmd_name>`)
3	The first input argument in `<args>`
4	The second input argument in `<args>`
⋮	⋮

(viii) Repeat steps (vi) and (vii) up the hierarchy until the criterion is matched. If the process reaches class `TclObject` and the criterion does not match, function `command` of class `TclObject` will report an error (e.g., no such method, requires additional args), and return `TCL_ERROR` (see file ~*tclcl*/Tcl.cc).

(xi) Return down the class hierarchy. When reaching C++ class `TcpAgent`, return to OTcl (instprocs `cmd` and `unknown{...}`, respectively) with a return value (e.g., `TCL_OK` or `TCL_ERROR`), and complete the command invocation.

An Alternative for OTcl Command Invocation

In the last subsection, we invoked an OTcl command by executing

    $tcp <cmd_name> <args>

which starts from position (1) in Fig. 3.4. Alternatively, we can also invoke a command using the following syntax:

    $tcp cmd <cmd_name> <args>

which starts from position (2) in Fig. 3.4.

The latter (position (2)) invocation method avoids the ambiguity when OTcl defines an instproc whose name is the same as the OTcl command name. Suppose an OTcl command <cmd_name> associated with an object "tcp" has an implementation in the C++ class TcpAgent. Suppose further that an instproc <cmd_name> (same name) is also defined in OTcl class Agent/TCP. When invoking "$tcp <cmd_name> <args>", NS2 will perform the actions specified in the OTcl instproc <cmd_name>. To invoke the OTcl command <cmd_name> whose implementation is in C++, we need to invoke "$tcp cmd <cmd_name> <args>". Since instproc cmd is defined solely in class SplitObject, this invocation avoids the ambiguity of OTcl command and instproc names.

OTcl Command Returning Mechanism

After performing the desired actions specified in C++, NS2 returns to OTcl with a certain return value. In file nsallinone-2.30/tcl8.4.13/generic/tcl.h, NS2 defines five following return values (as 0–5), as specified in Program 3.10, which inform the interpreter of the command invocation result.

Program 3.10 Return values in NS2.

```
    //nsallinone-2.30/tcl8.4.13/generic/tcl.h
1   #define TCL_OK        0
2   #define TCL_ERROR     1
3   #define TCL_RETURN    2
4   #define TCL_BREAK     3
5   #define TCL_CONTINUE  4
```

- TCL_OK: The command completes successfully.
- TCL_ERROR: The command does not complete successfully. The interpreter will explain the reason for the error.
- TCL_RETURN: After returning from C++, the interpreter exits (or returns from) the current instproc without performing the rest of instproc.
- TCL_BREAK: After returning from C++, the interpreter breaks the current loop. This is similar to executing C++ keyword break, but the results prevail to the OTcl domain.
- TCL_CONTINUE: After returning from C++, the interpreter continues to the next iteration. This is similar to executing C++ keyword continue, but the results prevail to the OTcl domain..

Among these five types, TCL_OK and TCL_ERROR are the most common ones. If C++ returns TCL_OK, the interperter may read the value passed from the C++ domain. Recalling from Section 3.2.3, the interpreter does not read the return value, but it reads the value specified in the statement. The return code `TCL_OK` only tells OTcl that the value stored by the statement `tcl.result(...)` is valid.

If an OTcl command returns `TCL_ERROR`, on the other hand, the interpreter will invoke procedure `tkerror` (defined in file `~tclcl/tcl-object.tcl`), which simply shows an error on the screen.

Exercise 3.11. What are the differences among a C++ function, an OTcl instproc, and an OTcl command?

3.5 Class TclClass

When a TclObject is created, NS2 automatically constructs a shadow compiled object. In Section 3.4.2, we have explained the TclObject creation mechanism. We have mentioned that class `TclClass` is responsible for the shadow object creation process. We now explain the details of class `TclClass` as well as the shadow object creation process.

3.5.1 An Overview of Class `TclClass`

Class `TclClass` is mainly responsible for creating a shadow object in the compiled hierarchy. It maps an OTcl class to a C++ static mapping variable, and provides a method to create a shadow object in the compiled hierarchy. As an example, Program 3.11 shows the details of class `TcpClass`, which maps class `Agent/TCP` in the interpreted hierarchy to the static mapping variable `class_tcp` in the compiled hierarchy.

Program 3.11 Declaration and implementation of class `TcpClass`.

```
    //~ns/tcp/tcp.cc
1   static class TcpClass : public TclClass {
2   public:
3       TcpClass() : TclClass("Agent/TCP") {}
4       TclObject* create(int , const char*const*) {
5           return (new TcpAgent());
6       }
7   } class_tcp;
```

Unlike other classes, a child class of class `TclClass` is declared, implemented, and instantiated (e.g., of variable `class_tcp` in Line 7) in the same place. From Program 3.11, a child class of class `TclClass` consists of only two functions: the constructor (`TcpClass` in Line 3) and function `create(...)` (Lines 4–6) which creates a shadow object. To construct a shadow object for an OTcl object of class `Agent/TCP`, we need to perform the following actions in the compiled hierarchy:

(i) Create a shadow compiled class (e.g., `TcpAgent`).
(ii) Derive a mapping class (e.g., `TcpClass`) from class `TclClass`.
(iii) Instantiate a static mapping variable (e.g., `class_tcp`).
(iv) Define the constructor of the mapping class (Line 3 in Program 3.11). Feed the OTcl class name (e.g., `Agent/TCP`) as an input argument to the base constructor (i.e., class `TclClass`).
(v) Define function "`create(...)`" to construct a shadow compiled object; Invoke "`new`" to create a shadow compiled object (e.g., `new TcpAgent`) and return the created object to the caller (Line 5 in Program 3.11).

3.5.2 TclObject Creation

We now explain the entire TclObject creation process. Once again, consider Fig. 3.3. The TclObject creation process proceeds as follows:

- Create an OTcl object as in Section 3.4.2.
- Invoke instproc `create-shadow` of class `TclClass` (see file ˜*tclcl*/Tcl.cc).
- From within function `create_shadow(...)`, invoke function `create(...)` of class `TcpClass`.
- In Program 3.11, function `create(...)` in Line 5 executes "`new TcpAgent`" and returns the created object to the caller.
- Construct a `TcpAgent` object, by calling the constructor of its parent classes (`Agent` and `TclObject`).
- Construct `Agent` object. This includes binding all variables to those in the interpreted hierarchy.[8]
- Return to class `TcpAgent`. Construct the `TcpAgent` object, and bind all variables to those in the interpreted hierarchy.
- Return the created shadow object to instproc `SplitObject::init{...}`, and proceed as specified in Section 3.4.2.

3.5.3 Naming Convention for Class `TclClass`

The convention to name a class derived from class `TclClass` and the corresponding static variable are described now. First, every class derives directly from class `TclClass`, irrespective of its class hierarchy. For example, class `RenoTcpAgent` derives from class `TcpAgent`. However, their mapping classes `RenoTcpClass` and `TcpClass` derive from class `TclClass`.

Secondly, the naming convention is very similar to the C++ variable naming convention. In most cases, we simply name the mapping class by attaching the word `Class` to the C++ class name. The static mapping variable is named by attaching the word "`class_`" to the front. Table 3.4 shows few examples of the above naming convention.

---

[8] Recall from Section 3.4.3 that NS2 binds variables of both hierarchies in the constructor.

Table 3.4. Examples of naming convention for class `TclClass`.

TclObject	SplitObject	Mapping class	Mapping variable
TcpAgent	Agent/TCP	TcpClass	class_tcp
RenoTcpAgent	Agent/TCP/Reno	RenoTcpClass	class_reno
DropTail	Queue/DropTail	DropTailClass	class_drop_tail

3.5.4 Instantiation of Mapping Variables

At the startup, NS2 instantiates all static mapping variables. Here, class `TclClass` stores the OTcl class names in its member variable `classname_` and stores all mapping variables to its linked list "`all_`". After all mapping variables are inserted into the linked list, function `TclClass::bind(...)` is invoked. Function `bind(...)` registers all mapping variables in "`all_`" into the system, and creates the interpreted class hierarchy. Function `bind(...)` also binds instprocs `create-shadow` and `delete-shadow` to functions `create_shadow(...)` and `delete_shadow(...)` of the mapping classes (e.g., `TcpClass`[9]), respectively. After this point, NS2 recognizes all OTcl class names. Creation of an OTcl object will follow the procedures specified in Sections 3.4.2 and 3.5.2.

Exercise 3.12. What are the major differences among classes `TclObject`, `TclClass`, and `InstVar`? Explain their roles during an object creating process.

3.6 Class TclCommand

As discussed in Section 3.4.4, OTcl command is a method to access the compiled hierarchy form the interpreted hierarchy. This section discusses another method called TclCommand to do the same. The main difference of OTcl command and TclCommands is as follows. Each OTcl command is associated with an OTcl/C++ class and cannot be invoked independently. Each TclCommand, on the other hand, is not bound to any class and is available globally. Since TclCommands violate the object oriented concept, it is not advisable to create this type of commands.

3.6.1 Invoking a TclCommand

A TclCommand is invoked as if it is a global OTcl procedure. We will explain how to invoke a TclCommand through Example 3.13.

Example 3.13. Consider the TclCommands `ns-version` and `ns-random`, specified in file `~ns`/common/misc.cc.

---

[9] In fact, class `TcpClass` inherits functions `create_shadow` and `delete_shadow` from class `TclClass`.

- TclCommand **ns-version** takes no argument and returns NS2 version.
- TclCommand **ns-random** returns a random number uniformly distributed in $[0, 2^{31} - 1]$ when no argument is specified. If an input argument is given, it will be used to set the seed of the random value generator.

These two TclCommands can be invoked globally. For example,

```
>>ns-version
2.30
>>ns-random
729236
>>ns-random
1193744747
### TERMINATE NS2 ###
>>ns-random
729236
>>ns-random
1193744747
### TERMINATE NS2 ###
>>ns-random 101
101
>>ns-random
72520690
>>ns-random
308637100
```

By executing **ns-version**, the version (2.30) of NS2 is shown on the screen. TclCommand **ns-random** with no argument returns a random number (e.g., 729236, 1193744747, $\cdots$). In NS2, a random number is generated by picking a number from a sequence of pseudo-random numbers. A random seed specifies the starting position in the sequence. By default, NS2 always sets random seed to be 0. *The results from multiple simulations would be the same unless the seeds are set differently.* In the above example, we do not feed the seed for the first two runs. Therefore, the generated random numbers are the same for the first two runs. In the third run, we set the seed to be 101, and obtain a different set of random values (i.e., 72520690, 308637100, $\cdots$). *An important note: you must set random seeds differently for different runs. Otherwise, NS2 will generate the same result.*

3.6.2 Creating a TclCommand

A TclCommand creation process is similar to those of a TclClass and function **command** of a TclObject. A TclCommand is defined in a class derived from class **TclCommand**. The name of a Tclcommand is provided as an input argument of class **TclCommand**, while the implementation is defined in function "**command(...)**". When NS2 starts, it binds all TclCommand names to function "**command(...)**" of the corresponding classes.

Program 3.12 Declaration and function command of class RandomCommand.

```
   //~ns/common/misc.cc
1  class RandomCommand : public TclCommand {
2  public:
3      RandomCommand() : TclCommand("ns-random") { }
4      virtual int command(int argc, const char*const* argv);
5  };

6  int RandomCommand::command(int argc, const char*const* argv)
7  {
8      Tcl& tcl = Tcl::instance();
9      if (argc == 1) {
10         sprintf(tcl.buffer(), "%u", Random::random());
11         tcl.result(tcl.buffer());
12     } else if (argc == 2) {
13         int seed = atoi(argv[1]);
14         if (seed == 0)
15             seed = Random::seed_heuristically();
16         else
17             Random::seed(seed);
18         tcl.resultf("%d", seed);
19     }
20     return (TCL_OK);
21 }
```

Program 3.12 shows the details of TclCommand ns-random, which is associated with class RandomCommand. Here ns-random is fed to the constructor of class TclCommand (Line 3). When invoking ns-random, NS2 invokes function command(...) of class RandomCommand, passing the command name as well as its input arguments to the function command(...). When invoking the command ns-random, Lines 10–11 generate a random number, and pass it to the interpreter. If the number of arguments is one, Lines 17-18 set the random seed to the input argument and pass the seed to the interpreter.

TclCommands ns-version and ns-random in Example 3.13 are defined in file ~ns/common/misc.cc. At the startup time, NS2 invokes function init_misc(...) (see Program 3.13) in file ~tclcl/TclAppInit.cc. This function simply instantiates all TclCommands by calling "new{...}" (e.g., Lines 3–4 in Program 3.13). After this point, every TclCommand invoked from the OTcl domain will refer to the corresponding instantiated TclCommand object.

3.6.3 Defining Your Own TclCommand

To create a TclCommand, you need to

(i) Derive a TclCommand class directly from class TclCommand,
(ii) Feed the TclCommand name to the constructor of class TclCommand,

---

Program 3.13 Function `misc_init`, which instantiates of TclCommands.

```
     //~ns/common/misc.cc
 1   void init_misc(void)
 2   {
 3         (void)new VersionCommand;
 4         (void)new RandomCommand;
 5         ...
 6   }
```

(iii) Provide implementation (i.e., desired actions) in the function `command` (...), and

(iv) Add an object instantiation statement in function `init_misc(...)`.

Example 3.14. Let the TclCommand `print-all-args` show all input arguments on the screen. We can implement this TclCommand by including the following codes to file ~*ns*/common/misc.cc:

```
class PrintAllArgsCommand : public TclCommand {
    public:
    PrintAllArgsCommand():TclCommand("print-all-args") {};
    int command(int argc, const char*const* argv);
}

int PrintAllArgsCommand::command(int argc,
                        const char*const* argv) {
    cout << "Input arguments: "
    for (int i = 1; i < argc; i++) {
        count << argv[i];
    }
    return (TCL_OK);
}

void init_misc(void)
{   ...
    (void)new PrintAllArgsCommand;
    ...
}
```

3.7 Class EmbeddedTcl

Although written in two languages, NS2 mainly operates in C++. At the compilation, NS2 translates all the OTcl script (e.g., all the script files in directory ˜ns/tcl/lib) into the C++ language using class `EmbeddedTcl`. The translation process consists of two main steps:

(i) During the compilation, NS2 translates the scripts into EmbeddedTcl objects (e.g., `et_ns_lib`, `et_tclobject`) by the following statement in file `Makefile`:

```
$(TCLSH) bin/tcl-expand.tcl tcl/lib/ns-lib.tcl \
                | $(TCL2C) et_ns_lib > gen/ns_tcl.cc
```

This statement creates an EmbeddedTcl object `et_ns_lib`.[10] There are two main parts in this statement: expanding file ˜ns/tcl/lib/ns-lib.tcl and creating an EmbeddedTcl object `et_ns_lib` as well as a C++ file ˜ns/gen/ns_tcl.cc. The details of these two parts are as follows:

(a) The first part, "`$(TCLSH) bin/tcl-expand.tcl tcl/lib/ns-lib.tcl`", runs the Tcl shell (`TCLSH`) to interpret the scripting file ˜ns/bin/tcl-expand.tcl, which takes the scripting file ˜ns/tcl/lib/ns-lib.tcl as an input argument.

Apart from containing its own codes, file ˜ns/tcl/lib/ns-lib.tcl has lines with format "`source <fn>`". These lines "source" another scripting file whose name is `<fn>`. By sourcing a Tcl file, we mean to include the file into the translation process. The script file ˜ns/bin/tcl-expand.tcl simply expands the file ˜ns/tcl/lib/ns-lib.tcl by replacing the source statement "`source <fn>`" with the content of the file `<fn>`.

(b) The second part, "`| $(TCL2C) et_ns_lib > gen/ns_tcl.cc`", translates the expanded file in the first part into an EmbeddedTcl object `et_ns_lib` (using a unix pipe "`|`") and redirects the printed result into file ˜ns/gen/ns_tcl.cc (using the unix redirect operator "`>`").

(ii) During NS2 startup, NS2 loads the translated EmbeddedTcl objects into NS2.

To incorporate a new scripting file "`file`" into NS2, we need to source the file by inserting the statement "`source file`" into file ˜ns/tcl/lib/ns-lib.tcl. At the compilation, a new scripting file will be included into NS2, and will be ready to use thereafter.

3.8 Chapter Summary

NS2 is written in OTcl (interpreted class hierarchy) and C++ (compiled class hierarchy). Loosely speaking, OTcl sets up a network (e.g., creating and connecting nodes), while C++ runs actual simulation (e.g., passing packets from one node to another). When an object is created from the interpreted hierarchy, a so-called *shadow* object is also created in the compiled hierarchy. The connection between the interpreted and compiled hierarchies is established through TclCL which consists of following C++ classes.

---

[10] Other EmbeddedTcl object (e.g., `et_tclobject`) are created similarly.

- Class `TclObject` is the main class where all classes in the compiled hierarchy derive. It corresponds to an OTcl class `SplitObject`, which is the base class for all classes in the interpreted hierarchy. `Class TclObject` has four main responsibilities. The first two responsibilities are to provide methods to create and destroy a C++ shadow object, when an OTcl object is created and destroyed, respectively. The third responsibility is to bind class variables in both hierarchies together so that a change in the variable in one hierarchy will result an automatic update in the bound variable in another hierarchy. The last responsibility is to provide method–namely OTcl command–to access C++ from OTcl domain.
- Class `TclClass` maps an OTcl class name to a C++ static mapping variable. While class `TclObject` initiates the shadow object creation process, the actual shadow object creation is performed by class `TclClass`.
- Class `InstVar` defines NS2 variable data types which can be bound in both the hierarchies.
- Class `Tcl` provides an access to the interpreted hierarchy from the compiled hierarchies.
- Similar to OTcl command, class `TclCommand` provides a global access to the compiled hierarchy from the interpreted hierarchy.
- Class `EmbeddedTcl` translates OTcl scripts into C++ codes.

4

Implementation of Discrete-Event Simulation in NS2

NS2 is a discrete-event simulator, where actions are associated with events rather than time. An event in a discrete-event simulator consists of execution time, a set of actions, and a reference to the next event (Fig. 4.1). These events connect to each other and form a *chain of events* on the *simulation timeline* (e.g., that in Fig. 4.1). Unlike a time-driven simulator, in an event-driven simulator, time between a pair of events does not need to be constant. When the simulation starts, events in the chain are executed from left to right (i.e., chronologically).[1] In the next section, we will discuss the simulation concept of NS2. In Sections 4.2, 4.3, and 4.4, we will explain the details of classes `Event` and `Handler`, class `Scheduler`, and class `Simulator`, respectively. Finally, we summarize this chapter in Section 4.6.

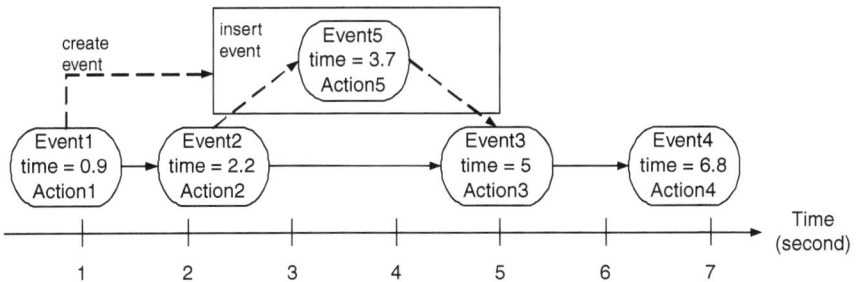

Fig. 4.1. A sample chain of events in a discrete-event simulation. Each event contains execution time and a reference to the next event. In this figure, Event1 creates and inserts Event5 after Event2 (the execution time of Event 5 is at 3.7 second).

---

[1] By execution, we mean taking actions associated with an event.

T. Issariyakul, E. Hossain, *Introduction to Network Simulator NS2*,
DOI: 10.1007/978-0-387-71760-9_4, © Springer Science+Business Media, LLC 2009

4.1 NS2 Simulation Concept

NS2 simulation consists of two major phases.

Phase I: Network Configuration Phase

In this phase, NS2 constructs a network and sets up an initial chain of events. The initial chain of events consists of events which are scheduled to occur at certain times (e.g., start FTP (File Transfer Protocol) traffic at 1 second.). These events are called *at-events* (see Section 4.2). This phase corresponds to every line in a Tcl simulation script before executing instproc run{} of the Simulator object.

Phase II: Simulation Phase

This part corresponds to a single line, which invokes instproc Simulator::run {}. Ironically, this single line contributes to most (e.g., 99%) of the simulation.

   In this part, NS2 moves along the chain of events and executes each event chronologically. Here, the instproc Simulator::run{} starts the simulation by *dispatching* the first event in the chain of events. In NS2, "dispatching an event" or "firing an event" means "taking actions corresponding to that event". An action is, for example, starting FTP traffic or creating another event and inserting the created event into the chain of events. In Fig. 4.1, at 0.9 s, Event1 creates Event5 which will be dispatched at 3.7 s, and inserts Event5 after Event2. After dispatching an event, NS2 moves down the chain and dispatches the next event. This process repeats until the last event corresponding to instproc halt{} of OTcl class Simulator is dispatched, signifying the end of simulation.

4.2 Events and Handlers

4.2.1 An Overview of Events and Handlers

As shown in Fig. 4.1, an event specifies an action to be taken at a certain time. In NS2, an event contains a *handler* which specifies the action, and the *firing time* or *dispatching time*. Program 4.1 shows declaration of classes Event and Handler. Class Event declares variables handler_ (whose class is Handler; Line 5) and time_ (Line 6) as its associated handler and firing time, respectively. To maintain the chain of events, each Event object contains pointers next_ (Line 3) and prev_ (Line 4) to the next and previous Event objects, respectively. Variable uid_ (Line 7) is an ID unique to every event.

   Lines 10–14 in Program 4.1 show the declaration of an abstract class Handler. Class Handler specifies the *default action* to be taken when an associated event is dispatched in its pure virtual function handle(e) (Line 13)[2].

---

[2] We call actions specified in the function handle(e) *default action*, since they are taken by default when the associated event is dispatched.

Program 4.1 Declaration of classes `Event` and `Handler`.

```
//~/ns/common/scheduler.h
1  class Event {
2  public:
3      Event* next_;        /* event list */
4      Event* prev_;
5      Handler* handler_;   /* handler to call when event ready */
6      double time_;        /* time at which event is ready */
7      scheduler_uid_t uid_;   /* unique ID */
8      Event() : time_(0), uid_(0) {}
9  };

10 class Handler {
11 public:
12     virtual ~Handler () {}
13     virtual void handle(Event* e) = 0;
14 };
```

This declaration forces all its instantiable derived classes to provide the action in function `handle(e)`. In the following, we will discuss few classes which derive from classes `Event` and `Handler`. These classes are `NsObject`, `Packet`, `AtEvent`, and `AtHandler`.

4.2.2 Class `NsObject`: A Child Class of Class `Handler`

Derived from class `Handler`, class `NsObject` is one of the main classes in NS2. It is a base class for most of the network components. We will discuss the details of this class in Chapter 5. Here, we only show the implementation of function `NsObject::handle(e)` in Program 4.2. Function `NsObject::handle(e)` casts an `Event` object associated with the input pointer (`e`) to a `Packet` object. Then it feeds the casted object to function `recv(p)` (Line 3). Usually, function `recv(p)`, where `p` is a pointer to a packet, indicates that an object has received a packet `p` (see Chapter 5). Unless function `handle(e)` is overridden, function `handle(e)` (i.e., an action associated with an event *p) of an NsObject simply indicates packet reception.

Program 4.2 Function `NsObject::handle`.

```
//~/ns/common/object.cc
1  void NsObject::handle(Event* e) 2  { 3      recv((Packet*)e); 4
}
```

4.2.3 Classes Packet and AtEvent: Child Classes of Class Event

Classes Packet and AtEvent are among key NS2 classes which derive from
class Event. These two classes can be placed on the chain of events so that
their associated handler will take actions at the firing time. While the details
of class AtEvent are discussed in this section, that of class Packet will be
discussed later in Chapter 8.

Program 4.3 Declaration of classes AtEvent and AtHandler, and function
AtHandler::handle.

```
    //~/ns/common/scheduler.cc
1   class AtEvent : public Event {
2   public:
3       AtEvent() : proc_(0) {
4       }
5       ~AtEvent() {
6           if (proc_) delete [] proc_;
7       }
8       char* proc_;
9   };

10  class AtHandler : public Handler {
11  public:
12      void handle(Event* event);
13  } at_handler;

14  void AtHandler::handle(Event* e)
15  {
16      AtEvent* at = (AtEvent*)e;
17      Tcl::instance().eval(at->proc_);
18      delete at;
19  }
```

Declared in Program 4.3, class AtEvent represents events whose action is
the execution of an OTcl statement. It contains one string variable proc_ (Line
8) which holds an OTcl statement string. At the firing time, its associated
handler, whose class is AtHandler, will retrieve and execute the OTcl string
from this variable.

Derived from class Handler, class AtHandler specifies the actions to be
taken at firing time in its function handle(e) (Lines 14–19). Here, Line 16
casts the input event into an AtEvent object. Then Line 17 extracts and
executes the OTcl statement from variable proc_ of the cast event.

In the OTcl domain, an AtEvent object is placed in a chain of events at
a certain firing time by instproc "at{time statement}" of class Simulator.
The syntax for the invocation is given below:

```
$ns at <time> <statement>
```

where **ns** is the **Simulator** object (see Section 4.4), **<time>** is the firing time, and **<statement>** is an OTcl statement string which will be executed when the simulation time is **<time>** second.

Program 4.4 Instance procedure **at** of class **Simulator** and command **at** of class **Scheduler**.

```
   //~/ns/tcl/lib/ns-lib.tcl
1  Simulator instproc at args {
2      $self instvar scheduler_
3      return [eval $scheduler_ at $args]
4  }

   //~/ns/common/scheduler.cc
5  if (strcmp(argv[1], "at") == 0) {
6      /* t < 0 means relative time: delay = -t */
7      double delay, t = atof(argv[2]);
8      const char* proc = argv[3];
9      AtEvent* e = new AtEvent;
10     int n = strlen(proc);
11     e->proc_ = new char[n + 1];
12     strcpy(e->proc_, proc);
13     delay = (t < 0) ? -t : t - clock();
14     if (delay < 0) {
15         tcl.result("can't schedule command in past");
16         return (TCL_ERROR);
17     }
18     schedule(&at_handler, e, delay);
19     sprintf(tcl.buffer(), UID_PRINTF_FORMAT, e->uid_);
20     tcl.result(tcl.buffer());
21     return (TCL_OK);
22 }
```

Program 4.4 shows the details of instproc at{...} of an OTcl class **Simulator** and an OTcl command **at** of class **Scheduler**. The instproc "at{...}" of class Simulator invokes an OTcl command "at" of the Scheduler object (See Lines 5–22).

Command **at** of class **Scheduler** stores the firing time in variable t (Line 7). Line 9 then creates an **AtEvent** object. Lines 8 and 10–12 store the input OTcl command in the variable **proc_** of the created **AtEvent** object. Line 13 converts the firing time to the **delay** time from the current time. Finally, Line 18 schedules the created **AtEvent** e at **delay** seconds in future, feeding the address of variable **at_handler** (see Program 4.3) as an input argument to function **schedule**(...).

4.3 The Scheduler

The scheduler maintains the chain of events and simulation (virtual) time. At runtime, it moves along the chain, and dispatches one event after another. Since there is only one chain of events in a simulation, there is exactly one `Scheduler` object in a simulation. Hereafter, we will refer to the `Scheduler` object simply as the Scheduler. Also, NS2 supports the four following types of schedulers: List Scheduler, Heap Scheduler, Calendar Scheduler (default), and Real-time Scheduler. For brevity, we do not discuss the differences among all these schedulers here. The details of these schedulers can be found in [15].

Program 4.5 Declaration of class `Scheduler`.

```
    //~ns/common/scheduler.h
1   class Scheduler : public TclObject {
2   public:
3       static Scheduler& instance() { return (*instance_); }
4       void schedule(Handler*, Event*, double delay);
5       virtual void run();
6       virtual void cancel(Event*) = 0;
7       virtual void insert(Event*) = 0;
8       virtual Event* lookup(scheduler_uid_t uid) = 0;
9       virtual Event* deque() = 0;
10      virtual const Event* head() = 0;
11      double clock() const { return (); }
12      virtual void reset();
13  protected:
14      void dispatch(Event*);
15      void dispatch(Event*, double);
16      Scheduler();
17      virtual ~Scheduler();
18      int command(int argc, const char*const* argv);
19      double clock_;
20      static Scheduler* instance_;
21      static scheduler_uid_t uid_;
22      int halted_;
22  };
```

4.3.1 Main Components of the Scheduler

Declared in Program 4.5, class `Scheduler` consists of a few main variables and functions. Variable `clock_` (Line 19) contains the current simulation time, and function `clock()` (Line 11) returns the value of the variable `clock_`. Variable `halted_` (Line 22) is initialized to 0, and is set to 1 when the simulation is stopped or paused. Variable `instance_` (Line 20) is the reference to the

Scheduler, and function `instance()` (Line 3) returns the variable `instance_`. Variable `uid_` is the event unique ID. In NS2, the Scheduler acts as a single point of unique ID management. When an event is inserted into the simulation timeline, the Scheduler creates a new unique ID, and assigns the ID to the event. Both the variables `instance_` and `uid_` are static, since there is only one scheduler and unique ID in a simulation.

4.3.2 Data Encapsulation and Polymorphism Concepts

Program 4.5 implements the concepts of *data encapsulation* and *polymorphism* (see Appendix B). It hides the chain of events from the outside world, and declares pure virtual functions `cancel(e)`, `insert(e)`, `lookup(uid)`, `deque()`, and `head()` in Lines 6–10 to manage the chain. Classes derived from class `Scheduler` provide implementation of the chain as well as all of the above functions. The beauty of this mechanism is the ease of modifying type of scheduler at runtime. NS2 implements most of the codes in relation to class `Scheduler`, not its derived classes (e.g., `CalendarScheduler`). At runtime (e.g., in a Tcl simulation script), we can select a scheduler to be of any derived class (e.g., `CalendarScheduler`) of class `Scheduler` without having to modify the codes for the base class (`Scheduler`).

4.3.3 Main Functions of the Scheduler

Three main functions of class `Scheduler` are `run()` (Program 4.6), `schedule(h,e,delay)` (Program 4.7) and `dispatch(p,t)` (Program 4.8). In Program 4.6, function `run()` first sets variable `instance_` to the address of the scheduler (`this`) in Line 3. Then, it keeps dispatching events (Line 6) in the chain until `halted_` $\neq 0$ or untill all the events are executed (Line 5).

Program 4.6 Function `run` of class `Scheduler`.

```
     //~ns/common/scheduler.cc
1    void scheduler::run()
2    {
3        instance_ = this;
4        Event *p;
5        while (!halted_ && (p - deque())) {
6            dispatch(p, p->time_);
7        }
8    }
```

Function `schedule(h,e,delay)` in Program 4.7 takes three input arguments: A `Handler` pointer(`h`), an `Event` pointer(`e`), and the delay(`delay`), respectively. It inserts the input `Event` object(`*e`) into the chain of events.

Lines 3–12 check for possible errors. Line 13 increments the unique ID of the
Scheduler and assigns it to the input Event object. Line 14 associates the
input Handler object with the input Event object. Line 15 converts input
delay time (delay) to the firing time (time_) of the Event object e. Line
17 inserts the configured Event object e in the chain of events via function insert(e). Since the scheduler increments its unique ID when invoking
function schedule(...), every scheduled event will have different unique ID.

Finally, the errors in Lines 3–12 include

1. Null handler (Line 3)
2. Positive Event unique ID (Lines 4-7; See Section 4.3.4)
3. Negative delay (Line 8)
4. Negative Scheduler unique ID[3]

Program 4.7 Function schedule of class Scheduler.

```
    //~ns/common/scheduler.cc
1   void Scheduler::schedule(Handler* h, Event* e, double delay)
2   {
3       if (!h) { /* error: Do not feed in NULL handler */ };
4       if (e->uid_ > 0) {
5           printf("Scheduler: Event UID not valid!\n\n");
6           abort();
7       }
8       if (delay < 0) { /* error: negative delay */ };
9       if (uid_ < 0) {
10          fprintf(stderr, "Scheduler: UID space exhausted!\n");
11          abort();
12      }
13      e->uid_ = uid_++;
14      e->handler_ = h;
15      double t = clock_ + delay;
16      e->time_ = t;
17      insert(e);
18  }
```

Function dispatch(p,t) in Program 4.8 is invoked by function run()
at the firing time. It takes a dispatching event (*p) and firing time (t) as
input arguments. Since the scheduler moves forward in the simulation time,
the firing time (t) cannot be less than the current simulation time (clock_).
Line 3 will show an error, if t < clock_. Line 4 sets the current simulation
virtual time to be the firing time of the event. Line 5 inverts the sign of the

---

[3] The unique ID of the Scheduler is always positive. Its negative value indicates
possible abnormality such as memory overflow or inadvertent memory access
violation.

uid_ of the event, indicating that the is event is being dispatched. Line 6 invokes function handle(p) of the associated handler handler_, feeding the event (p) as an input argument.

Program 4.8 Function dispatch of class Scheduler.

```
     //~ns/common/scheduler.cc
1    void Scheduler::dispatch(Event* p, double t)
2    {
3        if (t < clock_) { /* error */ };
4        clock_ = t;
5        p->uid_ = -p->uid_;       // being dispatched
6        p->handler_->handle(p); // dispatch
7    }
```

4.3.4 Dynamics of the Unique ID of an Event

The dynamics of the event's unique ID (uid_) is fairly subtle. In general, the scheduler maintains the unique ID, and assigns the unique ID to the event being scheduled. To make uid_ unique, the Scheduler increments uid_ and assigns the incremented uid_ to the scheduling event in its function schedule(...) (Line 13 in Program 4.7). When dispatching an event, the scheduler inverts the sign of uid_ of the dispatching event (Line 5 in Program 4.8). Figure 4.2 shows the dynamics of the unique ID caused by the above schedule(...) and dispatch(...) functions. The sign toggling mechanism of unique ID ensures that events will be scheduled and dispatched properly. If a scheduled event is not dispatched, or is dispatched twice, its unique ID will be positive, and an attempt to schedule this undispatched event will cause an error (Lines 5 and 6 in Program 4.7).

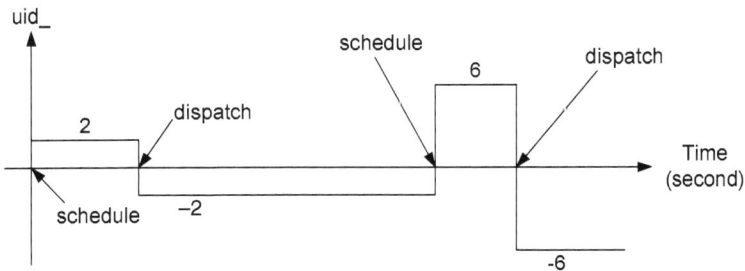

Fig. 4.2. Dynamics of Event unique ID (uid)_: Take a positive value from Scheduler::uid_ when being scheduled, and invert the sign when being dispatched. Increment upon schedule and inversion of sign upon dispatch

4.3.5 Scheduling-Dispatching Mechanism

We conclude this section through an example explaining the scheduling-dispatching mechanism. Consider the following script

```
set ns [new Simulator]
$ns at 10 [puts "An event is dispatched"]
$ns run
```

which prints out the message "An event is dispatched" 10 seconds after the simulation has started. Figure 4.3 shows the functions (shown in rectangles) and objects (shown in rounded rectangles) related to the scheduling-dispatching mechanism, whose names are shown in boldface font. Again, an AtEvent object is scheduled by the OTcl command "at" (in the upper-left rectangle), of class Scheduler. The Scheduler creates an AtEvent object e and stores input command (the fourth input argument str = puts "An event is dispatched") in e->proc_. Then, it schedules the event e with delay converted from time = 10 (the third input argument), feeding the address of AtHandler object (at_handler in the lower right rounded rectangle) as the corresponding handler.

The lower-left rectangle in Fig. 4.3 shows the details of function schedule(h,e,delay) of class Scheduler. Before inserting event e into the chain of events, function schedule(...) configures event e as follows: Update uid_ to be the same as that of Scheduler, store at_handler in the handler of event e, and set firing time to be clock_ (current time) + delay.

At the firing time, the scheduled AtEvent object is dispatched through function dispatch(p,t) (the upper-right rectangle in Fig. 4.3). When the scheduled Event object e[4] is dispatched, function dispatch(...) inverts the sign of its variable uid_, and invokes function handle(e) of the corresponding handler feeding Event object e as an input argument. Since the handler is at_handler (see the upper-left rectangle), the OTcl command puts "An event is dispatched" stored in e is executed.

4.3.6 Null Event and Dummy Event Scheduling

When being dispatched, an event p is fed to function handle(p) of the associated handler for a certain purpose. For example, the function handle(p) of class NsObject executes "recv(p), where "p" is a packet reception event. Here, the event *p must have been created and fed to function schedule(...) prior to the ongoing dispatching process.

In some cases, an event only indicates the time where the default action is taken but takes no part in such the action. For example, a queue unblocking event, informs the associated Queue object of the completion of the ongoing

---

[4] In Program 4.8, the first argument of function dispatch is p. Here, we use e as the first argument for the sake of explanation.

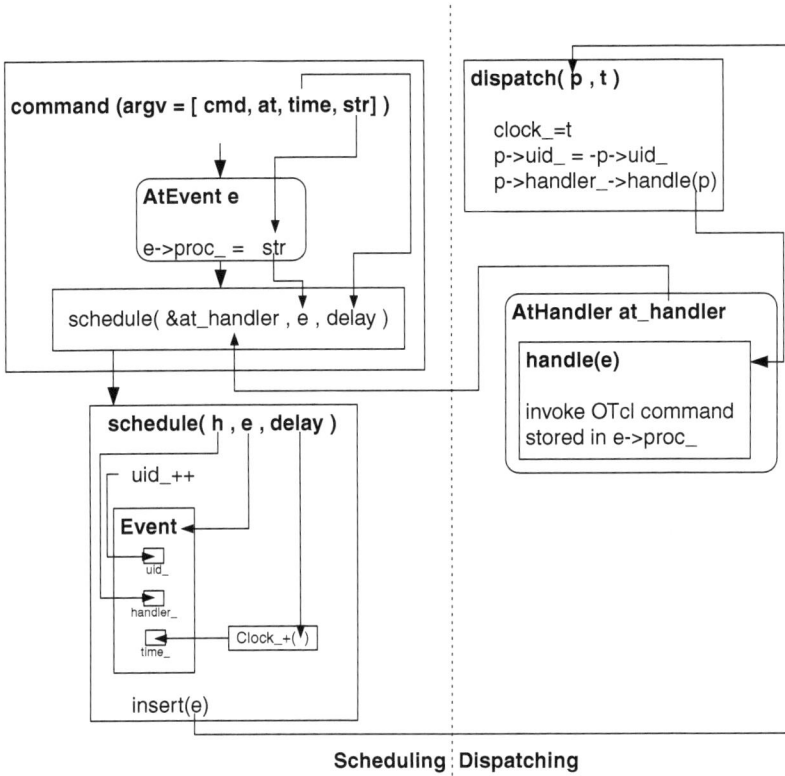

Fig. 4.3. Scheduling and dispatching mechanism of an AtEvent.

transmission (see Section 7.3). Function handle(p) of the associated handler in this case simply invokes function resume() which take no input argument of the associated Queue object. Clearly the queue unblocking event takes no role in the dispatching process. In this case, we do not need to explicity create an event. Instead, we can use a null event or a dummy event as an input to function schedule(...).

Scheduling of a Null Event

Function schedule(h,e,delay) takes a pointer to an event as its second input argument. A null event refers to a null pointer which is fed as the second input argument to the function schedule(...) (e.g., schedule(handler,0,delay)).

   Although simple to use, a null event could lead to runtime error which is difficult to be located. A null event is not an *actual* event. Its unique ID does not follow semantic in Fig. 4.3. The Scheduler ignores the unique ID when scheduling and dispatching a null event, and allows an undispatched event to be rescheduled. This breaks the scheduling-dispatching protection mechanism.

Using null events, the users are responsible for ensuring the proper sequence of scheduling-dispatching by themselves.

Scheduling of a Dummy Event

This is another approach to schedule and dispatch events which do not take part in default actions. A dummy event is usually declared as a member variable of a C++ class, and is used repeatedly in a scheduling-dispatching process.

Consider a packet departure event which is modeled by class `LinkDelay` (see Section 7.2) for example. During simulation, an NsObject informs a `LinkDelay` object to schedule packet departure events. At the firing time, the packet completely departs the NsObject, and the NsObject is allowed to fetch another packet for transmission. The packet departure event takes no part in the default action, since a new packet is fetched or created by another object.

As we shall see, a packet departure event is represented by a dummy event variable `intr_` of class `LinkDelay`, and the packet departure is scheduled through the variable `intr_` only. Since variable `intr_` is a dummy `Event`, its unique ID follows the semantic in Fig. 4.3. An attempt to schedule an undispatched event would immediately cause runtime error. Note that `intr_` is a variable of class `LinkDelay`. It is used over and over again to indicate packet departure from a `LinkDelay` object.

As a final note, under a simple configuration, it is recommended to use the null event scheduling approach. For a complicated configuration, on the other hand, the dummy event scheduling is preferable, since it provides a protection against scheduling of undispatched events.

4.4 The Simulator

OTcl and C++ classes `Simulator` are the main classes which supervise the entire simulation. Like the `Scheduler` object, there can be only one `Simulator` object throughout a simulation. This object will be referred to as *the Simulator* hereafter. The Simulator contains two types of key components: simulation objects and information-storing objects. While simulation object (e.g., the Scheduler) are the key components which derive the simulation, as well as the simulator are created during the Network Configuration Phase, and will be used in the Simulation Phase.

Information-storing objects (e.g., the reference to created nodes) contain information which is shared among several objects. For example, NS2 needs to know all created nodes and links in order to construct a routing table. These information-storing objects are created via various instprocs (e.g., `Simulator::node{}`) during the Network Configuration Phase. In the Simulation Phase, most objects access these information-storing objects via its instvar `ns_` (set by executing `set ns_ [Simulator instance]`), which is the reference to the Simulator.

4.4.1 Main Components of a Simulation

Interpreted Hierarchy

Created by various instprocs, the main OTcl simulation components are as follows:

- *The Scheduler* (`scheduler_` created by instproc `Simulator::init`) maintains the chain of events and executes the events chronologically.
- *The null agent* (`nullAgent_` created by instproc `Simulator::init`) provides the common packet dropping point.[5]
- *Node reference* (`Node_` created by instproc `Simulator::node`) is an associative array whose elements are the created nodes and indices are node IDs.
- *Link reference* (`link_` created by instprocs `simplex-link{...}` or `duplex-link{...}`) is an associative array. Associated with an index with format "sid:did", each element of `link_` is the created uni-directional link which carries packet from node "sid" to node "did".

Compiled Hierarchy

In the compiled hierarchy, class `Simulator` also contains variables and functions as shown in Program 4.9. Variable `instance_` (Line 18) is a pointer to the Simulator. It is a static variable, which means that there is only one variable `instance_` of class `Simulator` for the entire simulation. Variable `nodelist_` (Line 14) is the linked list containing the created nodes. The linked list can contain upto "`size_`" elements (Line 17), while the total number of nodes is "`nn_`" (Line 16). Variable `rtobject_` (Line 15) is a pointer to a `RouteLogic` object, which is responsible for the routing mechanism (see Chapter 6).

Function `populate_flat_classifiers{...}` (Line 7) pulls out the routing information stored in variable `*rtobject_` and installs the routing table in the created nodes and links (see Section 6.6). Function `add_node(...)` (Line 8) puts the input argument `node` into the linked list of nodes (`nodelist_`). Function `get_link_head(...)` returns the link head object (see Chapter 7) of the link with ID "nh" which connects to a `ParentNode` object `*node`. Function `node_id_by_addr(addr)` (Line 10) converts node address "addr" to node ID. Function `alloc(n)` (Line 11) allocates spaces in `nodelist_` which can accommodate up to "n" nodes, and clears all components of `nodelist_` to NULL. Function `check(n)` immediately returns if `n` is less than `size_`. Otherwise, it will create more space in `nodelist_`, which can accommodate upto "n" nodes. Static function `instance()` in Line 3 returns the variable `instance_` which is the pointer to the simulator.

---

[5] By "dropping a packet", we mean "removing a packet" from the simulation. We will discuss the dropping mechanism in Chapter 5. For the moment, it is sufficient to know that `nullAgent_` drops or removes all received packets from the simulation.

Program 4.9 Declaration of class `Simulator`.

```
   //~ns/common/simulator.h
1  class Simulator : public TclObject {
2  public:
3      static Simulator& instance() { return (*instance_); }
4      Simulator() : nodelist_(NULL),
                   rtobject_(NULL), nn_(0), size_(0) {}
5      ~Simulator() { delete []nodelist_;}
6      int command(int argc, const char*const* argv);
7      void populate_flat_classifiers();
8      void add_node(ParentNode *node, int id);
9      NsObject* get_link_head(ParentNode *node, int nh);
10     int node_id_by_addr(int address);
11     void alloc(int n);
12     void check(int n);
13 private:
14     ParentNode **nodelist_;
15     RouteLogic *rtobject_;
16     int nn_;
17     int size_;
18     static Simulator* instance_;
19 };
```

4.4.2 Retrieving the Instance of the Simulator

Program 4.10 Retrieving the instance of the Simulator using instproc
instance of class `Simulator`.

```
   //~ns/tcl/lib/ns-lib.tcl
1 Simulator proc instance {} {
2     set ns [Simulator info instances]
3     if { $ns != "" } {
4         return $ns
5     }
6     ...
7 }
```

From the interpreted hierarchy, we can also retrieve the simulator instance
by invoking instproc `instance{}` of class `Simulator`. This instproc executes
the OTcl built-in command "`info`" with an option "`instances`". This execution returns all the instances of a certain class. Since there is only one
`Simulator` instance, the statement "`Simulator info instances`" returns
the `Simulator` object as required.

4.4.3 Simulator Initialization

Simulator initialization refers to the process in the Network Configuration Phase, which creates the Simulator as well as its components. The Simulator is created by executing new Simulator. This command invokes the constructor (i.e., instproc init{...} of class Simulator) shown in Program 4.11.

Program 4.11 Instance procedures init and use-scheduler of class Simulator.

```
        //~ns/tcl/lib/ns-lib.tcl
 1   Simulator instproc init args {
 2        $self create_packetformat
 3        $self use-scheduler Calendar
 4        $self set nullAgent_ [new Agent/Null]
 5        $self set-address-format def
 6        eval $self next $args
 7   }

 8   Simulator instproc use-scheduler type {
 9        $self instvar scheduler_
10        if [info exists scheduler_] {
11            if { [$scheduler_ info class] == "Scheduler/$type" } {
12                return
13            } else {
14                delete $scheduler_
15            }
16        }
17        set scheduler_ [new Scheduler/$type]
18   }
```

The constructor first initializes the packet format in Line 2, and invokes instproc use-scheduler{type} in Line 3 to specify the type of the Scheduler. By default, the type of the Scheduler is Calendar. Line 4 creates a null agent (nullAgent). Line 5 sets the address format to the default format in Line 5. Instproc use-scheduler{type} (Lines 8–18) will delete the existing scheduler if it is different from that specified in the input argument type. Then it will create a scheduler with type = type, and store the created Scheduler object in instvar scheduler_.

4.4.4 Running Simulation

The Simulation Phase starts at the invocation of instproc Simulator::run{}. As shown in Program 4.12, the instproc Simulator::run{} first invokes instproc "configure{}" of class RouteLogic (Line 2). This instproc computes the optimal routes and creates the routing table (see Chapter 6). Lines

5–10 reset nodes and queues. Finally, Line 11 starts the Scheduler by invoking
the OTcl command run{} of class Scheduler, which in turn invokes the C++
function run{} of class Scheduler shown in Program 4.6. Again, this function
executes events in the chain of events one after another until the Simulator is
halted (i.e., varaible halted_ of class Scheduler is 1), or untill all the events
are executed.

Program 4.12 Instance procedure Simulator::run.

```
    //~/ns/tcl/lib/ns-lib.tcl
1   Simulator instproc run  {
2       [$self get-routelogic] configure
3       $self instvar scheduler_ Node_ link_ started_
4       set started_ 1
5       for each nn [array names Node_] {
6           $Node_($nn) reset
7       for each qn [array names link_] {
8           set q [$link_($qn) queue]
9           $q reset
10      }
11      return [$scheduler_ run]
12 }
```

4.5 Instprocs of OTcl Class Simulator

The list of useful instprocs of class Simulator is shown below.

now{} Retrieve the current simulation time.
nullagent{} Retrieve the shared null agent.
use-scheduler{type} Set the scheduler to be <type>.
at{time stm} Execute the statement <stm> at <time> second.
run{} Start the simulation.
halt{} Terminate the simulation.
cancel{e} Cancel the scheduled event <e>.

4.6 Chapter Summary

This chapter explains the details of discrete-event simulation in NS2. The
simulation is carried out by running a *Tcl simulation script*, which consists of
two parts. First, the *Network Configuration Phase* establishes a network, and
configures all simulation components. This phase also creates a chain of events
by connecting the created events chronologically. Secondly, the *Simulation*

Phase chronologically executes (or dispatches) the created events until the Simulator is halted, or untill all the events are executed.

There are four main classes involved in an NS2 simulation:

- Class `Simulator` supervises the simulation. It contains simulation components such as the Scheduler, the null agent, etc. It also contains information storing objects which are share by other (simulation) components.
- Class `Scheduler` maintains the chain of events and chronologically dispatches the events.
- Class `Event` consists of the firing time and the associated handler. Events are put together to form a chain of events, which are dispatched one by one by the Scheduler. Classes `Packet` and `AtEvent` are among the classes derived from class `Event`, which can be placed on the simulation timeline (i.e., in the chain of event). They are associated with different handlers, and take different actions at the firing time.
- Class `Handler`: Associated with an event, a handler specifies default actions to be taken when the associated event is dispatched. Classes `NsObject` and `AtHandler` are among classes derived from class `Handler`. They are always associated with `Packet` and `AtEvent` events, respectively. Their actions are to receive the `Packet` object and to execute an OTcl statement specified in the `AtEvent` object, respectively.

5

Network Objects: Creation, Configuration, and Packet Forwarding

NS2 is a simulation tool designed specifically for communication networks. The main functionalities of NS2 are to set up a network of connecting nodes and to pass packets from one node (which is a network object) to another.

A network object is one of the main NS2 components, which is responsible for packet forwarding. NS2 implements network objects by using the polymorphism concept in Object-Oriented Programming (OOP). Polymorphism allows network objects to take different actions ways under different contexts. For example, a `Connector` object immediately passes the received packet to the next network object, while a `Queue`[1] object enques the received packets and forwards only the head of the line packet.

This chapter first introduces the NS2 components by showing four major classes of NS2 components, namely, network objects, packet-related objects, simulation-related objects, and helper objects in Section 5.1. A part of the C++ class hierarchy, which is related to network objects, is also shown here. Section 5.2 presents class `NsObject` which acts as a template for all network objects. An example of network objects as well as packet forwarding mechanism are illustrated through class `Connector` in Section 5.3. Finally, the chapter summary is given in Section 5.4. Note that the readers who are not familiar with object-oriented programming are recommended to go through a review of the OOP polymorphism concept in Appendix B before proceeding further.

5.1 Overview of NS2 Components

5.1.1 Functionality-Based Classification of NS2 Modules

Based on the functionality, NS2 modules (or objects) can be classified into four following types:

---

[1] Class `Queue` is a child class of class `Connector`.

T. Issariyakul, E. Hossain, *Introduction to Network Simulator NS2*,
DOI: 10.1007/978-0-387-71760-9_5, © Springer Science+Business Media, LLC 2009

- *Network objects* are responsible for sending, receiving, creating, and destroying packet-related objects. Since these objects are those derived from class `NsObject`, they will be referred to hereafter as NsObjects.
- *Packet-related objects* are various types of packets which are passed around a network.
- *Simulation-related objects* control simulation timing, and supervise the entire simulation. As discussed in Chapter 4, examples of simulation-related objects are events, handlers, the Scheduler, and the Simulator.
- *Helper objects* do not explicitly participate in packet forwarding. However, they implicitly help to complete the simulation. For example, a routing module calculates routes from a source to a destination, while network address identifies each of the network objects.

In this chapter, we focus only on network objects. Note that, the simulation-related objects were discussed in Chapter 4. The packet-related objects will be discussed in Chapter 8. The main helper objects will be discussed in Chapter 12.

5.1.2 C++ Class Hierarchy

This section gives an overview of C++ class hierarchies. The entire hierarchy consists of over 100 C++ classes and `struct` data types. Here, we only show a part of the hierarchy (in Fig. 5.1). The readers are referred to [16] for the complete class hierarchy.

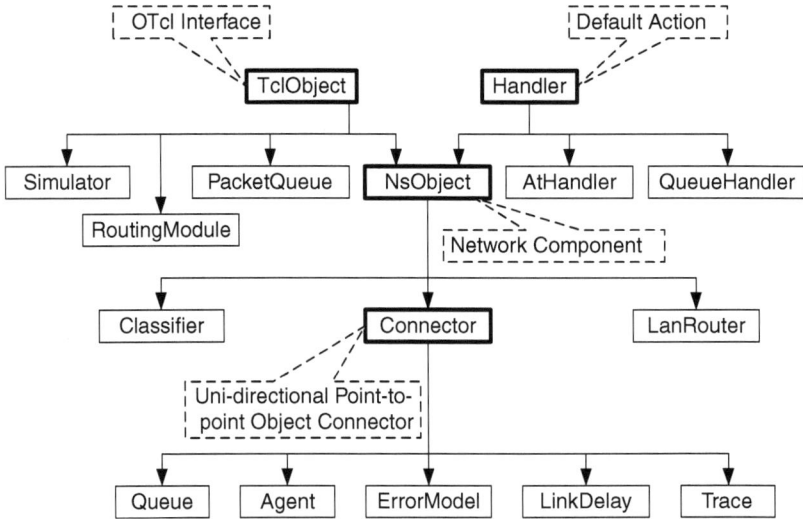

Fig. 5.1. A part of NS2 C++ class hierarchy (this chapter emphasizes on classes in boxes with *thick solid lines*).

As discussed in Chapter 3, all classes deriving from class `TclObject` form the compiled hierarchy. Classes in this hierarchy can be accessed from the OTcl domain. For example, they can be created by the global OTcl procedure "`new{...}`". Classes derived directly from class `TclObject` include network classes (e.g., `NsObject`), packet-related classes (e.g., `PacketQueue`), simulation-related classes (e.g., `Scheduler`), and helper classes (e.g., `Routing-Module`). Again, classes which do not need OTcl counterparts (e.g., classes derived from class `Handler`) form their own standalone hierarchies. These hierarchies are not a part of the compiled hierarchy nor the interpreted hierarchy.

As discussed in Chapter 4, class `Handler` specifies an action associated with an event. Again, class `Handler` contains a pure virtual function `handle(e)` (see Program 4.1). Therefore, its derived classes are responsible for providing the implementation of function `handle(e)`. For example, function `handle(e)` of class `NsObject` tells the NsObject to receive an incoming packet (Program 4.2), while that of class `QueueHandler` invokes function `resume()` of the associated `Queue` object (Lines 1–4 in Program 5.1; also see Section 7.3.2).

Program 5.1 Function `handle` of class `QueueHandler`.

```
    //~/ns/queue/queue.cc
1 void QueueHandler::handle(Event*)
2 {
3     queue_.resume();
4 }
```

Derived directly from class `TclObject` and `Handler` (see Program 5.2), class `NsObject` is the template class for all NS2 network objects. It inherits OTcl interfaces from class `TclObject` and the default action (i.e., function `handle(e)`) from class `Handler`. In addition, it defines a packet reception template, and forces all its derived classes to provide packet reception implementation. We will discuss the details of class `NsObject` in Section 5.2.

There are three main classes deriving from class `NsObject`: `Connector`, `Classifier`, and `LanRouter`. Connecting two NsObjects, a Connector object immediately forwards a received packet to the connecting NsObject (see Section 5.3). Connecting an NsObject to several NsObjects, a `Classifier` object classifies packets based on packet header (e.g., destination address, flow ID), and forwards the packets with the same classification to the same connecting NsObject (see Section 6.4). Class `LanRouter` also has multiple connecting NsObjects. However, it forwards every received packet to all connecting NsObjects.

5.2 NsObjects: A Network Object Template

5.2.1 Class NsObject

Representing NsObjects, class `NsObject` is the base class for all network objects in NS2 (see the declaration in Program 5.2). Again, the main responsibility of an NsObject is to forward packets. Therefore, class `NsObject` defines a pure virtual function `recv(p,h)` (see Line 5 in Program 5.2) as a uniform packet reception interface to force all its derived classes to implement this function.

Program 5.2 Declaration of class `NsObject`.

```
    //~/ns/common/object.h
 1  class NsObject : public TclObject, public Handler {
 2  public:
 3      NsObject();
 4      virtual ~NsObject();
 5      virtual void recv(Packet*, Handler* callback = 0) = 0;
 6      virtual int command(int argc, const char*const* argv);
 7  protected:
 8      virtual void reset();
 9      void handle(Event*);
10      int debug_;
11 };
```

Function `recv(p,h)` is in fact the very essence of packet forwarding mechanism in NS2. In NS2, an upstream object maintains a reference to the connecting downstream object. It passes a packet to the downstream object by invoking the function `recv(p,h)` of the downstream object and feeding the packet as an input argument. Since NS2 focuses mainly on forwarding packets in a downstream direction, NsObjects do not need to have a reference to its upstream objects. In most cases, NsObject configuration involves downstream (not upstream) objects only.

Function `recv(p,h)` takes two input arguments: a packet p to be received and a handler h. Most invocation of function `recv(p,h)` involves only packet "p", not the handler.[2] For example, a `Queue` object (see Section 7.3.3) puts the received packet in the buffer and transmits the packet at the head of the buffer. An ErrorModel object (see Section 12.3) imposes error probability on the received packet, and forwards the packet to the connecting object if the transmission is not in error.

---

[2] We will discuss the *callback* mechanism which involves a handler in Section 7.3.3.

Class NsObject derives from classes TclObject and Handler. Again, the functionality of class TclObject creates and binds the compiled shadow NsObject when an NsObject is created from the interpreted hierarchy. As a handler, an NsObject overrides function handle(e) which specifies the default action taken at the firing time of an associated event. Again, since the main responsibility of an NsObject is the packet forwarding, its function handle(e) (i.e., default action) is to receive a packet (cast from an event) through function recv(p,h) (see Program 4.2).

5.2.2 Packet Forwarding Mechanism of NsObjects

An NsObject forwards packets in two following ways:

- *Immediate packet forwarding*: To forward a packet to a downstream object, an upstream object needs to obtain a reference (e.g., a pointer) to the downstream object and invokes function recv(p,h) of the downstream object through the obtained reference. For example, a Connector (see Section 5.3) has a private pointer target_ to its downstream object. Therefore, it forwards a packet to its downstream object by invoking target_->recv(p,h).
- *Delayed packet forwarding*: To delay packet forwarding, a Packet object is cast to be an Event object, associated with a packet receiving NsObject, and placed on the simulation timeline at a given simulation time. At the firing time, function handle(e) of the NsObject will be invoked, and the packet will be received through function recv(p,h) (see an example of delayed packet forwarding in Section 5.3).

5.3 Connectors

As shown in Fig. 5.2, a Connector is a NsObject which connects three NsObjects in a uni-directional manner. It receives a from an upstream NsObject. By default, a Connector immediately forwards the received packet to its downstream NsObject. Alternatively, it can drop the packet by forwarding the packet to a packet dropping object.[3]

In NS2, each NsObject acts as a packet forwarder. Since it has no knowledge about its upstream objects, it does not have any interface to configure an upstream object. From Fig. 5.2, a Connector is interested in configuring its downstream NsObject and packet dropping NsObject only. The connection from an upstream object to a Connector, on the other hand, must be configured from within the scope of the upstream object.

---

[3] A packet dropping network object (e.g., a null agent) is responsible for destroying packets.

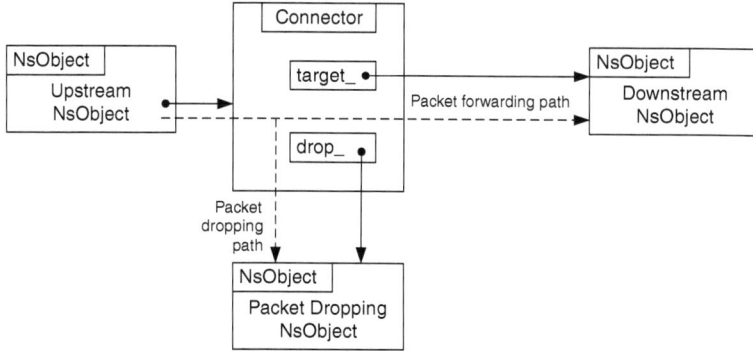

Fig. 5.2. Diagram of a connector. The *solid arrows* represent pointers, while the *dotted arrows* show packet forwarding and dropping paths.

Program 5.3 Declaration and function recv of class Connector.

```
   //~/ns/common/connector.h
1  class Connector : public NsObject {
2  public:
3      Connector();
4      inline NsObject* target() { return target_; }
5      void target (NsObject *target) { target_ = target; }
6      virtual void drop(Packet* p);
7      void setDropTarget(NsObject *dt) {drop_ = dt; }
8  protected:
9      virtual void drop(Packet* p, const char *s);
10     int command(int argc, const char*const* argv);
11     void recv(Packet*, Handler* callback = 0);
12     inline void send(Packet* p, Handler* h){target_->recv(p, h);}
13
14     NsObject* target_;
15     NsObject* drop_;    // drop target for this connector
16 };

   //~/ns/common/connector.cc
17 void Connector::recv(Packet* p, Handler* h){send(p, h);}
```

5.3.1 Class Declaration

Program 5.3 shows the declaration of class Connector. Class Connector contains two pointers (Lines 14–15 in Program 5.3) to NsObjects:[4] target_ and

---

[4] Since class Connector contains two pointers to abstract object (i.e., class NsObject), it can be regarded as an abstract user class for class composition discussed in Section B.8. We will discuss the details of how the class composition concept applies to a Connector in the next section.

drop_. From Fig. 5.2, target_ is the pointer to the connecting downstream object, while drop_ is the pointer to the packet dropping object.

Class Connector derives from the abstract class NsObject. It overrides the pure virtual function recv(p,h), by simply invoking function send(p,h) (see Line 12 in program 5.3). Function send(p,h) simply forwards the received packet to its downstream object by invoking function recv(p,h) of the downstream object (i.e., target_->recv(p,h) in Line 12).

Program 5.4 Function Connector::drop.

```
    //~/ns/common/connector.cc
1   void Connector::drop(Packet* p)
2   {
3       if (drop_ != 0)
4           drop_->recv(p);
5       else
6           Packet::free(p);
7   }
```

Program 5.4 shows the implementation of function drop(p), which drops or destroys a packet. Function drop(p) takes one input argument, which is a packet to be dropped. If the dropping NsObject exists (i.e., drop_$\neq$ 0), this function will forward the packet to the dropping NsObject by invoking drop_->recv(p,h). Otherwise, it will destroy the packet by invoking function Packet::free(p) (see Chapter 8). Note that function drop(p) is declared as virtual (Line 9). Hence, classes derived from class Connector may override this function without any function ambiguity[5].

5.3.2 OTcl Configuration Commands

As discussed in Section 4.1, NS2 simulation consists of two steps: Network Configuration Phase and Simulation Phase. In the Network Configuration Phase, a Connector is set up as shown in Fig. 5.2. Again, a Connector configures its downstream and packet dropping NsObjects only.

Suppose OTcl has instantiated three following objects: a Connector object (conn_obj), a downstream object (down_obj), and a dropping object (drop_obj). Then, the Connector is configured using the following two OTcl commands (see Program 5.5):

- OTcl command target with one input argument conforms to the following syntax:

    ```
    $conn_obj target $down_obj
    ```

---

[5] Function ambiguity is discussed in Appendix B.2

Program 5.5 OTcl commands target and drop-target of class Connector.

```
//~/ns/common/connector.cc
1  int Connector::command(int argc, const char*const* argv)
2  {
3      Tcl& tcl = Tcl::instance();
4      if (argc == 2) {
5          if (strcmp(argv[1], "target") == 0) {
6              if (target_ != 0)
7                  tcl.result(target_->name());
8              return (TCL_OK);
9          }
10         if (strcmp(argv[1], "drop-target") == 0) {
11             if (drop_ != 0)
12                 tcl.resultf("%s", drop_->name());
13             return (TCL_OK);
14         }
15     }
16     else if (argc == 3) {
17         if (strcmp(argv[1], "target") == 0) {
18             if (*argv[2] == '0') {
19                 target_ = 0;
20                 return (TCL_OK);
21             }
22             target_ = (NsObject*)TclObject::lookup(argv[2]);
23             if (target_ == 0) {
24                 tcl.resultf("no such object %s", argv[2]);
25                 return (TCL_ERROR);
26             }
27             return (TCL_OK);
28         }
29         if (strcmp(argv[1], "drop-target") == 0) {
30             drop_ = (NsObject*)TclObject::lookup(argv[2]);
31             if (drop_ == 0) {
32                 tcl.resultf("no object %s", argv[2]);
33                 return (TCL_ERROR);
34             }
35             return (TCL_OK);
36         }
37     }
38     return (NsObject::command(argc, argv));
39 }
```

This command casts the input argument `down_obj` to be of type `NsObject*` and stores it in variable `target_` (Line 22).

- OTcl command `target` with no input argument (e.g., `$conn_obj target`) returns OTcl instance corresponding to the C++ variable `target_` (Line 5–9). Note that function `name()` of class `TclObject` returns the OTcl reference string associated with the input argument.
- OTcl command `drop-target` with one input argument is very similar to that of OTcl command `target` but the input argument is cast and stored in the variable `drop_` instead of the variable `target_`.
- OTcl command `drop-target` with no input argument is very similar to that of OTcl command `target` but it returns the OTcl instance corresponding to the variable `drop_` instead of the variable `target_`.

Example 5.1. Consider the connector configuration in Fig. 5.3. Let the downstream object be of class `TcpAgent`, which corresponds to class `Agent/Tcp` in the OTcl domain. Also, let a `Agent/Null` object be a packet dropping NsObject. The following code shows how the network is set up from the OTcl domain:

```
set conn_obj [new Connector]
set tcp [new Agent/TCP]
set null [new Agent/Null]

$conn_obj target $tcp
$conn_obj drop-target $null
```

The first three lines create a Connector (`conn`), a TCP object (`tcp`), and a packet dropping object (`null`). The last two lines use the OTcl commands `target` and `drop-target` to set `tcp` and `null` as the downstream object and the dropping object of the Connector, respectively.

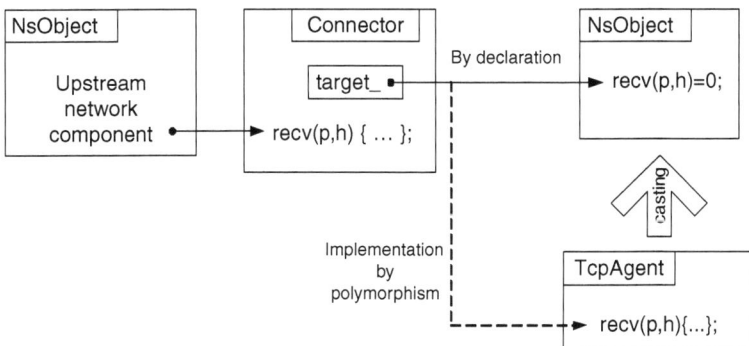

Fig. 5.3. A polymorphism implementation of a Connector. A Connector declares `target_` as an `NsObject` pointer. In the Network Configuration Phase, the OTcl command `target` is invoked to setup a downstream object of the Connector, and the NsObject `*target_` is cast to a `TcpAgent` object.

Connector configuration complies with the class composition programming concept discussed in Appendix B.5. Table 5.1 shows the components in Example 5.1 and the corresponding class composition. Classes `Agent/TCP` and `Agent/Null` are OTcl classes whose corresponding C++ classes derive from class `NsObject`. Class `Connector` stores pointers (i.e., `target_` and `drop_`) to NsObjects, and is therefore considered to be an abstract user class. Finally, as a user class, the Tcl Simulation Script instantiates NsObjects `conn`, `tcp`, and `null` from classes `Connector`, `Agent/Tcp`, and `Agent/Null`, respectively, and binds `tcp` and `null` to variables `target_` and `drop_`, respectively.

Table 5.1. Class composition of network components in Example 5.1.

Abstract class	`NsObject`
Derived class	`Agent/Tcp` and `Agent/Null`
Abstract user class	`Connector`
User class	A Tcl Simulation Script

When invoking "`target`" and "`drop-target`", `tcp` and `null` are first type-cast to `NsObject` pointers. Then they are assigned to `target_` and to `drop_`, respectively. Since a virtual function is unaffected by type casting, function `recv(p,h)` of both `tcp` and `null` are associated to class `Agent/TCP` and `Agent/Null`, respectively.

5.3.3 Packet Forwarding Mechanism

From Section 5.2.2, an NsObject forwards a packet in two ways: immediate and delayed packet forwarding. This section demonstrates both the packet forwarding mechanisms through a Connector.

Immediate Packet Forwarding

Immediate packet forwarding is carried out by invoking function `recv(p,h)` of a downstream object. In Example 5.1, the Connector forwards a packet to the TCP object by invoking function `recv(p,h)` of the TCP object (i.e., `target_->recv(p,h)`, where `target_` is configured to be a TCP object). C++ polymorphism is responsible for associating function `recv(p,h)` to class `Agent/TCP` (i.e., the construction type), not `NsObject` (i.e., the declaration type).

Delayed Packet Forwarding

Delayed packet forwarding is implemented with the aid of the Scheduler. Here, a packet is cast to an event, associated with a receiving NsObject, and placed on the simulation timeline. For example, to delay packet forwarding in Example 5.1 for "d" seconds, we may invoke the following statement instead of target_->recv(p,h).

```
Scheduler& s = Scheduler::instance();
s.schedule(target_, p, d);
```

Consider Fig. 5.4 and Program 5.6 altogether. Figure 5.4 shows the diagram of delayed packet forwarding, while Program 5.6 shows the details of functions schedule(h,e,delay) as well as dispatch(p,t) of class Scheduler. When "schedule(target_, p, d)" is invoked, function schedule (...) casts packet *p and the NsObject *target_ into Event and Handler objects, respectively (Line 1 of Program 5.6). Line 5 of Program 5.6 associates packet *p with the NsObject *target_. Lines 6-7 insert packet *p into the simulation timeline at the appropriate time. At the firing time, the event (*p) is dispatched (Lines 9-14). The Scheduler invokes function handle(p) of the handler associated with event *p. In this case, the associated handler is the NsObject *target_. Therefore, in Line 13, the default action handle(p) of target_, invokes function recv(p,h) to receive the scheduled packet.

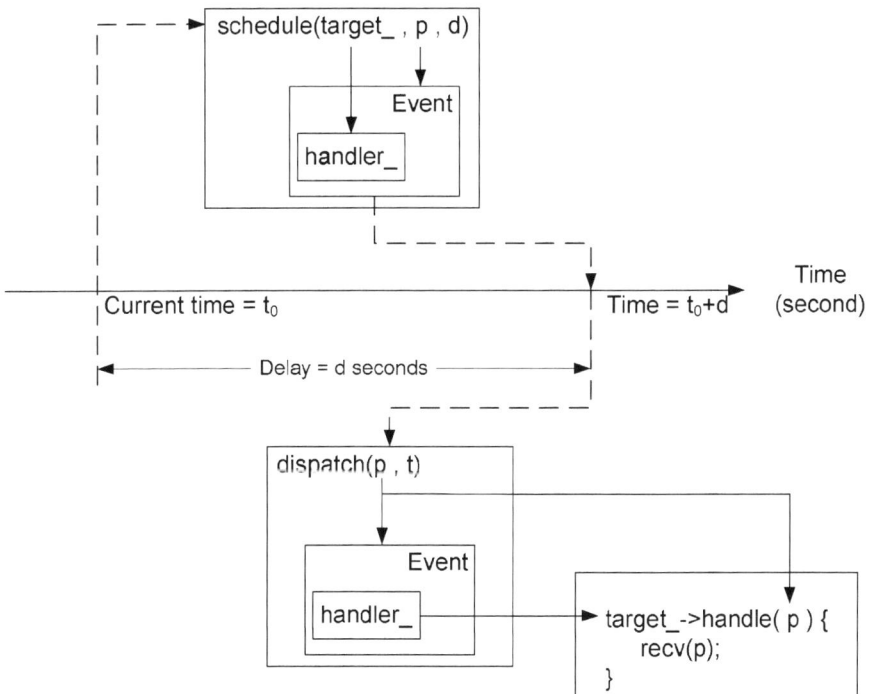

Fig. 5.4. Delayed packet forwarding.

Program 5.6 Functions `schedule` and `dispatch` of class `Scheduler`.

```
//~/ns/common/scheduler.cc
1   void Scheduler::schedule(Handler* h, Event* e, double delay)
2   {
3       ...
4       e->uid_ = uid_++;
5       e->handler_ = h;
6       e->time_ = clock_ + delay;
7       insert(e);
8   }

9   void Scheduler::dispatch(Event* p, double t)
10      ...
11      clock_ = t;
12      p->uid_ = -p->uid_; // being dispatched
13      p->handler_->handle(p); // dispatch
14  }
```

5.4 Chapter Summary

Referred to as an NsObject, a network object is responsible for sending, receiving, creating, and destroying packets. As an object of class `NsObject`, it derives OTcl interfaces from class `TclObject` and the default action (i.e., function `handle(e)`) from class `Handler`. It defines a pure virtual function `recv(p,h)` as a uniform packet reception interface for all its derived classes. Based on the polymorphism concept, all its derived classes must provide their own implementation of how to receive a packet.

In NS2, an NsObject needs to create a connection to its downstream object only. Normally, an NsObject forwards a packet to a downstream object by invoking function `recv(p,h)` of its downstream object. In addition, an NsObject can defer packet forwarding by associating a packet to the downstream object and inserting the packet on the simulation timeline. At the firing time, the scheduler dispatches the packet, and the default action of the downstream object is invoked to receive the packet.

As an example, we show the details of class `Connector`, one of the main NsObject classes in NS2. Class `Connector` contains two pointers to NsObjects: `target_` pointing to a downstream object and `drop_` pointing to a packet dropping object. To configure a Connector, an object whose class derives from class `NsObject` can be set as downstream and dropping objects via OTcl command `target` and `drop-target`, respectively. These two OTcl commands cast the downstream and dropping objects to NsObjects, and assign them to C++ variables `*target_` and `*drop_`, respectively.

6

Nodes as Routers or Computer Hosts

This chapter focuses on a basic network component, *Node*. In NS2, a Node acts as a computer host (e.g., a source or a destination) and a router (e.g., an intermediate node). It receives packets from an attached application or an upstream object, and forwards them to the attached links specified in the routing table (as a router) or delivers them to the ports specified in the packet header (as a host).

In the following, we first give an overview of Nodes and routing mechanism in NS2 in Sections 6.1 and 6.2, respectively. Sections 6.3, 6.4, and 6.5 discuss three main routing components: Route logic, classifiers, and routing modules, respectively. In Section 6.6 we show how the aforementioned Node components are assembled to compose a Node. Finally, the chapter summary is provided in Section 6.7.

6.1 An Overview of Nodes in NS2

A Node plays two important roles in NS2. As a router, it forwards packets to the connecting link based on a routing table. As a host, it delivers packets to the transport layer agent attached to the port specified in the packet header. NS2 configures the connection to its downstream NsObjects only. A Node does not need to have a connection to its upstream NsObject (e.g., a sending transport agent or an upstream link). Instead, its upstream NsObject will create a connection to the Node.

6.1.1 Architecture of a Node

In the OTcl domain, a Node is defined in a C++ class `Node` which is bound to an OTcl class with the same name. Unless specified otherwise, this chapter deals with the OTcl class only. A Node is a composite object whose architecture is shown in Fig. 6.1. It provides a single point of packet entrance,

T. Issariyakul, E. Hossain, *Introduction to Network Simulator NS2*,
DOI: 10.1007/978-0-387-71760-9_6, © Springer Science+Business Media, LLC 2009

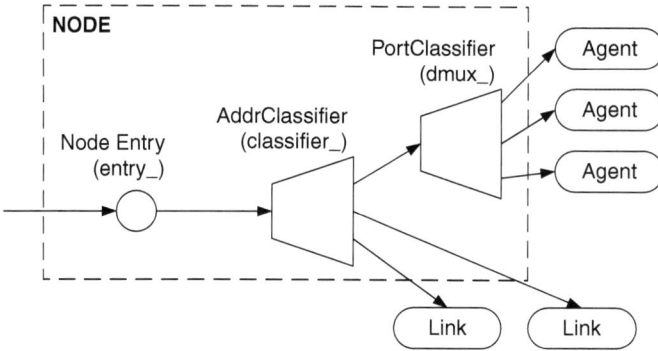

Fig. 6.1. Node architecture.

entry_ (which is a `Connector` object). After entering the Node entry, the packet enters an address classifier (an instvar `classifier_`). If the Node is not the final destination, the address classifier will forward the packet to the link specified in the routing table. Otherwise, it will forward the packet to the demultiplexer or port classifier (an instvar `dmux_`), which forwards the packet to the agent attached to the port specified in the packet header.

Apart from the above packet forwarding components, a Node also has other components. The list of major Node OTcl components is given below.

id_	Node ID
agents_	List of attached transport layer agents
nn_	Total number of Nodes (a static class instvar belonging to an OTcl class `Node`)
neighbor_	List of neighboring nodes
nodetype_	Node type (e.g., regular node or mobile node)
ns_	Simulator
dmux_	Demultiplexer or port classifier
module_list_	List of enabled routing modules
reg_module_	List of registered routing modules
rtnotif_	List of routing modules which will be notified of route updates
ptnotif_	List of routing modules which will be notified of port attachment/detachment
hook_assoc_	Sequence of the chain of classifiers
mod_assoc_	Association of classifiers and routing modules, whose indexes and values are classifiers and the associated routing modules, respectively.

6.1.2 Related Instproc of Class Node

An OTcl class Node defines the following main instprocs, which can be classified into three categories.

Initialization Instprocs

enable-module{mod_name}	Appends "mod_name" to the module list "module_list_".
disable-module{mod_name}	Removes "mod_name" from the module list "module_list_".
register-module{mod}	Inserts an input routing module "mod" into an entry of the instance associative array "reg_module_" whose index in the module name.
unregister-module{mod}	Removes an entry of an instance associative array "reg_module_" whose index matches with the name of the input routing module "mod".
route-notify{module}	Inserts an input routing module "module" into the route notification list "rtnotif_".
unreg-route-notify{... module}	Removes a routing module "module" from the route notification list "rtnotif_"
port-notify{module}	Inserts an input routing module "module" into the agent attachment list "ptnotif_".
unreg-port-notify{... module}	Removes an input routing module "module" from the agent attachment list "ptnotif_".

Route Adding/Deleting and Agent Attachment/Detachment Instprocs

add-route{... dst target}	Recursively adds a routing entry (dst,target), where "dst" and "target" are a destination node and a forwarding NsObject, respectively, for all routing modules in the link list "rtnotif_".
delcto-route{args}	Recursively removes a routing entry specifies in the input arguments from all routing modules in the linked list "rtnotif_".
alloc-port{... nullagent}	Returns a free port of the demultiplexer "dmux_" of the Node.
agent{port}	Returns the agent whose port is "port".
add-target{... agent port}	Recursively attaches the input agent "agent" to the port "port" of the demultiplexer "dmux_" associated with all routing modules in the instvar "ptnotif_"

attach{agent port}   Attaches an input agent "`agent`" to the port
"`port`" of the Node; Sets up necessary instvars and
invoke instprocs "`add-target`" to install the input
agent "`agent`" in slot "`port`" of the demultiplexer
"`dmux_`" associated with all routing modules in the
instvar "`ptnotif_`".

detach{agent ...   Recursively detaches an input agent "`agent`" from
nullagent}   the demultiplexer "`dmux_`" associated with all
routing modules in "`ptnotif_`". Replaces the
"`agent`" installed in the demultiplexer with the
input null agent "`nullagent`".

Classifier Manipulation Instprocs

insert-entry{module...   Inserts an input classifier "`clsfr`" as the
clsfr hook}   head (i.e., the first) classifier connecting from
the Node entry, and installs the existing (if
any) head classifier in the slot "`hook`" of the
classifier "`clsfr`". Also, updates the instvars
"`hook_assoc_`" and "`mod_assoc_`"
accordingly.

install-entry{module...   Does what the instproc "`insert-entry`"
clsfr hook}   does. Also destroy the existing head classifier
if any.

install-demux{demux...   Replaces the existing demultiplexer "`dmux_`"
port}   with the input demultiplexer "`demux`". If
"`port`" is an integer, installs the existing
demultiplexer "`dmux_`" in the slot "`port`" of
"`demux`".

mk-default-classifier{}   Creates classifiers and routing modules specified in the instvar "`module_list_`", and asso-
ciates them to the Node.

6.1.3 Default Nodes and Node Configuration Interface

A default NS2 Node is based on flat-addressing and static routing. With flat-addressing, an address of every new node is incremented by one from that of the previously created node. Static routing assumes no change in topology. The routing table is computed once at the beginning of the Simulation phase and does not change thereafter. By default, NS2 employs the Dijkstra shortest path algorithm [17] to compute optimal routes for all pairs of Nodes. The details about other routing protocols as well as hierarchical addressing can be found in the NS manual [15].

To provide a default Node with more functionalities such as link layer or Medium Access Control (MAC) protocol functionalities, we may use instproc `node-config` of class `Simulator` whose syntax is as follows:

```
$ns node-config -<option> [<value>]
```

where $ns is the Simulator object. This instproc does not immediately configure the Nodes as specified in the <option>. Instead, it stores <value> in the instvars of the Simulator corresponding to <option>. This stored configuration will be used during a Node construction process. Therefore, this instproc must be executed prior to the Node construction.

An example use of the instproc node-config{args} for the default setting is shown below:

```
$ns_ node-config -addressType    flat
                 -adhocRouting
                 -llType
                 -macType
                 -propType
                 -ifqType
                 -ifqLen
                 -phyType
                 -antType
                 -channel
                 -channelType
                 -topologyInstance
```

By default, almost every option is specified as NULL with the exception of addressType, which is set to be flat addressing. Another important option reset is used to restore default parameter setting:

```
$ns node-config -reset
```

The details of instproc node-config (e.g., other options) can be found in the file ˜ns/tcl/lib/ns-lib.tcl and [15].

6.2 Routing Mechanism in NS2

In general, a Node may connect to several downstream NsObjects (i.e., targets). As a router, it needs to select one of the downstream NsObjects as a forwarding NsObject for each incoming packet. In most cases, this process is carried out using a so-called routing table each row of which is called a routing entry. A routing entry specifies a forwarding NsObject for a packet which matches a predefined criterion. For example, (dst,target) specifies that, a packet whose destination address is dst, must be forwarded to a forwarding NsObject target.

The routing mechanism in NS2 consists of four main components:

- *Routing agent*: collects information (e.g., the network topology) needed to compute a routing table.

- *Route logic*: uses the information collected by the routing agent, and compute the routing table.
- *Classifier*: employs the computed routing table for packet forwarding.
- *Routing module*: acts as a single point of management of a group of classifiers in a Node. It takes configuration commands from a routing agent, a route logic, and a Node, and propagates them to relevant classifiers.

In this book we focus on static routing, where routing agents are not involved in the routing process. Therefore, we omit the details of routing agents hereafter (the details of which can be found in [15]).

Figure 6.2 shows the routing components in NS2. Each box in this figure represents an object whose type is indicated on the top, while each word in a box represents an instproc of the corresponding object. The arrow shows the sequence of instproc invocation (details of the instprocs will be shown later in this chapter). For example, the instproc `new{...}` of the Node invokes the insproc `register{proto args}` of the routing module.

Depending on their functionality, the above four routing components are stored in different simulation objects. A route logic computes the routing table for every node. It is shared by several simulation objects, and is therefore stored in the Simulator. Acting as a routing table, an address classifier is specific to and is hence stored in a Node. A routing module is an interface to all the routing components of a Node. Hence, it is stored as an instvar of a Node.

Next, we will discuss the details of route logic, classifiers, and routing module in Sections 6.3, 6.4, and 6.5, respectively. Then, in Section 6.6, we will revisit NS2 routing mechanism, and discuss how the above routing components are configured in a Node.

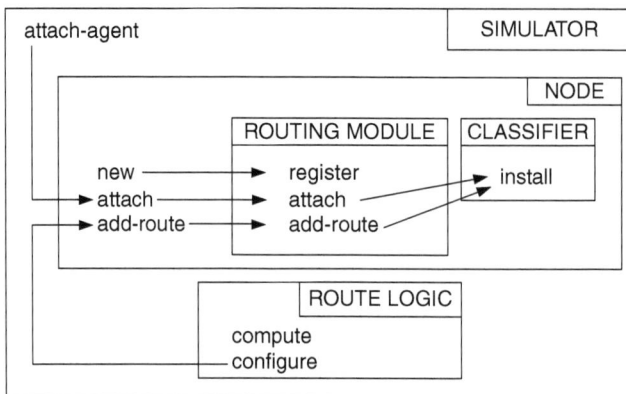

Fig. 6.2. Configuration of routing components in NS2.

6.3 Route Logic

The main responsibility of a route logic object is to compute the routing table. Route logic is implemented in a C++ class `RouteLogic` which is bound to the OTcl class with the same name (see Program 6.1). Class `RouteLogic` has two key variables: "`adj_`", which is the adjacency matrix used to compute the routing table, and "`route_`", which is the routing table. It has the following three main functions:

Program 6.1 Declaration of class `RouteLogic` and the corresponding OTcl mapping class.

```
   //~/ns/routing/route.h
1  class RouteLogic : public TclObject {
2  public:
3      RouteLogic();
4      ~RouteLogic();
5      int command(int argc, const char*const* argv);
7      virtual int lookup_flat(int sid, int did);
8  protected:
9      void reset(int src, int dst);
10     void reset_all();
11     void compute_routes();
12     void insert(int src, int dst, double cost);
13     void insert(int src, int dst, double cost, void* entry);
14     adj_entry *adj_;
15     route_entry *route_;
16 };

   //~/ns/routing/route.cc
17 class RouteLogicClass : public TclClass {
18 public:
19     RouteLogicClass() : TclClass("RouteLogic") {}
20     TclObject* create(int, const char*const*) {
21         return (new RouteLogic());
22     }
23 } routelogic_class;
```

insert(src,... Inserts a new entry including a source ID (src), a
    dst,cost) destination ID (dst), and the corresponding routing cost
      (cost) into the adjacency matrix.
compute_route() Uses the adjacency matrix adj_ to compute the optimal
    routes for all source-destination pairs and store the com-
    puted routes in the variable route_.
lookup_flat(... Searches within variable route_ for an entry with
    sid,did) matching source ID (sid) and destination ID (did), and
    returns the index of the forwarding object (e.g.,
    connecting link).

Program 6.2 Instprocs register, configure and lookup of class
RouteLogic.

```
    //~/ns/tcl/lib/ns-route.tcl
1   RouteLogic instproc register {proto args} {
2       $self instvar rtprotos_ node_rtprotos_ default_node_rtprotos_
3       if [info exists rtprotos_($proto)] {
4           eval lappend rtprotos_($proto) $args
5       } else {
6           set rtprotos_($proto) $args
7       }
8   }

9  RouteLogic instproc configure {} {
10     $self instvar rtprotos_
11     if [info exists rtprotos_] {
12         foreach proto [array names rtprotos_] {
13             eval Agent/rtProto/$proto init-all $rtprotos_($proto)
14         }
15     } else {
16         Agent/rtProto/Static init-all
17     }
18 }

19 RouteLogic instproc lookup { nodeid destid } {
20     if { $nodeid == $destid } {
21         return $nodeid
22     }
23     set ns [Simulator instance]
24     set node [$ns get-node-by-id $nodeid]
25     $self cmd lookup $nodeid $destid
26 }
```

In the interpreted hierarchy, the OTcl class `RouteLogic` has two major instprocs to configure the route logic and one major instproc to query the routing information (see Program 6.2).

register{... proto,args}	Stores a routing agent `<args>` as an element of the instance associative array `rtprotos_` whose index is `<proto>`.
configure	Reads instvar `rtprotos_` and invokes instproc `init-all` of all registered routing agents to create routing tables.
lookup{... nodeid destid}	Looks in the routing table for the forwarding object corresponding to the input source and destination pair (`nodeid`,`destid`). Returns `nodeid` (Line 11) if `nodeid=destid`. Otherwise, returns the forwarding object returned from the function `lookup_flat` of the C++ class `RouteLogic`.

6.4 Classifiers: Multi-target Packet Forwarders

A classifier is a packet forwarding object with multiple connecting target. It forwards incoming packets whose header matches with a certain criterion (e.g., same destination host) to the same forwarding NsObject. Similar to a Connector, a classifier identifies each target using a pointer. It installs each of these pointers so-called *slots*. Based on a predefined criterion, a classifier selects a slot for each incoming packet, and forwards the packet to the NsObject whose pointer is installed in that slot. In this section, we will explain the packet forwarding mechanism, the internal variables and functions, and the configuration interface of the classifiers. The process of assembling classifiers and composing a Node will be discussed in Section 6.6.

6.4.1 Class `Classifier` and Its Main Components

NS2 implements classifiers in a C++ class `Classifier` (see the declaration in Program 6.3), which is bound to an OTcl class with the same name. The main components of a classifier include the following.

C++ Variables

The C++ class `Classifier` has two key variables: `slot_` and `default_target_`. The variable `slot_` (Line 13 in Program 6.3) is a linked list of pointers whose entries are a pointer a to downstream NsObjects. Each of these NsObjects corresponds to a predefined criterion. Packets matched with a predefined criterion are forwarded to the corresponding NsObject. Class `Classifier` also define another pointer to an NsObject, `default_target_`. The variable `default_target_` points to a receiving NsObject for packets which do not match with any predefined criterion.

Program 6.3 Declaration of class `Classifier`.

```
    //~/ns/classifier/classifier.h
1   class Classifier : public NsObject {
2   public:
3       Classifier();
4       virtual ~Classifier();
5       virtual void recv(Packet* p, Handler* h);
6       virtual NsObject* find(Packet*);
7       virtual int classify(Packet *);
8       virtual void clear(int slot);
9       virtual void install(int slot, NsObject*);
10      inline int mshift(int val) {return((val >> shift_) & mask_);}
11  protected:
12      virtual int command(int argc, const char*const* argv);
13      NsObject** slot_;
14      NsObject *default_target_;
15      int shift_;
16      int mask_;
17  };
```

The class `Classifier` also have two supplementary variables: `shift_` (Line 15) and `mask_` (Line 16). These two variables are used in function `mshift(val)` (Line 10) to reformat the address (see also Section 12.4).

C++ Functions

The main C++ functions of class `Classifier` can be classified into packet forwarding functions (i.e., `recv(p,h)`, `find(p)`, and `classify(p)`) and configuration functions (i.e., `install(slot,p)`, `install_next(node)`, `do_install(dst,target)`, and `clear(slot)`).

`recv(p,h)`	Receives a packet `*p` and handler `*h`.
`find(p)`	Returns a forwarding NsObject pointer for an incoming packet `*p`.
`classify(p)`	Returns a slot number of an entry which match with the header of an incoming packet `*p`.
`install(slot,p)`	Stores the input NsObject pointer "p" in the slot number "slot" of the variable `slot_`.
`install_next(node)`	Installs the NsObject pointer "node" in the next available slot.
`do_install(...dst,target)`	Installs an input NsObject pointer `target` in the slot number `dst`.
`clear(slot)`	Removes the NsObject pointer installed in the slot number "slot".

mshift(val) Shifts val to the left by "shift_" bits. Masks the shifted value
by using a logical AND (&) operation with "mask_".

As an NsObject, a classifier receives a packet by having its upstream object
invoke its function recv(p,h), passing the packet "*p" and a handler "*h"
as input arguments. In Program 6.4, Line 3 retrieves for an NsObject pointer
"node" for an incoming packet "*p" by invoking function find(*p). Then,
Line 8 passes the packet "*p" and the handler "*h" to its forwarding NsObject
*node by executing node->recv(p,h).

Program 6.4 Functions recv and find of class Classifier.

```
      //~/ns/classifier/classifier.cc
 1    void Classifier::recv(Packet* p, Handler* h)
 2    {
 3        NsObject* node = find(p);
 4        if (node == NULL) {
 5            Packet::free(p);
 6            return;
 7        }
 8        node->recv(p,h);
 9    }

10    NsObject* Classifier::find(Packet* p)
11    {
12        NsObject* node = NULL;
13        int cl = classify(p);
14        if (cl < 0 || cl >= nslot_ || (node = slot_[cl]) == 0) {
15            /*There is no potential target in the slot;*/
16        }
17        return (node);
18    }
```

Function find(p) (Lines 10–18 in Program 6.4) examines the incoming
packet *p, and retrieves the matched NsObject pointer installed in the variable
slot_. Line 13 invokes function classify(p) to retrieve the slot number (cl)
corresponding to the packet *p. Then, Lines 14 and 17 return the NsObject
pointer (i.e., node) stored in slot cl of variable slot_.

Function classify(p) is perhaps the most important function of a classifier. This is the place where the classification criterion is defined. The function
classify(p) returns an NsObject pointer installed in the slot whose criterion matches with the input packet *p. Since classification criteria could be
different for different types of classifiers, the function classify(p) is usually
overridden in the derived classes of class Classifier. In Sections 6.4.2 and

6.4.3, we will show two example implementations of function classify(p) in classes HashClassifier and PortClassifier, respectively.

Program 6.5 Functions clear, install, and install_next of class Classifier.

```
//~ns/classifier/classifier.cc
1   void Classifier::install(int slot, NsObject* p)
2   {
3       if (slot >= nslot_)
4           alloc(slot);
5       slot_[slot] = p;
6       if (slot >= maxslot_)
7           maxslot_ = slot;
8   }

9   int Classifier::install_next(NsObject *node) {
10      int slot = maxslot_ + 1;
11      install(slot, node);
14
12      return (slot);
13  }

14  void Classifier::clear(int slot)
15  {
16      slot_[slot] = 0;
17      if (slot == maxslot_)
18          while (--maxslot_ >= 0 && slot_[maxslot_] == 0);
19  }

    //~ns/classifier/classifier.h
20  virtual void do_install(char* dst, NsObject *target) {
21      int slot = atoi(dst);
22      install(slot, target);
23  }
```

Consider Program 6.5. Function install(slot,p) stores the input NsObject pointer "p" in the slot number "slot" of the variable "slot_" (Line 5), and updates the variable maxslot_ (the total number of slots) if necessary. Function install_next(node) installs the input NsObject pointer "node" in the next available slot (Lines 10–11). Function do_install(dst,target) converts dst to be an integer variable (Line 21), and installs the NsObject pointer target in the slot corresponding to dst (Line 22). Finally, function clear(slot) removes the installed NsObject pointer from the slot number "slot" of the variable slot_ (Line 6).

Defined in Line 10 of Program 6.3, function mshift(val) simply returns val. The constructor of class Classifier sets the default values of shift_ and mask_ to be zero and 0xffffffff. The function mshift(val) shifts the input argument val by zero bit. Also, the logical AND with 0xffffffff leaves the input argument unchanged. Hence, function mshift(val) of class Classifier has no effect on the input argument val.

OTcl Commands

Class Classifier also defines the following key OTcl commands in a C++ function command of class Classifier. These OTcl command can be invoked from the OTcl domain.

slot{index} Returns the NsObject stored in the slot number index
clear{slot} Clears the NsObject pointer installed in the slot number slot.
install{index object} Installs object in the slot number index.
installNext{object} Installs object in the next available slot.

6.4.2 Hash Classifiers

An Overview of Hash Classifiers

Hash table is a data structure which facilitate a key-value lookup process[1]. It eliminates the need to sequentially search for a matched key and retrieve the corresponding value. A hash table uses a hash function to transform a hash key into a hash index, and stores the corresponding hash value in an array entry (i.e., a record of the hash table) whose index corresponding to the hash index. Given a hash key, the search process transforms the hash key into a hash index using a hash function, and directly accesses the array entry corresponding to the hash index. Since a hash function has low complexity, the search time when using hash table is usually much smaller than that when using a sequential search.

In NS2, a hash classifier classifies packets based on a hash table. Table 6.1 shows an example of hash tables used for a hash classifier. Here, each row of the hash table is called a *hash record*. A *hash value* is the slot number. A *hash key* has three components: Flow ID, source address, and destination address. A hash classifier examines the header of an incoming packet, searches in the hash table for a hash entry whose key matches with information provided in the packet header, and returns the hash value (i.e., slot number) of the matched entry. From Table 6.1, the hash classifier returns the slot number 1 for a packet

---

[1] Suppose we have a table which associates keys and values. The objective of a key-value lookup process is as follows. Given a key, search in the table for the matched key and return the corresponding value.

Table 6.1. An example of hash table.

Slot number	Flow ID	Source address	Destination address
1	1	1	1
2	1	1	2
⋮	⋮	⋮	⋮

with (flow ID, source address, destination address) = (1,1,1), and returns 2 for a packet with (flow ID, source address, destination address) = (1,1,2).

Implementation of Hash Classifier in NS2

Hash classifier is declared in a C++ class HashClassifier in the compiled hierarchy (Program 6.6), and mapped to an OTcl class Classifier/Hash in the interpreted hierarchy.

Program 6.6 Declaration of class HashClassifier.

```
   //~ns/classifier/classifier-hash.h
 1 class HashClassifier : public Classifier {
 2 public:
 3     HashClassifier(int keylen): default_(-1), keylen_(keylen);
 4     ~HashClassifier();
 5     virtual int classify(Packet *p);
 6     virtual long lookup(Packet* p) ;
 7     void set_default(int slot) { default_ = slot; }
 8 protected:
 9     long lookup(nsaddr_t src, nsaddr_t dst, int fid);
10     void reset();
11     int set_hash(nsaddr_t src, nsaddr_t dst, int fid, long slot);
12     long get_hash(nsaddr_t src, nsaddr_t dst, int fid);
13     virtual int command(int argc, const char*const* argv);
14     virtual const char* hashkey(nsaddr_t, nsaddr_t, int)=0;
15     int default_;
16     Tcl_HashTable ht_;
17     int keylen_;
18 };
```

Declared in Program 6.6, the class HashClassifier has three main variables. First, variable default_ (Line 15) contains the default slot for a packet which does not match with any entry in the table. Secondly, variable ht_ (Line 16) is the hash table. Finally, variable keylen_ (Line 17) is the total number of hash keys. By default, the hash keys include flow ID, source address, and destination address, and the variable keylen_ is 3.

Apart from function `classify(p)` derived from class `Classifier`, class `HashClassifier` defines the following functions (see the function declaration in Program 6.6):

lookup(p) Returns the slot number of the entry which matches with the incoming packet "p" (Line 6).

lookup(src,... Returns the slot number of the entry whose source
dst,fid) address, destination address, and flow ID are "src", "dst", and "fid", respectively. (Line 9).

set_hash(src,... Inserts an entry with source address "src", destination
dst,fid,slot) address "dst", and flow ID "fid" to the hash table, and associates the entry to slot number "slot" (Line 11).

get_hash(src,... Returns the slot number which matches with the
dst,fid) values returned from function `hashkey` (...) (Line 12).

hashkey(src,... Returns an identifier for a hash entry corresponding to
dst,fid) the input hash key (src,dst,fid). This function is pure virtual and should be overridden by child classes of `HashClassifier`.

Program 6.7 Functions `lookup` and `get_hash` of class `HashClassifier`.

```
    //~ns/classifier/classifier-hash.cc
1   long HashClassifier::lookup(Packet* p) {
2       hdr_ip* h = hdr_ip::access(p);
3       return get_hash(mshift(h->saddr()),mshift(h->daddr()),
                                               h->flowid());
4   }

5   long HashClassifier::get_hash(nsaddr_t src,
                                  nsaddr_t dst, int fid) {
6       Tcl_HashEntry *ep= Tcl_FindHashEntry(&ht_,
                                  hashkey(src, dst, fid));
7       if (ep)
8           return (long)Tcl_GetHashValue(ep);
9       return -1;
10  }
```

Program 6.7 shows the details of functions `lookup(p)` and `get_hash(src, dst,fid)` of class `HashClassifier`. Function `lookup(p)` returns the slot number of an entry whose source address, destination address, and flow ID match with those indicated in the header of an incoming packet `*p`[2] (by invoking function `get_hash(...)` in Line 3). To retrieve an entry, the function `get_hash(...)` invokes function `Tcl_FindHashEntry` (...) to get the input

---

[2] See the details of IP packet header in Section 8.3.3.

entry from the hash table `ht_` in Line 6. If the entry exists, Line 8 will retrieve the slot number by invoking function `Tcl_GetHashValue(ep)`. Declared as pure virtual in class `HashClassifier`, function `hashkey(...)` (invoked in Line 6), which computes a hash index from a hash key, should be overridden by the child classes of class `HashClassifier`.

Program 6.8 Declaration of class `DestHashClassifier`.

```
     //~ns/classifier/classifier-hash.h
 1   class DestHashClassifier : public HashClassifier {
 2   public:
 3       DestHashClassifier() : HashClassifier(TCL_ONE_WORD_KEYS) {}
 4       virtual int command(int argc, const char*const* argv);
 5       int classify(Packet *p);
 6       virtual void do_install(char *dst, NsObject *target);
 7   protected:
 8       const char* hashkey(nsaddr_t, nsaddr_t dst, int) {
 9           long key = mshift(dst);
10           return (const char*) key;
11       }
12   };
```

As an example, consider class `DestHashClassifier` (Program 6.8), a child class of class `HashClassifier`, which classifies incoming packets by the destination address only. Class `DestHashClassifier` overrides functions `classify(p)`, `do_install(dst,target)`, and `hashkey(...)`, and uses other functions (e.g., `lookup(p)`) of class `HashClassifier` (i.e., its parent class).

Program 6.9 shows the implementation of function `classify(p)` of class `DestHashClassifier`. This function obtains a matching slot number "slot" by invoking `lookup(p)` (Line 2; See also Fig. 6.3), and returns "slot" if it is valid (Line 4). Otherwise, Line 6 will return variable "default_" if "slot" is invalid. If neither `slot` nor `default_` is valid, Line 7 will return −1, indicating no matching entry in the hash table. Function `do_install(dst,target)` installs (Line 12) an `NsObject` pointer `target` in the next available slot, and registers this installation in the hash table (Line 13). Defined in class `Classifier`, function `getnxt(target)` in Line 11 returns the slot where `target` is installed or the next available slot if `target` is not found. Again, the statement `set_hash(0,d,0,slot)` inserts an entry with source address "0", destination address "d", and flow ID "0" into the hash table, and associates the entry with a slot number "slot".

Figure 6.3 shows a process when a `DestHashClassifier` object invokes function `lookup(p)`. In this figure, the function name is indicated at the top of each box, while the corresponding class is shown in the right of a block arrow. The process follows what we discussed earlier. The important point here is the function `hashkey(...)`. From Lines 8–11 in Program 6.8,

Program 6.9 Functions `classify` and `do_install` of class `DestHashClassifier`.

```
    //~ns/classifier/classifier-hash.cc
1   int DestHashClassifier::classify(Packet * p) {
2       int slot = lookup(p);
3       if (slot >= 0 && slot <=maxslot_)
4           return (slot);
5       else if (default_ >= 0)
6           return (default_);
7       else return (-1);
8   }

9   void DestHashClassifier::do_install(char* dst, NsObject *target) {
10      nsaddr_t d = atoi(dst);
11      int slot = getnxt(target);
12      install(slot, target);
13      if (set_hash(0, d, 0, slot) < 0)
14          /* show error */
15  }
```

class `DestHashClassifier` overrides function `hashkey(...)` by returning the destination address (see the detail of function `mshift(val)` in Line 10 of Program 6.3). In Fig. 6.3, functions `lookup(p)` and `get_hash(...)` belong to class `HashClassifier`, while function `hashkey(...)` is attributed to class

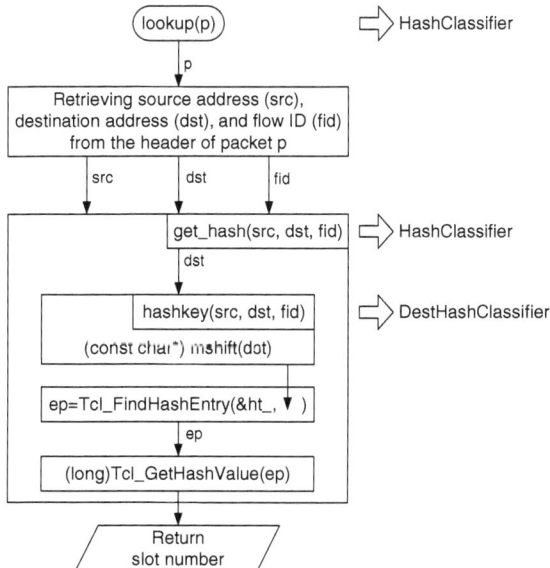

Fig. 6.3. Flowchart of function `lookup(p)` invoked from class `DestHashClassifier`.

DestHashClassifier. This is a beauty of OOP, since we only need to override one function for a derived class (e.g., class DestHashClassifier), and are able reuse the rest of the code from the parent class (e.g., class HashClassifier).

Apart from class DestHashClassifier, class HashClassifier has three other major child classes (class names on the left and right are compiled and interpreted classes, respectively):

- SrcDestHashClassifier ⇔ Classifier/Hash/SrcDest: classifies packets based on source and destination addresses.
- FidHashClassifier ⇔ Classifier/Hash/Fid: classifies packets based on a flow ID.
- SrcDestFidHashClassifier ⇔ Classifier/Hash/SrcDestFid: classifies packets based on source address, destination address, and flow ID.

6.4.3 Port Classifiers

A port classifier classifies packets based on the destination port. From Line 5 in Program 6.10, function classify(p) returns the destination port number of the IP header of the incoming packet p.

Program 6.10 Function classify of class PortClassifier.

```
   //~ns/classifier/classifier-port.cc
 1 int PortClassifier::classify(Packet *p)
 2 {
 3   hdr_ip* iph = hdr_ip::access(p);
 4   return iph->dport();
 5 }
```

A port classifier is used as a demultiplexer (e.g., dmux_ in Fig. 6.1) which bridges a node to a receiving transport agent. When function recv(p,h) of dmux_ (i.e., a PortClassifier object) is invoked, the packet is forwarded to an NsObject associated with slot_[cl], where cl is the destination port number specified in the packet header. By installing a receiving agent in slot_[cl], the classifier forwards packets whose destination port is "cl" to the receiving agent.

6.4.4 Installing Classifiers in a Node

This section discusses the how classifiers are installed in a Node. As shown in Fig. 6.1, a Node can have more than one classifier. These classifiers are inter-connected and form a so-called *chain of classifiers*.

Class Node has three instvars related to classifier installation: classifier_, hook_assoc_, and mod_assoc_. Instvar classifier_ is the head of the chain

Table 6.2. An example of hook_assoc_ for a chain of classifiers

index	_o2	_o3	_o4
hook_assoc_(index)	_o1	_o2	_o3

of classifiers, which connects from the node entry. Instvar hook_assoc_ is an associative array whose index is a classifier and its value is the downstream classifier in the chain. For example, let us install classifiers _o1, _o2, _o3, and _o4 in sequence into a Node. Then, the instvar classifier_ would be _o4. The value of hook_assoc_ in this case is shown in Table 6.2. Finally, instvar mod_assoc_ is an associative array whose index is a classifier and its value is the associated routing module.

As discussed in Section 6.1, class Node provides three instprocs to configure classifiers. First, as shown in Program 6.11, instproc insert-entry{module clsfr hook} takes three input arguments: a routing module module, a classifier clsfr, and an optional argument hook. Line 4 updates the instvar hook_assoc_. Line 8 installs the current head classifier in the slot number "hook" of the input classifier clsfr. Line 11 associates clsfr with the input routing module module. Line 12 replaces the head classifier classifier_ with the input classifier clsfr. Note that clsfr does not need to be a classifier. If clsfr is an NsObject, it can be inserted into the head of the chain. In this case, hook must be specified as "target" so that Line 6 will set the target of clsfr to be the head classifier.

Program 6.11 Instproc insert-entry of class Node

```
   //~ns/tcl/lib/ns-node.tcl
1  Node instproc insert-entry { module clsfr {hook ""} } {
2      $self instvar classifier_ mod_assoc_ hook_assoc_
3      if { $hook != "" } {
4          set hook_assoc_($clsfr) $classifier_
5          if { $hook == "target" } {
6              $clsfr target $classifier_
7          } elseif { $hook != "" } {
8              $clsfr install $hook $classifier_
9          }
10     }
11     set mod_assoc_($clsfr) $module
12     set classifier_ $clsfr
13 }
```

The second classifier configuration instproc install-entry{module clsfr hook} is shown in Program 6.12. It is very similar to instproc insert-entry. The only difference is, it also destroys the existing head classifier, if any.

Program 6.12 Instproc `install-entry` of class `Node`.

```
   //~ns/tcl/lib/ns-node.tcl
 1 Node instproc install-entry { module clsfr {hook ""} } {
 2     $self instvar classifier_ mod_assoc_ hook_assoc_
 3     if [info exists classifier_] {
 4         if [info exists mod_assoc_($classifier_)] {
 5             $self unregister-module $mod_assoc_($classifier_)
 6             unset mod_assoc_($classifier_)
 7         }
 8         if [info exists hook_assoc_($classifier_)] {
 9             if { $hook == "target" } {
10                 $clsfr target $hook_assoc($classifier_)
11             } elseif { $hook != "" } {
12                 $clsfr install $hook $hook_assoc_($classifier_)
13             }
14             set hook_assoc_($clsfr) $hook_assoc_($classifier_)
15             unset hook_assoc_($classifier_)
16         }
17     }
18     set mod_assoc_($clsfr) $module
19     set classifier_ $clsfr
20 }
```

Finally, Program 6.13 shows the details of instproc **install-demux**{demux port}. This instproc takes two input arguments: **demux** (mandatory) and **port** (optional). It replaces the existing demultiplexer[3] **dmux_** with the input demultiplexer **demux** (Line 2, 9 and 10). If **port** exists, the current demultiplexer **dmux_** will be installed in the slot number "**port**" of the input demultiplexer **demux** (Lines 5–7).

6.5 Routing Modules

6.5.1 An Overview of Routing Modules

The main functionality of a routing module is to facilitate classifier management. Since a Node maintains only the head of the chain of classifiers, access to a classifier in a long chain could be difficult. In addition, it is fairly inconvenient to (possibly selectively) propagate a configuration command to *several* classifiers. Such the difficulty is shown in Figure 6.4, where 10 address classifiers are connected from the head classifier. As the network topology changes, all the address classifiers need to be reconfigured. NS2 employs routing modules to facilitate the classifier configuration process.

---

[3] A demultiplexer classifies packets based on port number specified in the packet header (see Section 6.4.3 for more details).

Program 6.13 Instproc Node::install-demux.

```
//~ns/tcl/lib/ns-node.tcl
1  Node instproc install-demux {demux {port ""} } {
2      $self instvar dmux_ address_
3      if { $dmux_ != "" } {
4          $self delete-route $dmux_
5          if { $port != "" } {
6              $demux install $port $dmux_
7          }
8      }
9      set dmux_ $demux
10     $self add-route $address_ $dmux_
11 }
```

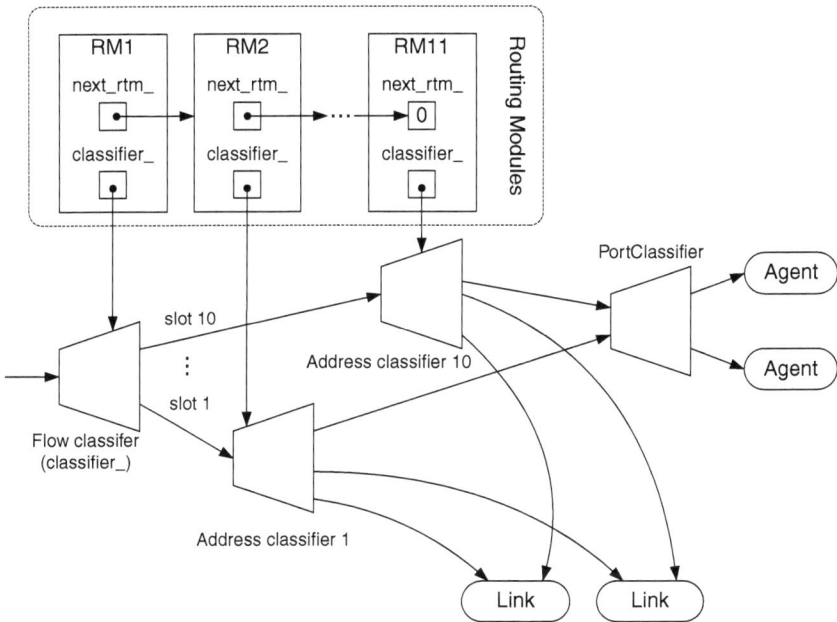

Fig. 6.4. The relationship among routing modules and classifiers in a Node.

Routing modules provide a single point of management for a group of classifiers. Here, each routing module is associated with a classifier, and has a pointer to another routing module (see Fig. 6.4). Together, they form a linked list of routing modules for a group of classifiers. The head of the linked list acts as an interface to propagate configuration commands to classifiers in the linked list. For example, to add a route, we only need to keep the reference of the head routing module (as opposed to keeping the references of 10 address classifiers). Then, the new routing information is entered through this head routing module which will propagate the information to all routing modules in the linked list. Each routing module determines whether the information is relevant to the associated classifier. If so, it will (re)configure the classifier according to the received information. From this point of view, the routing agents and the route logic interact only to the head routing module to deliver classifier configuration commands (e.g., adding or deleting routes) to the relevant classifiers. Note that a classifier can also be configured directly, if the reference is available. Routing modules only facilitate the configuration process of a group of classifiers.

Routing modules are implemented in a C++ class `RoutingModule`, which are bound to an OTcl class `RtModule` (see Program 6.14). Again, these two classes are the base classes from which more specific classes derive (see the built-in routing module classes in Table 6.3). In the following, we will discuss the base class routing module (classes `RoutingModule` and `RtModule`) and the base routing modules (classes `BaseRoutingModule` and `RtModule/Base`) only.

Table 6.3. Built-in routing modules in NS2.

Routing module	C++ class	OTcl class
Routing Module	`RoutingModule`	`RtModule`
Base Routing Module	`BaseRoutingModule`	`RtModule/Base`
Multicast Routing Module	`McastRoutingModule`	`RtModule/Mcast`
Hierarchical Routing Module	`HierRoutingModule`	`RtModule/Hier`
Manual Routing Module	`ManualRoutingModule`	`RtModule/Manual`
Source Routing Module	`SourceRoutingModule`	`RtModule/Source`
Quick Start for TCP/IP Routing Module (Determine initial congestion window)	`QSRoutingModule`	`RtModule/QS`
Virtual Classifier Routing Module	`VCRoutingModule`	`RtModule/VC`
Pragmatic General Multicast Routing Module (Reliable multicast)	`PgmRoutingModule`	`RtModule/PGM`
Light-Weight Multicast Services Routing Module (Reliable multicast)	`LmsRoutingModule`	`RtModule/LMS`

Hereafter, we define a term *name* of a routing module as the suffix (which follows `RtModule/`) of the OTcl class name (see Table 6.3). For example,

Program 6.14 Declaration and the constructor of a C++ class RoutingModule which is bound to an OTcl class RtModule.

```
   //~ns/routing/rtmodule.h
 1 class RoutingModule : public TclObject {
 2 public:
 3     RoutingModule();
 4     inline Node* node() { return n_; }
 5     virtual int attach(Node *n) { n_ = n; return TCL_OK; }
 6     virtual int command(int argc, const char*const* argv);
 7     virtual const char* module_name() const { return NULL; }
 8     void route_notify(RoutingModule *rtm);
 9     void unreg_route_notify(RoutingModule *rtm);
10     virtual void add_route(char *dst, NsObject *target);
11     virtual void delete_route(char *dst, NsObject *nullagent);
12     RoutingModule *next_rtm_;
13 protected:
14     Node *n_;
15     Classifier *classifier_;
16 };

17 static class RoutingModuleClass : public TclClass {
18 public:
19     RoutingModuleClass() : TclClass("RtModule") {}
20     TclObject* create(int, const char*const*) {
21         return (new RoutingModule);
22     }
23 } class_routing_module;

24 RoutingModule::RoutingModule() :
25             next_rtm_(NULL), n_(NULL), classifier_(NULL) {
26     bind("classifier_", (TclObject**)&classifier_);
27 }
```

the name of classes RtModule/Base and RtModule/Hier are Base and Hier, respectively.

6.5.2 C++ Class RoutingModule

Program 6.14 shows the declaration of class RoutingModule, which has three main variables. Variable classifier_ in Line 15 is a pointer to a Classifier object. To provide a single pointer of management for a group of classifiers, routing modules form a linked list using their pointers next_rtm_ (Line 12) to another RoutingModule object. Another important variable is n_ (Line 14), which is a pointer to the associated Node object. These three variables are initialized to NULL in the constructor of class RoutingModule (Line 25).

Also, variable `classifier_` is bound to an OTcl instvar with the same name (Line 26).

The key functions of class `RoutingModule` include (see Program 6.15).

`node()`	Returns the attached `Node` object `n_`.
`attach(n)`	Stores an input `Node` object "n" in the variable `n_`.
`module_name()`	Returns the name of the routing module.
`route_notify(rtm)`	Adds an input `RoutingModule *rtm` to the end of the linked list.
`unreg_route_notify(rtm)`	Removes an input `RoutingModule` pointer `*rtm` from the linked list.
`add_route(dst,target)`	Informs every classifier in the link list to add a routing entry (`dst`,`target`).
`delete_route(...` `dst,nullagent)`	Informs every classifier in the linked list to delete a routing entry with destination `dst`.

Class `RoutingModule` is usually not instantiated from the OTcl domain. Therefore, its name is defined as `NULL` in function `module_name()` (Line 7 in Program 6.14). Its derived classes override this function by returning their own name to the caller (for class `BaseRoutingModule` see Line 4 in Program 6.6).

Program 6.15 shows the details of functions `route_notify(rtm)` and `unreg_route_notify(rtm)`. Function `route_notify(rtm)` recursively invokes itself until it reaches the last routing module in the linked list, where `next_rtm_` is `NULL`. Then, it attaches the input routing module `*rtm` as the last component of the linked list (Line 5). Function `unreg_route_notify` recursively searches down the linked list (Line 13) until it finds the input routing module pointer `rtm` (Line 9), and removes it from the linked list (Line 10).

Lines 17–30 in Program 6.15 show the details of functions `add_route(dst, target)` and `delete_route(dst,nullagent)`. Function `add_route(dst, target)` takes a destination node `dst` and a forwarding `NsObject` pointer `target` as input arguments. It installs the pointer `target` in all the associated classifiers (Line 20). Again, this entry is propagated down the linked list (Line 22), until reaching the last element of the linked list (Line 14). Function `delete_route(dst,nullagent)` does the opposite of the function `add_route(dst,target)` does. It recursively installs a null agent "nullagent" (i.e., a packet dropping point) as a target for packets destined for a destination node `dst` in all the classifiers, essentially removing the entry with the destination `dst` from all the classifiers.

Class `RoutingModule` also defines three OTcl commands – namely `node`, `attach-node`, and `module-name` – which simply invoke the functions `node()`, `attach(n)`, and `module_name()`, respectively.

Program 6.15 Functions route_notify, unreg_route_notify, add_route, and delete_route of class RoutingModule.

```
   //~ns/routing/rtmodule.cc
1  void RoutingModule::route_notify(RoutingModule *rtm) {
2      if (next_rtm_ != NULL)
3          next_rtm_->route_notify(rtm);
4      else
5          next_rtm_ = rtm;
6  }

7  void RoutingModule::unreg_route_notify(RoutingModule *rtm) {
8      if (next_rtm_) {
9          if (next_rtm_ == rtm) {
10             next_rtm_ = next_rtm_->next_rtm_;
11         }
12         else {
13             next_rtm_->unreg_route_notify(rtm);
14         }
15     }
16 }

17 void RoutingModule::add_route(char *dst, NsObject *target)
18 {
19     if (classifier_)
20         classifier_->do_install(dst,target);
21     if (next_rtm_ != NULL)
22         next_rtm_->add_route(dst, target);
23 }

24 void RoutingModule::delete_route(char *dst, NsObject *nullagent)
25 {
26     if (classifier_)
27         classifier_->do_install(dst, nullagent);
28     if (next_rtm_)
29         next_rtm_->add_route(dst, nullagent);
30 }
```

6.5.3 OTcl Class `RtModule`

In the OTcl domain, the routing module is defined in class `RtModule`. Class `RtModule` has two instvars: `classifier_` and `next_rtm_`. Bound to the compiled variable with the same name, instvar `classifier_` stores a reference to the associated classifier. Instvar `next_rtm_` provides a support to create a linked list of routing module. This instvar has no relationship with variable `next_rtm_` of the compiled class, since the bond is not created in the constructor of the C++ class `RoutingModule` (see Lines 24–27 of Program 6.14).

The OTcl class `RtModule` also defines the following instprocs which can be classified into two categories. For brevity, we do not show the details of these instprocs here. The readers may find the details of these instprocs in file `~ns`/tcl/lib/ns-rtmodule.tcl.

Initialization Instprocs

`register{node}`	Associates the input Node `node` with the routing module, and updates instvars `rtnotif_` and `ptnotif_` of the input Node `node`.
`unregister{}`	Removes the classifier of the routing module. Also removes the routing module from instvars `rtnotif_` and `ptnotif_` of the associated Node.
`route-notify{... module}`	Moves down the linked list in the OTcl domain (via instvar `next_rtm_`) and stores the input routing module "`module`" as the last element of the link-list.
`unreg-route-notify{... module}`	Looks for the input routing module "`module`" and removes it from the linked list of routing modules in the OTcl domain.

Instprocs for Route Addition/Deletion and Agent Attachment/Detachment

`add-route{dst target}`	Adds a routing entry with a destination "`dst`" and a forwarding NsObject "`target`" in all the classifiers in the linked list of routing modules.
`delete-route{... dst nullagent}`	Removes the routing entry with destination `dst` from all the classifiers in the linked list of routing module. Replaces the target of the classifiers with the null agent "`nullagent`".
`attach{agent port}`	Attaches the input agent "`agent`" to the associated Node. Set the target of the input (sending) agent "`agent`" to be the entry of the Node. Also, installs the input (receiving) agent "`agent`" in the slot number "`port`" of the demultiplexer "`dmux_`" of the Node.

6.5.4 C++ Class `BaseRoutingModule` and OTcl class `RtModule/Base`

Derived from the C++ class `RoutingModule`, class `BaseRoutingModule` is declared in Program 6.16, and is bound to an OTcl class `RtModule/Base`. It overrides function `module_name()`, by setting its name to be "Base" (Line 4). A base routing module classifies packets based on its destination address only. Therefore, the type of the variable `classifier_` is defined as a `DestHashClassifier` pointer.

Program 6.16 Declaration of class `BaseRoutingModule` which is bound to the OTcl class `RtModule/Base`.

```
    //~ns/routing/rtmodule.h
1   class BaseRoutingModule : public RoutingModule {
2   public:
3       BaseRoutingModule() : RoutingModule() {}
4       virtual const char* module_name() const { return "Base"; }
5       virtual int command(int argc, const char*const* argv);
6   protected:
7       DestHashClassifier *classifier_;
8   };

    //~ns/routing/rtmodule.cc
9   static class BaseRoutingModuleClass : public TclClass {
10  public:
11      BaseRoutingModuleClass() : TclClass("RtModule/Base") {}
12      TclObject* create(int, const char*const*) {
13          return (new BaseRoutingModule);
14      }
15  } class_base_routing_module;
```

In the OTcl domain, class `RtModule/Base` also overrides instproc `register` { node} of class `RtModule`. We will discuss the details of this instproc later in Section 6.6.4.

6.6 Node Object Configuration

Having discussed the key Node components, we now show how these components are assembled to compose a Node. In Section 6.6.1 we first show the relationship among few closely related Node components. We show the instprocs to add/delete routes in Section 6.6.2, and the instprocs to attach/detach agents in Section 6.6.3. We show the Node construction process (via procedure `new{...}`) in Section 6.6.4. As we will see, the main Node component (e.g., routing module, classifiers, demultiplexer) are assembled during this process.

Finally, the route configuration process (i.e., configuring classifiers) is shown in Section 6.6.5.

6.6.1 Relationship Among Instvars module_list_, reg_module_, rtnotif_, and ptnotif_

As shown in Fig. 6.5, the following five instvars of an OTcl Node are closely related: `module_list_`, `reg_module_`, `rtnotif_`, `ptnotif_`, and `mod_assoc_`. Instvar `module_list_` is a list of strings, each of which represents the name of enabled routing module. Instvar `reg_module_` is an associative array whose index and value are the name of the routing module and the routing module instance. Instvars `rtnotif_` and `ptnotif_` are the objects which should be notified of a route change and an agent attachment/detachment, respectively. While `rtnotif_` is the head of the linked list of the routing modules, `ptnotif_` is simply an OTcl list whose elements contain the routing modules. Finally, instvar `mod_assoc_` is an associative array whose indexes and values are classifiers and the associated routing modules, respectively.

The relationship among `module_list_`, `reg_module_`, `rtnotif_`, and `ptnotif_` is shown in Fig. 6.5. The instvars are shown in boxes, while the instprocs of class `Node` are encircled with ellipses. The arrow from an instproc to an instvar indicates that the instvar is configured from within the instproc. Here, instprocs `enable-module{mod_name}` and `disable-module{mod_name}`

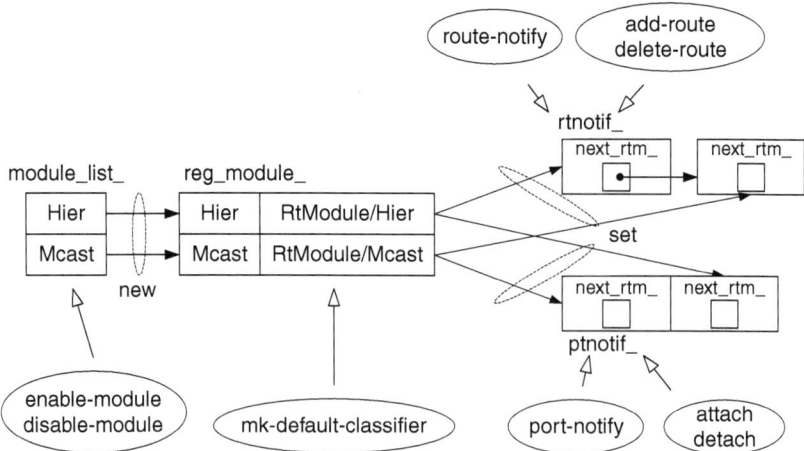

Fig. 6.5. Relationship among instvars module_list, reg_module_, rtnotif_, and ptnotif_ of class Node.

place and remove the name of a routing module `mod_name` in and from instvar `module_list_`, respectively. When instproc `mk-default-classifier` is invoked, the names in `module_list_` are used to instantiate routing module instances. The instantiated objects are stored in the associative array `reg_module_` whose indexes are the corresponding names. Instproc `mk-default-classifier` also invokes the instprocs `route-notify{module}` and `port-notify{module}` to add all the instantiated objects into the list of routing modules `rtnotif_` and `ptnotif_`, respectively. Note that instvar `ptnotif_` is an OTcl list, and its pointer `next_rtm_` is not used. In Fig. 6.5, instprocs `port-notify{...}`, `attach{agent port}`, and `detach{agent nullag ent}` (see file ~*ns*/tcl/lib/ns-node.tcl) can directly access any component of `ptnotif_`. However, instprocs `route-notify{...}`, `add-route{dst target}`, and `delete-route{dst nullagent}` must access a routing module through the head of the linked list (i.e., `rtnotif_`) only.

6.6.2 Adding/Deleting a Routing Entry

A routing entry consists of a destination node address `dst` and a forwarding NsObject `target`. It can be added to a `Node` object by using instproc `add-route{dst target}` of class `Node`. In Program 6.17, instproc `add-route{dst target}` of class `Node` invokes the same instproc of the routing module `rtnotif_` which is of class `RtModule` (Line 4). Line 10 installs the routing entry in the `classifier_` of the routing module. Lines 11–13 recursively invoke instproc `add-route{dst target}` of all the routing modules in the linked list to install the routing entry in the `classifier_` associated with each routing module.

The mechanism for deleting a route entry is similar to that for adding a route entry, and is omitted for brevity. The readers may find the details of route entry deletion in instproc `delete-route{dst nullagent}` of classes `Node` and `RtModule` (see file ~*ns*/tcl/lib/ns-node.tcl and file ~*ns*/tcl/lib/ns-rtmodule.tcl).

6.6.3 Agent Attachment/Detachment

To attach an agent to a Node, we use instproc `attach-agent{node agent}` of class `Simulator` whose syntax is

    $ns attach-agent $node $agent

Here, $ns, $node, and $agent are `Simulator`, `Node`, and `Agent` objects, respectively. Program 6.18 shows the instprocs related to an agent attachment process. The process proceeds as follows:

- `Simulator::attach-agent{node agent}`: Invoke "$node attach $agent" (Line 2).

Program 6.17 Instprocs `add-route` of classes `Node` and `RtModule`.

```
    //~ns/tcl/lib/ns-node.tcl
1   Node instproc add-route { dst target } {
2       $self instvar rtnotif_
3       if {$rtnotif_ != ""} {
4           $rtnotif_ add-route $dst $target
5       }
6       $self incr-rtgtable-size
7   }

    //~ns/tcl/lib/ns-rtmodule.tcl
8   RtModule instproc add-route { dst target } {
9       $self instvar next_rtm_
10      [$self set classifier_] install $dst $target
11      if {$next_rtm_ != ""} {
12          $next_rtm_ add-route $dst $target
13      }
14  }
```

- `Node::attach`{agent port}: Update instvar "`agent`" (Lines 6-8 and Line 16), create "`dmux_`" if necessary (Lines 9-15), and invoke "`$self add-target $agent $port`" (Line 17).
- `Node::add-target`{agent port}: For each routing module "`m`" stored in the instvar `ptnotif_`, execute "`$m attach $agent $port`" (Lines 21-23).
- `RtModule::attach`{agent port}: As a sending agent, set the node entry to be the target of "`agent`" (Line 26). As a receiving agent, install "`agent`" in the slot number "`port`" of demultiplexer "`dmux_`" (Line 27). Note that although an agent can be *either* a sending agent or a receiving agent, this instproc assigns both roles to an agent. This does not cause any problem at runtime due to the following reasons. A sending agent is attached to a source node, and always transmits packets destined to a destination node. It takes no action when receiving a packet from a demultiplexer. A receiving agent, on the other hand, does not generate a packet. Therefore, it can never send a packet to the node entry.

6.6.4 Node Construction

As has already been mentioned before, a `Node` object is created in the OTcl domain by executing "`$ns node`", where `$ns` is the Simulator instance. Instproc "`node`" of class `Simulator` (see Line 4 in Program 6.19) employs instproc "`new{...}`" to create a `Node` object (Line 4 where `node_factory_` is set to `Node` in Line 1). It also updates instvars of the Simulator so that they can be later used by other simulation objects throughout the simulation.

The main steps in the node construction process are shown in Table 6.4.

Program 6.18 Instprocs `attach` and `add-target` of classes `Node`, and instproc `attach` of class RtModule.

```
//~ns/tcl/lib/ns-lib.tcl
1  Simulator instproc attach-agent { node agent } {
2      $node attach $agent
3  }

//~ns/tcl/lib/ns-node.tcl
4  Node instproc attach { agent { port "" } } {
5      $self instvar agents_ address_ dmux_
6      lappend agents_ $agent
7      $agent set node_ $self
8      $agent set agent_addr_ [AddrParams addr2id $address_]
9      if { $dmux_ == "" } {
10         set dmux_ [new Classifier/Port]
11         $self add-route $address_ $dmux_
12     }
13     if { $port == "" } {
14         set port [$dmux_ alloc-port [[Simulator
                                      instance] nullagent]]
15     }
16     $agent set agent_port_ $port
17     $self add-target $agent $port
18 }

19 Node instproc add-target { agent port } {
20     $self instvar ptnotif_
21     foreach m [$self set ptnotif_] {
22         $m attach $agent $port
23     }
24 }

//~ns/tcl/lib/ns-rtmodule.tcl
25 RtModule instproc attach { agent port } {
26     $agent target [[$self node] entry]
27     [[$self node] demux] install $port $agent
28 }
```

Program 6.19 Default value of instvar node_factory_ and instproc node of class Simulator.

```
//~ns/tcl/lib/ns-node.tcl
1  Simulator set node_factory_ Node

//~ns/tcl/lib/ns-node.tcl
2  Simulator instproc node args {
3      $self instvar Node_ routingAgent_
4      set node [eval new [Simulator set node_factory_] $args]
5      set Node_([$node id]) $node
6      $self add-node $node [$node id]
7      $node nodeid [$node id]
8      $node set ns_ $self
9      return $node
10 }
```

Table 6.4. Main steps in the Node construction process.

Step	Class	Instproc	Key statement(s)
1	Node	init	$self mk-default-classifier
2	Node	mk-default-classifier	$self register-module [... new RtModule/Base]
3	Node	register-module{mod}	$mod register $self set reg_module([$mod ... module-name]) $mod
4	RtModule/Base	register{node}	$self next $node $self set classifier_ [... new Classifier/Hash/Dest] 32 $node install-entry $classifier_
5	RtModule	register{node}	$self attach-node $node $node route-notify $self $node port-notify $self

Step 1: Constructor of the OTcl class Node

Instproc init{...} sets up instvars of class Node, and invokes instproc mk-default-classifier{} of the created Node object (Line 22 in Program 6.20).

Program 6.20 Constructor of class Node.

```
     //~/ns/tcl/lib/ns-node.tcl
 1  Node set module_list_ { Base }

 2  Node instproc init args {
 3      eval $self next $args
 4      $self instvar id_ agents_ dmux_ neighbor_ rtsize_ address_ \
 5              nodetype_ multiPath_ ns_ rtnotif_ ptnotif_
 6      set ns_ [Simulator instance]
 7      set id_ [Node getid]
 8      $self nodeid $id_    ;# Propagate id_ into c++ space
 9      if {[llength $args] != 0} {
10          set address_ [lindex $args 0]
11      } else {
12          set address_ $id_
13      }
14      $self cmd addr $address_; # Propagate address_ into C++ space
15      set neighbor_ ""
16      set agents_ ""
17      set dmux_ ""
18      set rtsize_ 0
19      set ptnotif_ {}
20      set rtnotif_ {}
21      set nodetype_ [$ns_ get-nodetype]
22      $self mk-default-classifier
23      set multiPath_ [$class set multiPath_]
24  }

25  Node instproc mk-default-classifier {} {
26      Node instvar module_list_
27      foreach modname [Node set module_list_] {
28          $self register-module [new RtModule/$modname]
29      }
30  }

31  Node instproc register-module { mod } {
32      $self instvar reg_module_
33      $mod register $self
34      set reg_module_([$mod module-name]) $mod
35  }
```

Step 2: Instproc mk-default-classifier{}

Instproc mk-default-classifier{} creates (using new{...}) and registers (using register-module{mod}) routing modules whose names are stored in the instvar module_list_ (Lines 27–29 in Program 6.20). By default, only "Base" routing module is stored in instvar module_list_ (Line 1 in Program 6.20). To enable/disable other routing module, the following two instprocs of class RtModule must be invoked prior to the execution of "$ns node":

    enable-module{name}
    disable-module{name}

where <name> is the name of the routing module, which is to be enabled/ disabled.

Step 3: Instproc register-module{mod} of class Node

This instproc invokes instproc register{node} of the input routing module mod and stores the registered module in the instvar reg_module_.

Step 4: Instproc register{node} of class RtModule/Base

This instproc first invokes instproc register{node} of its parent class (by the statement $self next $node in Line 7 of Program 6.21). Then, Lines 9–12 create (using new{...}) and configure (using install-entry{...}) the head classifier (i.e., classifier_) of the Node.

Step 5: Instproc register{node} of class RtModule

This instproc attaches input Node object "node"to the routing module. It also invokes instproc route-notify{module} and port-notify{module} of the associated Node to include the routing module into the route notification list rtnotif_ and port notification list ptnotif_ of the associated Node (see Program 6.22).

The details of instprocs route-notify{module} and port-notify {module} are shown in Program 6.22. The instproc route-notify{module} takes one input routing module. It stores the module in the last instvar next_rtm_ down the linked list of routing modules (see Lines 6 and 10-17). It also invokes the OTcl command route-notify of the input routing module (Line 8). The OTcl command route-notify invokes the C++ function route_notify(rtm) associated with the attached Node (see Lines 18-24) to store the routing module as the last routing module in the linked list (see Lines 25-30).

As shown in Lines 31-34 of Program 6.22, the instproc port-notify{ module} takes a routing module as an input argument, and appends the input argument module to the end of the link-list.

Program 6.21 Instprocs `register` of classes `RtModule` and `RtModule/Base`.

```
   //~/ns/tcl/lib/ns-rtmodule.tcl
 1 RtModule instproc register { node } {
 2     $self attach-node $node
 3     $node route-notify $self
 4     $node port-notify $self
 5 }

 6 RtModule/Base instproc register { node } {
 7     $self next $node
 8     $self instvar classifier_
 9     set classifier_ [new Classifier/Hash/Dest 32]
10     $classifier_ set mask_ [AddrParams NodeMask 1]
11     $classifier_ set shift_ [AddrParams NodeShift 1]
12     $node install-entry $self $classifier_
13 }
```

6.6.5 Route Configuration

At the beginning of the Simulation Phase, NS2 computes the optimal routes for all source-destination nodes, using the Dijkstra shortest path algorithm [17]. It installs the computed routing information in all the Nodes. This phase commences by the execution of instproc run{} of the Simulator. Table 6.5 shows the main steps in the instproc run{} which are related to the route configuration process.

Table 6.5. Main steps in the route configuration process.

Step	Class	Instproc	Invocation
1	Simulator	run	[$self get-routelogic] configure
2	RouteLogic	configure	Agent/rtProto/Static init-all
3	Agent/... rtProto/Static	init-all	[Simulator instance] ... compute-routes
4	Simulator	compute-routes	$self compute-flat-routes
5	Simulator	compute-flat- routes	set r [$self get-routelogic] $r compute set n [Node set nn_] $self ... populate-flat-classifiers $n

Program 6.22 Instprocs and functions which are related to instprocs route-notify and port-notify of the OTcl class Node.

```
//~/ns/tcl/lib/ns-node.tcl
1  Node instproc route-notify { module } {
2      $self instvar rtnotif_
3      if {$rtnotif_ == ""} {
4          set rtnotif_ $module
5      } else {
6          $rtnotif_ route-notify $module
7      }
8      $module cmd route-notify $self
9  }


   //~/ns/tcl/lib/ns-rtmodule.tcl
10 RtModule instproc route-notify { module } {
11     $self instvar next_rtm_
12     if {$next_rtm_ == ""} {
13         set next_rtm_ $module
14     } else {
15         $next_rtm_ route-notify $module
16     }
17 }


   //~ns/routing/rtmodule.cc
18 int BaseRoutingModule::command(int argc, const char*const* argv) {
19     Tcl& tcl = Tcl::instance();
20     if (argc == 3) {
21         if (strcmp(argv[1] , "route-notify") == 0) {
22             n_->route_notify(this);
23         }
24 }


   //~ns/common/node.cc
25 void Node::route_notify(RoutingModule *rtm) {
26     if (rtnotif_ == NULL)
27         rtnotif_ = rtm;
28     else
29         rtnotif_->route_notify(rtm);
30 }


   //~/ns/tcl/lib/ns-node.tcl
31 Node instproc port-notify { module } {
32     $self instvar ptnotif_
33     lappend ptnotif_ $module
34 }
```

Step 1: Instproc run{} of class Simulator

Shown in Line 2 of Program 4.12, instproc run{} of class Simulator retrieves the RouteLogic object using its instproc get-routelogic{} and invokes instproc configure{} associated with the retrieved RouteLogic object.

Step 2: Instproc configure{} of class RouteLogic

Defined in file ~ns/tcl/lib/ns-route.tcl, instproc configure{} of class Route Logic configures the routing table for all the Nodes by invoking instproc init-all{} of class Agent/rtProto/Static.

Step 3: Instproc init-all{} of class Agent/rtProto/Static

Defined in file ~ns/tcl/rtglib/route-proto.tcl, instproc init-all{} of class Agent /rtProto/Static invokes the instproc compute-routes{} of the Simulator.

Step 4: Instproc compute-routes{} of class Simulator

By default, NS2 uses flat addressing. Therefore, instproc compute-routes{} of class Simulator invokes instproc compute-flat-routes{} to compute and setup the routing table (see file ~ns/tcl/lib/ns-route.tcl).

Step 5: Instproc compute-flat-routes{} of class Simulator

Defined in file ~ns/tcl/lib/ns-route.tcl, instproc compute-flat-routes{} of class Simulator retrieves the associated route logic object (using instproc get-routelogic{}), computes the optimal route using the retrieved object (using instproc compute{}), and configures the classifiers in all the Nodes according to the computed route (using the command populate-flat-classifiers{n}).

Program 6.23 shows the details of OTcl command populate-flat-cla ssifiers{n}. This OTcl command stores the input number of nodes "n" in the variable nn_ (Line 4), and invokes function populate_flat_classifiers() (Line 5) to install the computed route in all the classifiers.

As shown in Lines 10–25 of Program 6.23, function populate_flat_class ifiers() is run for all pairs (i,j) of nn_ nodes. For each pair, Line 16 retrieves the next hop (i.e., forwarding) referencing point nh of a forwarding object for a packet traveling from Node "i" to Node "j", and Line 18 retrieves the link entry point l_head corresponding to the variable nh. Lines 19–20 add a new routing entry for the node i (i.e., nodelist_[i]). The entry specifies the link entry l_head as a forwarding target for packet destined for a destination node j. The entry is included to the Node "i" via its function add_route(dst,target).

Program 6.23 An OTcl command populate-flat-classifiers, a function populate_flat_classifiers of class Simulator, and a function add_route of class Node.

```
//~ns/common/simulator.cc
1  int Simulator::command(int argc, const char*const* argv) {
2       ...
3      if (strcmp(argv[1], "populate-flat-classifiers") == 0) {
4          nn_ = atoi(argv[2]);
5          populate_flat_classifiers();
6          return TCL_OK;
7      }
8       ...
9  }

10 void Simulator::populate_flat_classifiers() {
11     ...
12     for (int i=0; i<nn_; i++) {
13         for (int j=0; j<nn_; j++) {
14             if (i != j) {
15                 int nh = -1;
16                 nh = rtobject_->lookup_flat(i, j);
17                 if (nh >= 0) {
18                     NsObject *l_head=get_link_head(nodelist_[i],nh);
19                     sprintf(tmp, "%d", j);
20                     nodelist_[i]->add_route(tmp, l_head);
21                 }
22             }
23         }
23     }
25 }

//~ns/common/node.cc
26 void Node::add_route(char *dst, NsObject *target) {
27     if (rtnotif_)
28         rtnotif_->add_route(dst, target);
29 }
```

In Lines 26–29 of Program 6.23, function `add_route(dst,target)` simply invokes function `add_route(dst,target)` of the associated `RoutingModule` object `rtnotif_`. Defined in Program 6.15, function `add_route(dst,target)` of class `RoutingModule` recursively installs the input routing entry down the linked list of routing modules, by executing `do_install(dst,target)` of the variable `classifier_` associated with each routing module. The function `do_install(...)` installs NsObject `target` in slot `dst` of the classifier such that packets destined for the destination `dst` are forwarded to NsObject `target`.

6.7 Chapter Summary

A Node is a basic component which acts as a router and a computer host. Its main responsibilities are to forward packets according to a routing table and to bridge the high-layer protocols to a low-level network. A Node consists of two key components: classifiers and routing modules. A classifier is a multi-target packet forwarder. It is used in a Node to forward packets, which are destined to different destinations, to different forwarding NsObjects. It is also used as a demultiplexer, which forwards packets with different destination ports to different attached transport-layer agents.

As another main component, a routing module acts as a single point of management for a group of classifiers in a Node. When receiving a configuration command, it propagates the command to the related classifiers. It acts as an interface to other routing components such as route logic (which is responsible for computing the optimal routes), and to the agent attachment/detachment instprocs of class `Node`. Routing modules alleviate the need to configure every classifier separately, and therefore, greatly facilitate the classifier configuration process especially for a highly-complicated node configuration with numerous classifiers.

During the Network Configuration Phase, a Node is created by executing `$ns node` where `$ns` is the `object`. At the construction, address classifiers and routing modules are installed in the Node. However, the routing mechanism of the address classifiers are not configured here. The transport layer connections, on the other hand, are created in this phase using instproc **attach-agent** of class `Simulator`. At the beginning of the Simulation Phase, NS2 computes the optimal routes for all pairs of nodes, and installs the computed routing information in relevant classifiers.

Link and Buffer Management

A Link is an OTcl object which connects two nodes and carries packets from the beginning node to the terminating node. This chapter focuses on a class of most widely-used Link object, namely, `SimpleLink` objects. Conveying packets from one node to another, a `SimpleLink` object models packet transmission time, link propagation delay, and packet buffering. Here, packet transmission time refers to the time required by a transmitter to send out a packet. It is determined by the link bandwidth and packet size. Link propagation delay is the time needed to convey each bit from the beginning to the end of a link. In presence of bursty traffic, a transmitter may receive packets while transmitting a packet. The packets entering a busy transmitter could be placed in a buffer for future transmission. Unlike the real implementation, NS2 implements packet buffering in a Link, not a Node.

In the following, we first give an introduction to classes `Link` and `SimpleLink` in Section 7.1. Then, we show how NS2 models packet transmission time and propagation delay in Section 7.2. Next, the packet buffering, queue blocking, and callback mechanisms are discussed in Section 7.3. Section 7.4 shows a network construction and packet flow example. Finally, the chapter summary is provided in Section 7.5.

7.1 Introduction to SimpleLink Objects

NS2 models a link using classes derived from OTcl class `Link` object, among which OTcl class `SimpleLink` is the simplest one which can be used to connect two Nodes.

7.1.1 Main Components of a SimpleLink

Figure 7.1 shows the composition of class `SimpleLink`, which consists of the following basic objects and tracing objects in the interpreted hierarchy:

T. Issariyakul, E. Hossain, *Introduction to Network Simulator NS2*,
DOI: 10.1007/978-0-387-71760-9_7, © Springer Science+Business Media, LLC 2009

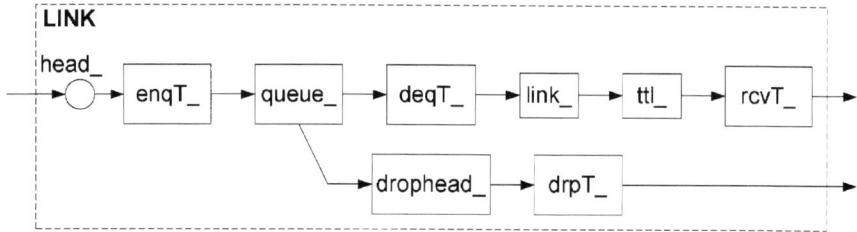

Fig. 7.1. Architecture of a `SimpleLink` object.

Basic Objects

head_ The entry point of a `SimpleLink` object.

queue_ As a `Queue` object, `queue_` models packet buffering of a "real" router (see Section 7.3).

link_ A `DelayLink` object, which models packet transmission time and link propagation delay (see Section 7.2).

ttl_ A *time to live* checker object whose class is `TTLChecker`. It decrements the time to live field of an incoming packet. After the decrement, if the time to live field is still positive, the packet will be forwarded to the next element in the link. Otherwise, it will be removed from the simulation (see file ~*ns*/common/ttl.h,cc).

drophead_ The common packet dropping point for the link. The dropped packets are forwarded to this object. It is usually connected to the null agent of the Simulator so that all `SimpleLink` objects share the same dropping point.

Tracing Objects

These objects will be inserted only if instvar `$traceAllFile_` of the is defined. We will describe the details of tracing objects in detail in Chapter 13. These objects are

enqT_ Trace packets entering `queue_`.
deqT_ Trace packets leaving `queue_`.
drpT_ Trace packets dropped from `queue_`.
rcvT_ Trace packets leaving the link or equivalently received by the next node.

7.1.2 Instprocs for Configuring a `SimpleLink` Object

In the OTcl domain, a `SimpleLink` object is created using the instprocs `simplex-link{..}` and `duplex-link{...}` of class `Simulator` whose syntax is as follows:

```
$ns simplex-link $n1 $n2 <bandwidth> <delay> <queue_type>
$ns duplex-link $n1 $n2 <bandwidth> <delay> <queue_type>
```

where $ns is the Simulator object, and $n1 and $n2 are Node objects.

Instproc simplex-link{...} above creates a uni-directional SimpleLink object connecting Node $n1 to Node $n2 (Program 7.1). The speed and the propagation delay of the link are given as <bandwidth> (in bps) and <delay> (in seconds), respectively. Again, as opposed to a "real" router, NS2 incorporates a queue in a SimpleLink object, not in a Node object. The type of the queue in the link is specified by <queue_type>.

Program 7.1 Instproc simplex-link of class Simulator.

```
    //~ns/tcl/lib/ns-lib.tcl
1   Simulator instproc simplex-link { n1 n2 bw delay qtype args } {
2       $self instvar link_ queueMap_ nullAgent_ useasim_
3       switch -exact $qtype {
4           /* See the detail in ~ns/tcl/lib/ns-lib.tcl */
5           default {
6               set q [new Queue/$qtype $args]
7           }
8       }
9       switch -exact $qtypeOrig {
10          /* See the detail in ~ns/tcl/lib/ns-lib.tcl */
11          default {
12              set link_($sid:$did) [new SimpleLink     \
                                        $n1 $n2 $bw $delay $q]
13          }
14      }
15  }
```

Program 7.1 shows details of instproc Simulator::simplex-link{...}. Line 6 creates an object of class Queue/qtype. Line 12 constructs a SimpleLink object, connecting node n1 to n2. It specifies delay, bandwidth, and Queue object of the link to be bw, delay, and q, respectively. The Simulator stores the created SimpleLink object in its instance associative array link_ ($sid:$did), where sid is the source node ID, and $did is the destination node ID (see Chapter 4).

Instproc duplex-link{...} creates two SimpleLink objects: one connecting Node $n1 to Node $n2 and another connecting Node $n2 to Node n1. For brevity, we do not show the detail here. The readers are encouraged to find the details of instproc duplex-link{...} in file ~ns/tcl/lib/ns-lib.tcl.

7.1.3 The Constructor of Class `SimpleLink`

Program 7.2 shows the details of instproc init{...} (i.e., the constructor) of class SimpleLink, which constructs and connects objects according to Fig. 7.1. Lines 3, 5, 11, 12, and 18 create instvars drophead_, head_, queue_, link_, and ttl_, whose OTcl classes are Connector, Connector, Queue, DelayLink, and TTLChecker, respectively. Note that the bandwidth and delay of instvar link_ are configured in Lines 13–14.

Program 7.2 The constructor of the OTcl class `SimpleLink`.

```
   //~ns/tcl/lib/ns-link.tcl
1  SimpleLink instproc init { src dst bw delay q {
                                  lltype "DelayLink"} } {
2    set ns [Simulator instance]
3    set drophead_ [new Connector]
4    $drophead_ target [$ns set nullAgent_]
5    set head_ [new Connector]
6    if { [[$q info class] info heritage ErrModule] ==
                                    "ErrorModule" } {
7        $head_ target [$q classifier]
8    } else {
9        $head_ target $q
10   }
11   set queue_ $q
12   set link_ [new $lltype]
13   $link_ set bandwidth_ $bw
14   $link_ set delay_ $delay
15   $queue_ target $link_
16   $link_ target [$dst entry]
17   $queue_ drop-target $drophead_
18   set ttl_ [new TTLChecker]
19   $ttl_ target [$link_ target]
20   $self ttl-drop-trace
21   $link_ target $ttl_
22 }
```

Apart from creating the above objects, the constructor also connects the created objects as in Fig. 7.1. Derived from class Connector, each of the created objects uses command **target** and **drop-target** to specify the next downstream object and the dropping point, respectively (see Chapter 5). Line 9 sets the target of head_ to be q. Line 15 sets the target of queue_ (which is set to "q" in Lines 11) to be link_. Line 16 sets the target of link_ to be the entry of the next node. Lines 19 and 21 insert ttl_ between link_ and the entry of the next node. Line 17 sets the dropping point of queue_ to be drophead_. Finally, Line 4 sets the target of drophead_ to be the null agent of the Simulator.

7.2 Modeling Packet Departure

7.2.1 Packet Departure Mechanism

NS2 models packet departure by using a C++ class Linkdelay (see Program 7.3), which is bound to an OTcl class DelayLink object. Again, the OTcl class DelayLink is used to instantiate the instvar SimpleLink::link_ which models the packet departure process.

Program 7.3 Declaration of class LinkDelay.

```
    //~ns/link/delay.h
1   class LinkDelay : public Connector {
2       public:
3           LinkDelay(): dynamic_(0), latest_time_(0), itq_(0){
4               bind_bw("bandwidth_", &bandwidth_);
5               bind_time("delay_", &delay_);
6           }
7           void recv(Packet* p, Handler*);
8           void send(Packet* p, Handler*);
9           void handle(Event* e);
10          inline double txtime(Packet* p) { /* Packet TXT Time */
11              return (8. * hdr_cmn::access(p)->size() / bandwidth_);
12          }
13      protected:
14          int command(int argc, const char*const* argv);
15          double bandwidth_;
16          double delay_;
17          PacketQueue* itq_;
18          Event intr_; /* In transit */
19  };

    //~ns/link/delay.cc
20  static class LinkDelayClass : public TclClass {
21  public:
22      LinkDelayClass() : TclClass("DelayLink") {}
23      TclObject* create(int argc , const char*const*  argv ) {
24          return (new LinkDelay);
25      }
26  } class_delay_link;
```

A packet departure process consists of packet transmission time and link propagation delay. While the former defines the time a packet stays in an upstream node, the summation of the former and the latter determines the time needed to deliver an entire packet to the connecting downstream node. Conceptually, when a LinkDelay object receives a packet, it places these two events on the simulation timeline:

(i) *Packet departure* from an upstream object: Define *packet transmission time* $= \frac{\text{packet size}}{\text{bandwidth}}$ as time needed to transmit a packet over a link. After a period of packet transmission time, the packet completely leaves (or departs) the transmitter, and the transmitter is allowed to transmit another packet. Upon a packet reception, a LinkDelay object waits for a period of packet transmission time, and informs its upstream object that it is ready to receive another packet.

(ii) *Packet arrival* at a downstream node: Define *propagation delay* as the time needed to deliver a data bit from the beginning to the end of the link. Again, an entire packet needs a period of "packet transmission time + propagation delay" to reach the destination. A LinkDelay object, therefore, schedules a packet reception event at the downstream node after this period.

7.2.2 C++ Class LinkDelay

Program 7.3 shows the declaration of C++ class LinkDelay, which is mapped to the OTcl class DelayLink. Class LinkDelay has the following four main variables. Variables bandwidth_ (Line 15) and delay_ (Line 16) store the link bandwidth and propagation delay, respectively. In Lines 4–5, these two variables are bound to OTcl instvars with the same name. In a link with large bandwidth-delay product, a transmitter can send a new packet before the previous packet reaches the destination. Class LinkDelay stores all packets *in-transit* in its buffer itq_ (Line 17), which is a pointer to a PacketQueue object (See Section 7.3.1). Finally, variable intr_ is a dummy Event object, which represent a packet departure (from the transmitting node) event. As discussed in Section 4.3.6, the packet departure is scheduled using variable intr_ which does not take part in event dispatching[1].

The main functions of class LinkDelay are recv(p,h), send(p,h), handle (e), and txttime(p). Function txttime(p) calculates the packet transmission time of packet *p (Lines 10–12 in Program 7.3). Function send(p,h) sends packet *p to the connecting downstream object (see Line 12 in Program 5.3). Function handle(e) is invoked when the Scheduler dispatches an event corresponding to the LinkDelay object (see Chapter 4). Function recv(p,h) (Program 7.4) takes a packet *p and a handler *h as input arguments, and schedules packet departure and packet arrival events.

---

[1] As a dummy Event object, variable intr_ ensures that an error message will be shown on the screen, if an undispatched event is rescheduled.

Program 7.4 Function `recv` of class `LinkDelay`.

```
      //~ns/link/delay.cc
1     void LinkDelay::recv(Packet* p, Handler* h)
2     {
3         double txt = txtime(p);
4         Scheduler& s = Scheduler::instance();
5         if (dynamic_) { /* See ~ns/link/delay.cc */ }
6         else if (avoidReordering_) { /* See ~ns/link/delay.cc */ }
7         else {
8             s.schedule(target_, p, txt + delay_);
9         }
10        s.schedule(h, &intr_, txt);
11    }
```

(i) *Packet departure event*: Since a packet spends "packet transmission time" (`txt` in Line 3) at the upstream object, function `recv(p,h)` schedules a packet departure event at `txt` seconds after the `LinkDelay` object receives the packet. To do so, Line 10 invokes function `schedule(h,&intr,txt)` of class `Scheduler`, where the first, second, and third input arguments are handler pointer, dummy event pointer, and delay, respectively (see Chapter 4). After `txt` seconds, the Scheduler dispatches this event by invoking function `handle(e)` associated with the handler `*h` to inform the upstream object of a packet departure. In most cases, the upstream object responds by transmitting another packet, if available (see Section 7.3.3 for the callback mechanism).

(ii) *Packet arrival*: Class `LinkDelay` also passes the packet to its downstream object (`*target_`). Line 8 schedules an event cast from the input packet `*p` with delay `txt+delay_` seconds, where `txt` is the packet transmission time and `delay_` is the link propagation delay. Here, `*target_` is passed to function `schedule(...)` as a handler pointer. After "`txt+delay_`" seconds, `h.handle(p)` will invoke function `recv(p)` (see Program 4.2), and packet `*p` will be passed to `*target_` after `txt+delay_` seconds.

The major difference between scheduling packet departure and arrival events is as follows. While a node can hold only one (head of the line) packet, a link can contain more than one packet. Correspondingly, at an instance, a link can schedule only one packet departure event (using `intr_`), and more than one packet arrival event (using `*p` which represents a packet). Every time a `LinkDelay` object receives a packet, it schedules the packet departure event using the same variable `intr_`. If variable `intr_` has not been dispatched, such a scheduling will cause runtime error, because it attempts to place a packet in the head of the buffer which is currently occupied by another packet. A packet arrival event, on the other hand, is tied to incoming packet. A `LinkDelay` object schedules a new packet arrival event for every received packet (see Line 8 in Program 7.4). Therefore, a link can schedule a packet arrival event, even if the previous arrival event has not been dispatched. This is essentially the case in a link with large bandwidth-delay product which can contain several packets.

7.3 Buffer Management

Another major component of a `SimpleLink` object is class `Queue`. Class `Queue` models the buffering mechanism in a network router. It stores a received packet in the buffer, and forwards a (in most case the head of the line) packet in the buffer to its downstream object when the ongoing transmission is complete. As shown in Program 7.5, class `Queue` derives from class `Connector`, and can be used to connect two NsObjects. It employs a `PacketQueue` object (see Section 7.3.1), `*pq_` in Line 20, for packet buffering. The buffer size is specified in variable `qlim_` (Line 16). The variables `blocked_` (Line 16), `unblock_on_resume_` (Line 17), and `qh_` (Line 18) are related to the so-called *callback mechanism*, and shall be discussed later in Section 7.3.3.

Program 7.5 Declaration of class `Queue`.

```
      //~ns/queue/queue.h
1   class Queue : public Connector {
2    public:
3        virtual void enque(Packet*) = 0;
4        virtual Packet* deque() = 0;
5        virtual void recv(Packet*, Handler*);
6        void resume();
7        int blocked() const { return (blocked_ == 1); }
8        void unblock() { blocked_ = 0; }
9        void block() { blocked_ = 1; }
10       int limit() { return qlim_; }
11       int length() { return pq_->length(); }
12       virtual ~Queue();
13   protected:
14       Queue();
15       void reset();
16       int qlim_;
17       int blocked_;
18       int unblock_on_resume_;
19       QueueHandler qh_;
20       PacketQueue *pq_;
21   };
```

There are a number of important functions of class `Queue`. Function `enque(p)` and `deque()` (Lines 3–4) place and take, respectively, a packet from the `PacketQueue` object `*pq_`. They are declared as pure virtual, and must be implemented by instantiable derived classes of class `Queue`. Derived from class `NsObject`, function `recv(p,h)` (Line 5) is the main packet reception function. Function `blocked()` in Line 7 indicates whether the `Queue` object is in a blocked state. Functions `resume()` (Line 6), `unblock()` (Line 8), and `block()` (Line 9) are used in the callback mechanism which will be discussed in Section 7.3.3. Finally, functions `limit()` and `length()` return the buffer size and current buffer occupancy, respectively.

7.3.1 Class `PacketQueue`: A Model for Packet Buffering

Declared in Program 7.6, class `PacketQueue` models low-level operations of the buffer including storing, enqueuing, and dequeuing packet. Class `PacketQueue` is a linked list of `Packets`, whose member variables are as follows. Variable `head_` in Line 11 is the pointer to the beginning of the linked list. Variable `tail_` in Line 12 is the pointer to the end of the linked list. The variable `len_` in Line 13 is the number of packets in the buffer. Function `enque(p)` in Line 5 puts the input packet `*p` to the end of the buffer. Function `deque()` in Line 6 returns the head of the line `packet` pointer or returns NULL when the buffer is non-empty or empty, respectively. Function `remove(p)` in Line 7 searches for a matching packet `*p` and removes it from the buffer (if found). Note that packet admitting/dropping is the functionality of class `Queue`, not of class `PacketQueue`. We will show an example of packet admitting/dropping of class `DropTail` in Section 7.3.4.

Program 7.6 Declaration of class `PacketQueue`.

```
     //~ns/queue/queue.h
 1   class PacketQueue : public TclObject {
 2      public:
 3         PacketQueue() : head_(0), tail_(0), len_(0), bytes_(0) {}
 4         virtual int length() const { return (len_); }
 5         virtual Packet* enque(Packet* p);
 6         virtual Packet* deque();
 7         virtual void remove(Packet*);
 8         Packet* head() { return head_; }
 9         Packet* tail() { return tail_; }
10      protected:
11         Packet* head_;
12         Packet* tail_;
13         int len_;
14   };
```

7.3.2 Queue Handler

Derived from class `Handler` (see Line 1 in Program 7.7), class `QueueHandler` is closely related to the (event) Scheduler. Again, a `QueueHandler` object defines its default actions in its function `handle(e)`. These default actions will be taken when an associated event is dispatched. As shown in Lines 8–11 of Program 7.7, the default action of a `QueueHandler` object is to execute function `resume()` of the associated `Queue` object `queue_`. We will discuss the details of function `resume()` in Section 7.3.3. In the rest of this section, we will demonstrate how a connection between `QueueHandler` and `Queue` objects is created.

Program 7.7 Declaration and function `handle` of class `QueueHandler`, and the constructor of class `Queue`.

```
   //~ns/queue/queue.h
1  class QueueHandler : public Handler {
2  public:
3      inline QueueHandler(Queue& q) : queue_(q) {}
4      void handle(Event*);
5  private:
6      Queue& queue_;
7  };

   //~ns/queue/queue.cc
8  void QueueHandler::handle(Event*)
9  {
10     queue_.resume();
11 }

12 Queue::Queue() : Connector(), blocked_(0),
                    unblock_on_resume_(1), qh_(*this),pq_(0)
13 { /* See the detail in ~ns/queue/queue.cc */ }
```

To associate a `Queue` object with a `QueueHandler` object, classes `Queue` and `QueueHandler` declare their member variables `qh_` (Line 19 in Program 7.5) and `queue_` (Line 6 in Program 7.7), as a `QueueHandler` pointer and a `Queue` reference, respectively. These two variables are initialized when a `Queue` object is instantiated (Line 12 in Program 7.7). The constructor of class `Queue` invokes the constructor of class `QueueHandler`, feeding itself as an input argument (i.e., `qh_(*this)` in Line 12 of Program 7.7). The constructor of `qh_` then sets its member variable `queue_` to share the same address as the input `Queue` object (i.e. `queue_(q)` in Line 3 of Program 7.7), hence creating a two-way connection between the `Queue` and `QueueHandler` objects. After this point, the `Queue` and the `QueueHandler` objects refer to each other by the variables `qh_` and `queue_`, respectively.

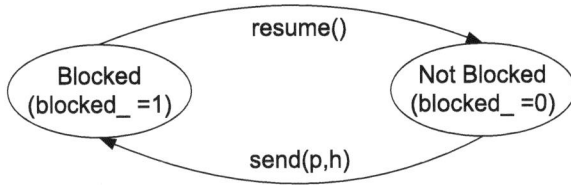

Fig. 7.2. State diagram of the queue blocking mechanism.

7.3.3 Queue Blocking and Callback Mechanism

Queue Blocking

NS2 uses the concept of *queue blocking*[2] to indicate whether a queue is currently transmitting a packet. By default, a queue can transmit one packet at a time. It is not allowed (i.e., blocked) to transmit another packet until the ongoing transmission is complete. A queue is said to be blocked or unblocked (i.e., blocked_ = 1 or blocked_ = 0), when it is transmitting a packet or is not transmitting a packet, respectively.

Figure 7.2 shows the state diagram of the queue blocking mechanism. When in the "Not Blocked" state, a queue is allowed to transmit a packet by executing "target_->recv(p,&qh_)", after which it enters the "Blocked" state. Here, a queue waits until the ongoing transmission is complete where the function resume() is invoked. After this point, the queue enters the "Not Blocked" state and the process repeats.

Callback Mechanism

As discussed in Chapter 5, a node in NS2 passes packets to a downstream node by executing function recv(p,h), where *p denotes a packet and *h denotes a handler. A callback mechanism refers to a process where a downstream object invokes an upstream object along the downstream path for a certain purpose. In a queue blocking process, a callback mechanism is used to unblock a Queue object by invoking function resume() of the upstream Queue object.

We now explain the callback mechanism process for queue unblocking via an example network in Fig. 7.3. Here, we assume that the following objects are sequentially connected: an upstream NsObject, a Queue object, a LinkDelay object, and a downstream object. Again, an NsObject passes a packet *p by invoking function recv(p,h) of its downstream object, where *h is a handler. In most cases, the input handler *h is passed along with the packet *p as input argument of function recv(p,h). However, this mechanism is different for the Queue object.

---

[2] Queue blocking has no relation to packet blocking when the buffer is full.

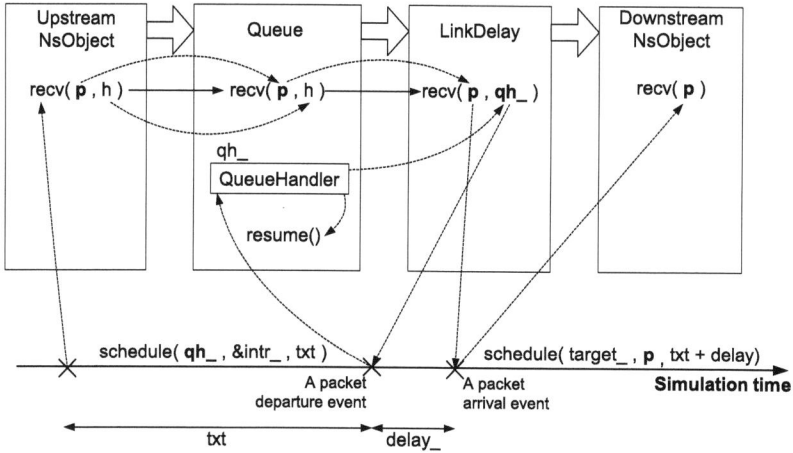

Fig. 7.3. Diagram of callback mechanism for a queue unblocking process.

Consider function `recv(p,h)` of class `Queue` in Program 7.8. Instead of immediately passing the incoming packet `*p` to its downstream object, Line 3 places the packet in the buffer (i.e., `pq_`). Again, a `Queue` object is allowed to transmit a packet only when it is not blocked (Line 4). In this case, Line 5 retrieves a packet from the buffer. If the packet is valid (Line 6), Line 7 will set the state of the `Queue` object to be "blocked", and Line 8 will forward the packet to its downstream object (i.e., `*target_`). The `Queue` object passes its `QueueHandler` pointer `qh_` (instead of the incoming handler pointer) to its downstream object. This `QueueHandler` pointer acts as a reference point for a queue blocking callback mechanism.

Program 7.8 Function `recv` of class `Queue`.

```
//~ns/queue/queue.cc
1   void Queue::recv(Packet* p, Handler*)
2   {
3       enque(p);
4       if (!blocked_) {
5           p = deque();
6           if (p != 0) {
7               blocked_ = 1;
8               target_->recv(p, &qh_);
9           }
10      }
11 }
```

From Fig. 7.3, the downstream object of the Queue object is a LinkDelay object. Upon receiving a packet, it schedules two events: packet departure and arrival events (see Lines 10 and 8 in Program 7.4). A packet arrival event is associated with the downstream object (i.e., *target_). At the firing time, the function handle(p) of the downstream object will invoke function recv(p) to receive packet *p (see Program 4.2).

Function recv(p) of class LinkDelay also schedules a packet departure event. Since the input handler pointer is a QueueHandler pointer, the departure event is associated with the QueueHandler object qh_. At the firing time, the Scheduler invokes function handle(p) of the associated QueueHandler object. In Program 7.7, this function in turn invokes function resume() to unblock the associated Queue object. Literally the LinkDelay object schedules an event which *calls back* to unblock the upstream Queue object.

Program 7.9 Function resume of class Queue.

```
    //~ns/queue/queue.cc
1   void Queue::resume()
2   {
3       Packet* p = deque();
4       if (p != 0)
5           target_->recv(p, &qh_);
6       else
7           if (unblock_on_resume_)
8               blocked_ = 0;
9           else
10              blocked_ = 1;
11  }
```

Program 7.9 shows the details of function resume(). Function resume() first retrieves the head of the line packet from the buffer (Line 3). If the buffer is non-empty (Line 4), Line 5 will send the packet to the downstream object of the queue regardless of the blocked status. In this case, the variable blocked_ would remain unchanged. If the Queue object is in a "Blocked" state, it will remained blocked after packet transmission, hence complying with the state diagram in Fig. 7.2. If the queue is idle (i.e., the buffer is empty), variable blocked_ will be set to zero and one in case that the flag unblock_on_resume_ is one and zero, respectively.

7.3.4 Class DropTail: A Child Class of Class Queue

Consider class DropTail, a child class of class Queue, which is bound to the OTcl class Queue/DropTail in Program 7.10. The constructor of class DropTail creates a pointer q_ (Line 13) to a PacketQueue object, and sets pq_ derived from class Queue to be the same as q_ (Line 5). Throughout

the implementation, class DropTail refers to its buffer by q_ instead of pq_.
Class DropTail overrides function enque(p) (Line 11 and Program 7.11) and
deque() (Line 12) of class Queue. It also allows packet dropping at the front
of the buffer, if the flag drop_front_ (Line 14) is set to 1. Class DropTail
does not override function recv(p,h). Therefore, it receives a packet through
the function recv(p,h) of class Queue.

Program 7.10 Declaration of class DropTail.

```
    //~ns/queue/drop-tail.h
1   class DropTail : public Queue {
2     public:
3       DropTail() {
4           q_ = new PacketQueue;
5           pq_ = q_;
6           bind_bool("drop_front_", &drop_front_);
7       };
8       ~DropTail() { delete q_; };
9     protected:
10      int command(int argc, const char*const* argv);
11      void enque(Packet*);
12      Packet* deque();
13      PacketQueue *q_;
14      int drop_front_;
15  };

    //~ns/queue/drop-tail.cc
16  static class DropTailClass : public TclClass {
17      public:
18      DropTailClass() : TclClass("Queue/DropTail") {}
19      TclObject* create(int, const char*const*) {
20          return (new DropTail);
21      }
22  } class_drop_tail;
```

In Program 7.11, function enque(p) first checks whether the incoming
packet will cause buffer overflow (Line 3). If so, it will drop the packet either
from the front (Lines 5–7) or from the tail (Line 9), where function drop(p)
(Lines 7 and 9) belongs to class Connector (see Program 5.4). If the buffer
has enough space, Line 10 will enqueue packet (p) to its buffer (q_).

7.4 A Sample Two-Node Network

We have introduced two basic NS2 components: nodes and links. Based
on these two components, we now create a two-node network with a uni-

Program 7.11 Function enque of class DropTail.

```
//~ns/queue/drop-tail.cc
1   void DropTail::enque(Packet* p)
2   {
3       if ((q_->length() + 1) >= qlim_)
4           if (drop_front_) {
5               q_->enque(p);
6               Packet *pp = q_->deque();
7               drop(pp);
8           } else
9               drop(p);
10      else
11          q_->enque(p);
12  }
```

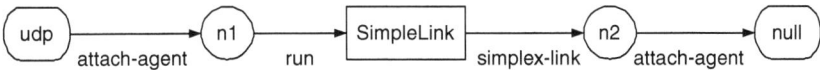

Fig. 7.4. A two-node network with a uni-directional link and the instprocs of class Simulator.

directional link and show the packet flow mechanism within this network in Fig. 7.4.

7.4.1 Network Construction

The network in Fig. 7.4 consists of a beginning node (n1), a termination node (n2), a SimpleLink connecting n1–n2, a source transport layer agent (udp), and a sink transport layer agent (null). This network can be created using the following Tcl simulation script:

```
set ns [new Simulator]
set n1 [$ns node]
set n2 [$ns node]
$ns simplex-link $n1 $n2 <bw> <delay> DropTail
set udp [new Agent/UDP]
set null [new Agent/Null]
$ns attach-agent $n1 $udp
$ns attach-agent $n2 $null
```

Here, command $ns node creates a Node object. The internal mechanism of the node construction process was described in Section 6.6. The statement $ns simplex-link $n1 $n2 <bw> <delay> DropTail creates a unidirectional SimpleLink object, which connects node n1 to node n2. The link

bandwidth and delay are <bw> bps and <delay> seconds, respectively. The
buffer in the link is of class DropTail. From Section 6.6.3, the commands $ns
attach-agent $n1 $udp and $ns attach-agent $n2 $null set the target
of agent udp to be the entry of Node n1, and installs agent null in the demultiplexer of Node n2.

7.4.2 Packet Flow Mechanism

To deliver a packet "*p" from agent udp to null,

(i) Agent udp sends packet *p to the entry of Node n1.[3]
(ii) Packet *p is sent to the head classifier classifier_ (which is of class
DestHashClassifier) of Node n1.
(iii) The DestHashClassifier object classifier_ examines the header of
packet *p. In this case, the packet is destined to the Node n2. Therefore,
it forwards the packet to the link head of the connecting SimpleLink
object.
(iv) The link head forwards the packet to the connecting Queue object.
(v) The Queue object enqueues the packet. If not blocked, it will forward
the head of the line packet to the connecting LinkDelay object and set
its status to blocked.
(vi) Upon receiving a packet, the LinkDelay object schedules the two following events:
(a) Packet departure event, which indicates that packet transmission is
complete. This event unblocks the associated Queue object.
(b) Packet arrival event, which indicates the packet arrival at the connecting TTLChecker object.
(vii) The TTLChecker object receives the packet, and decrements the TTL
field of the packet header. If the TTL field of the packet is non-positive,
the TTLChecker object will drop the packet. Otherwise, it will forward
the packet to the entry of Node n2 (see file ˜ns/common/ttl.cc).
(viii) Node n2 forwards the packet to the head classifier (classifier_). Since
the packet is destined to itself, the packet is forwarded to the demultiplexer (dmux_).
(ix) The demultiplexer forwards the packet to the agent null installed in
the demultiplexer.

7.5 Chapter Summary

This chapter focuses on class SimpleLink, a basic link class which can be used
to connect two nodes. The connection between two nodes n1 and n2 can be
created by the following instprocs:

---

[3] Note that, each object sends a packet *p to its downstream object by invoking
target_-> recv(p,h), where target_ is a pointer to the downstream object.

```
$ns simplex-link $n1 $n2 <bw> <delay> <queue_type>
$ns duplex-link $n1 $n2 <bw> <delay> <queue_type>
```

where the bandwidth and delay of the `SimpleLink` object are `<bw>` bps and `<delay>` seconds, respectively. Also the type of queue implemented in the `SimpleLink` object is `<queue_type>`.

A `SimpleLink` object models packet transmission time, link propagation delay, and packet buffering. Here, packet transmission time is the time required to transmit a packet, and is computed by $\frac{\text{packet size}}{\text{bandwidth}}$, while the link propagation time is the time required to deliver a data bit from the beginning to the end of the `SimpleLink` object. As shown in Fig. 7.1, an OTcl `SimpleLink` object consists of instvars `head_`, `drophead_`, `queue_`, `link_`, and `ttl_`, whose classes are `Connector`, `Connector`, `Queue`, `DelayLink`, and `TTLChecker`, respectively.

- Instvars `head_` and `drophead_` act as an entry point and a dropping point, respectively, of a `SimpleLink` object.
- Instvar `link_` models packet transmission time and link propagation delay of a link. When receiving a packet, it schedules two events: packet departure from the beginning node and packet arrival at the terminating node.
- Instvar `queue_` models packet buffering mechanism in a `SimpleLink` object. It operates very closely with the instvar `link_`. Upon receiving a packet, the instvar `queue_` enques the packet. If not blocked, it will block itself and forward the packet as well as the associated queue handler to the instvar `link_`. When the packet departure event (scheduled by the instvar `link_`) is dispatched, instvar `queue_` is unblocked (i.e., being called back) and allowed to transmit another packet.
- Instvar `ttl_` is a packet time to live checker, which drops packets which stay in the network for longer than a specified period of time.

Packets, Packet Headers, and Header Format

Generally, a packet consists of packet header and data payload. Packet header stores packet attributes (e.g., source and destination IP addresses) necessary for packet delivery, while data payload contains user information. Although this concept is typical in practice, NS2 models packets differently.

In most cases, NS2 extracts information from data payload and stores the information into packet header. This idea removes the need to process data payload at runtime. For example, instead of counting the number of bits in a packet, NS2 stores packet size in variable `hdr_cmn::size_` (see Section 8.3.5), and accesses this variable at runtime.[1]

This chapter discusses how NS2 models packets. Section 8.1 gives an overview on NS2 packet modeling. Section 8.2 discusses the packet allocation and deallocation processes. Sections 8.3 and 8.4 show the details of packet header and data payload, respectively. We give a guideline of how to customize packets (i.e., to define a new packet type and activate/deactivate new and existing protocols) in Section 8.5. Finally, the chapter summary is given in Section 8.6.

8.1 An Overview of Packet Modeling Principle

8.1.1 Packet Architecture

Figure 8.1 shows the architecture of an NS2 packet model. From Fig. 8.1, a packet model consists of four main parts: actual packet, class `Packet`, protocol specific headers, and packet header manager.

- **Actual Packet:** An actual packet refers to the portion of memory which stores packet header and data payload. NS2 does not directly access either the packet header or the data payload. Rather, it uses pointers

---

[1] For example, class `LinkDelay` determines packet size from a variable `hdr_cmn::size_` when computing packet transmission time (see Line 11 of Program 7.3).

T. Issariyakul, E. Hossain, *Introduction to Network Simulator NS2*,
DOI: 10.1007/978-0-387-71760-9_8, © Springer Science+Business Media, LLC 2009

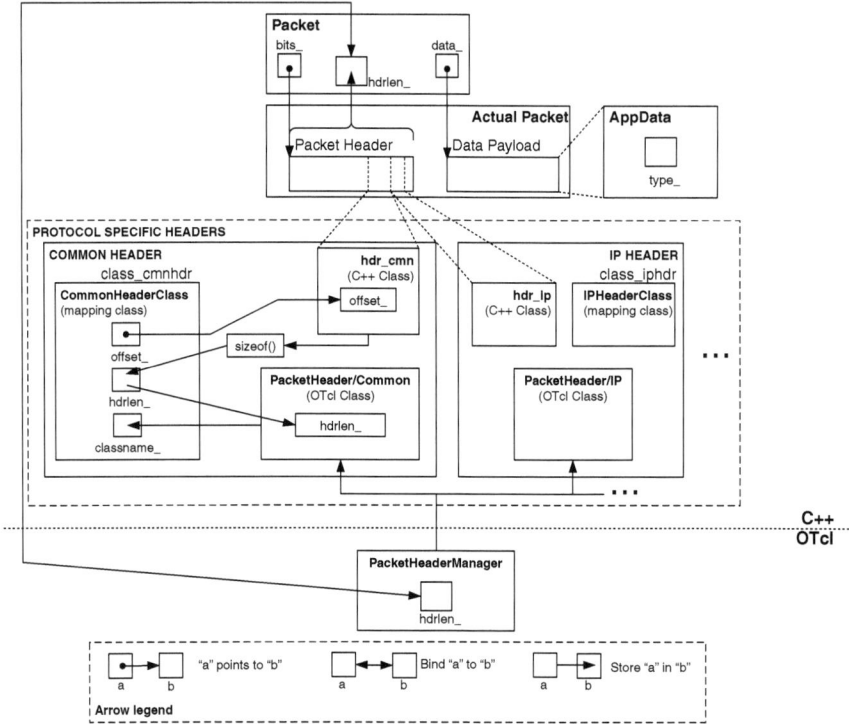

Fig. 8.1. Packet modeling in NS2.

`bits_` and `data_` of class `Packet` to access packet header and data payload, respectively. The details of packet header and data payload will be given in Sections 8.3 and 8.4, respectively.

- **Class `Packet`:** Declared in Program 8.1, class `Packet` is the C++ main class which represents packets. It contains the following variables and functions:

C++ variables of class `Packet`

`bits_`	A string which contains packet header
`data_`	Pointer to an `AppData` object which contains data payload
`fflag_`	Set to `true` if the packet is currently referred to by other objects and `false` otherwise
`free_`	Pointer to the head of the packet free list
`ref_count_`	Number of objects which currently refer to the packet
`next_`	Pointer to the next packet in the linked list of packets
`hdr_len_`	Length of packet header

Program 8.1 Declaration of class `Packet`.

```
   //~/ns/common/packet.h
1  class Packet : public Event {
2  private:
3      unsigned char* bits_;
4      AppData* data_;
5      static void init(Packet*) {bzero(p->bits_, hdrlen_);}
6      bool fflag_;
7  protected:
8      static Packet* free_;
9      int     ref_count_;
10 public:
11     Packet* next_;
12     static int hdrlen_;

       //Packet Allocation and Deallocation
13     Packet() : bits_(0), data_(0), ref_count_(0), next_(0) { }
14     inline unsigned char* const bits() { return (bits_); }
15     inline Packet* copy() const;
16     inline Packet* refcopy() { ++ref_count_; return this; }
17     inline int& ref_count() { return (ref_count_); }
18     static inline Packet* alloc();
19     static inline Packet* alloc(int);
20     inline void allocdata(int);
21     static inline void free(Packet*);

       //Packet Access
22     inline unsigned char* access(int off){return &bits_[off]);};
23 }
```

C++ functions of class `Packet`

init(p)   Clears the packet header `bits_` of the input packet `p`.
copy()   Returns a pointer to a duplicated packet.
refcopy() Increases the number of objects, which refer to the packet, by one.
alloc()   Creates a new packet and returns a pointer to the created packet.
alloc(n) Creates a new packet with "n" bytes of data payload and returns a pointer to the created packet.
allocdata(n) Allocates "n" bytes of data payload to the variable `data_`.
free(p)   Deallocates packet `p`.
access(off) Retrieves a reference to a certain point (specified by the offset "off") of the variable `bits_` (i.e., packet header).

- **Protocol Specific Header:** From Fig. 8.1, packet header consists of several protocol specific headers. Each protocol specific header uses a contiguous portion of packet header to store its packet attributes. In common with most TclObjects, there are three classes related to each protocol specific header: a C++ class, an OTcl class, and a mapping class.

- A C++ class (e.g., `hdr_cmn` or `hdr_ip`): provides a sturcture to store packet attributes.
- An OTcl class (e.g., `PacketHeader/Common` or `PacketHeader/IP`): acts as an interface to the OTcl domain. NS2 uses this class to configure packet header from the OTcl domain.
- A mapping class (e.g., `CommonHeaderClass` or `IPHeaderClass`): binds a C++ class to an OTcl class.

We will discuss the details of protocol specific header later in Section 8.3.5.

- **Packet Header Manager:** A packet header manager maintains a list of active protocols, and configures all active protocol specific headers to setup a packet header. It has an instvar `hdrlen_` which indicates the length of packet header consisting of protocol specific headers. Instvar `hdrlen_` is bound to a variable `hdrlen_` of class `Packet`. Any change in one of these two variables will result in an automatic change in another.
- **Data Payload:** From Line 4 in Program 8.1, the pointer `data_` points to data payload, which is of class `AppData`. We will discuss the details of data payload in Section 8.4.

8.1.2 A Packet as an Event: A Delayed Packet Reception Event

Derived from class `Event` (Line 1 in Program 8.1), class `Packet` can be placed on the simulation time line (see the details in Chapter 4). In Section 4.2, we mentioned two main classes derived from class `Event`: class `AtEvent` and class `Packet`. We also mentioned that an `AtEvent` object is an event created by a user from a Tcl simulation script. This chapter discusses the details of another derived class of class `Event`: class `Packet`.

As discussed in Section 5.2.2, NS2 models a delayed action by placing an event corresponding to the action on the simulation timeline at a certain delayed time. Derived from class `Event`, class `Packet` can be placed on the simulation timeline to signify a delayed packet reception. For example, the following statement (see Line 8 in Program 7.4) schedules a packet reception event, where the NsObject `*target_` receives a packet `*p` at `txt+delay_` seconds in future:

```
s.schedule(target_, p, txt + delay_)
```

where function `schedule(...)` of class `Scheduler` defined in Program 4.7 takes an `Event` pointer as its second input argument. A `Packet` pointer is cast to be an `Event` pointer before being fed as the second input argument.

At the firing time, the Scheduler dispatches the scheduled event (i.e., `*p`) and invokes `target->handle(p)`, which executes "`target_->recv(p)`" to forward packet `*p` to the NsObject pointer `*target_`.

8.1.3 A Linked List of Packets

Apart from the above 4 main packet components, a Packet object contains a pointer next_ (Line 11 in Program 8.1), which helps formulating a linked list of Packet objects (e.g., Packet List in Fig. 8.2). Program 8.2 shows the implementation of functions enque(p) and deque() of class PacketQueue. Function enque(p) (Lines 3–13) puts a Packet object *p to the end of the queue. If the PacketQueue is empty, NS2 sets head_, tail_, and p to point to the same place[2] (Line 5). Otherwise, Lines 7–8 set p as the last packet in the PacketQueue, and shift variable tail_ to the last packet pointer p. Since the pointer tail_ is the last pointer of PacketQueue, Line 10 sets the pointer tail_->next_ to 0 (i.e, points to NULL).

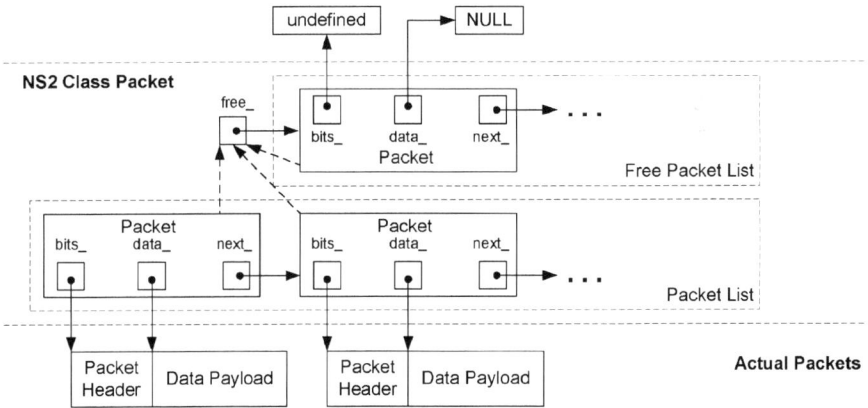

Fig. 8.2. A linked list of packets and a free packet list.

Function deque() (Lines 14–21) retrieves a pointer to the packet at the head of the buffer. If there is no packet in the buffer, the function deque() will return a NULL pointer (Line 15). If the buffer is not empty, Line 17 will shift the pointer head_ to the next packet, Line 19 will decrease the length of PacketQueue object by one, and Line 20 will return the packet pointer p which was set to the pointer head_ in Line 16.

8.1.4 Free Packet List

Unlike most NS2 objects, a Packet object, once created, will not be destroyed until the simulation terminates. NS2 keeps Packet objects which are no longer

---

[2] Note that, head_ and tail_ are pointers to the first and the last Packet objects, respectively, in a PacketQueue object.

Program 8.2 Functions enque and deque of class PacketQueue.

```
     //~/ns/common/queue.h
 1   class PacketQueue : public TclObject {
 2       ...
 3       virtual Packet* enque(Packet* p) { // Returns previous tail
 4           Packet* pt = tail_;
 5           if (!tail_) head_= tail_= p; // if the PacketQueue
               is empty
 6           else {
 7               tail_->next_= p;
 8               tail_= p;
 9           }
10           tail_->next_= 0;
11           ++len_;
12           return pt;
13       }

14       virtual Packet* deque() {
15           if (!head_) return 0;
16           Packet* p = head_;
17           head_= p->next_; // 0 if p == tail_
18           if (p == tail_) head_= tail_= 0;
19           --len_;
20           return p;
21       }
22       ...
23 };
```

in use in a *free packet list* (see Fig. 8.2). When NS2 needs a new packet, it first checks whether the free packet list is empty. If not, it will take a Packet object from the list. Otherwise, it will create another Packet object. We will discuss the details of how to *allocate* and *deallocate* a Packet object later in Section 8.2.

There are two variables which are closely related to the packet allocation/deallocation process: fflag_ and free_. Each Packet object uses a variable fflag_ (Line 6 in Program 8.1) to indicate whether it is in use. Variable fflag_ is set to true, when the Packet object is in use, and set to false otherwise. Shared by all the Packet objects, a static pointer free_ (Line 8 in Program 8.1) is a pointer to the first packet on the free packet list. Each packet on the free packet list uses its variable next_ to help form a link list of free Packet objects. This linked list of free packets is referred to as a *free packet list*. Although NS2 does not return memory allocated to a Packet object to the system, it does return the memory used by packet header (i.e., bits_) and data payload (i.e., data_) to the system (see Section 8.2.2), when the packet is deallocated. Since most memory required to store a Packet object is consumed by packet header and data payload, maintaining a free packet list does not result in a significant waste of memory.

8.2 Packet Allocation and Deallocation

Unlike most of the NS2 objects,[3] a `Packet` object is allocated and deallocated using static functions `alloc()` and `free(p)` of class `Packet`, respectively. If possible, function `alloc()` takes a `Packet` object from the free packet list. Only when the free packet list is empty, does the function `alloc()` creates a new `Packet` object using `new`. Function `free(p)` deallocates a `Packet` object, by returning the memory allocated for packet header and data payload to the system and storing the not-in-use `Packet` pointer `p` in the free packet list for future reuse. The details of packet allocation and deallocation will be discussed in the next two sections.

8.2.1 Packet Allocation

Program 8.3 shows the details of function `alloc()` of class `Packet`, the packet allocation function. Function `alloc()` returns a pointer to an allocated `Packet` object to the caller. This function consists of two parts: packet allocation in Lines 3–15, and packet initialization in Lines 16–24.

Consider the packet allocation in Lines 3–15. Line 3 declares `p` as a pointer to a `Packet` object, and sets the pointer `p` to point to the first packet on the free packet list[4]. If the free packet list is empty (i.e., `p = 0`), NS2 will create a new `Packet` object (in Line 11), and allocate memory space with size "`hdrlen_`" bytes for the packet header in Line 12. Variable `hdrlen_` is not configured during the construction of a `Packet` object. Rather, it is set up in the Network Configuration Phase (see Section 8.3.8), and is used by the function `alloc()` to create packet header.

Function `alloc()` does not allocate memory space for data payload. When necessary, NS2 creates data payload by using the function `allocdata(n)` (see Lines 8–14 in Program 8.4), which will be discussed in detail later in this section.

If the free packet list is non-empty, function `alloc()` will execute Lines 5–9 in Program 8.3 (see also the diagram in Fig. 8.3). In this case, function `alloc()` first makes sure that nobody is using the `Packet` object `p`, by asserting that `fflag_` is `false` (Line 5).[5] Then, Line 6 shifts the pointer `free_` by one position. Lines 8–9 initialize two variables (`uid_` and `time_`) of class `Event` (i.e., the mother class of class `Packet`) to be zero. Line 23 removes the packet from the free list by setting `p->next_` to zero.

After the packet allocation process is complete, Lines 16–24 initialize the allocated `Packet` object. Line 16 invokes function `init(p)`, which initializes

---

[3] Generally, NS2 creates and destroys most objects by using procedures `new` and `delete`, respectively.

[4] Again, `free_` is the pointer to the first packet on the free packet list.

[5] The C++ function `assert(cond)` can be used for an integrity check. It does nothing if the input argument `cond` is `true`. Otherwise, it will initiate an error handling process (e.g., showing an error on the screen).

Program 8.3 Function alloc of class Packet.

```
//~/ns/common/packet.h
1   inline Packet* Packet::alloc()
2   {
        //Packet Allocation
3       Packet* p = free_;
4       if (p != 0) {
5           assert(p->fflag_ == FALSE);
6           free_ = p->next_;
7           assert(p->data_ == 0);
8           p->uid_ = 0;
9           p->time_ = 0;
10      } else {
11          p = new Packet;
12          p->bits_ = new unsigned char[hdrlen_];
13          if (p == 0 || p->bits_ == 0)
14              abort();
15      }

        //Packet Initialization
16      init(p); // Initialize bits_[]
17      (HDR_CMN(p))->next_hop_ = -2; // -1 reserved for IP_BROADCAST
18      (HDR_CMN(p))->last_hop_ = -2; // -1 reserved for IP_BROADCAST
19      p->fflag_ = TRUE;
20      (HDR_CMN(p))->direction() = hdr_cmn::DOWN;
21      /* setting all direction of pkts to be downstream as default;
22         until channel changes it to +1 (upstream) */
23      p->next_ = 0;
24      return (p);
25  }
```

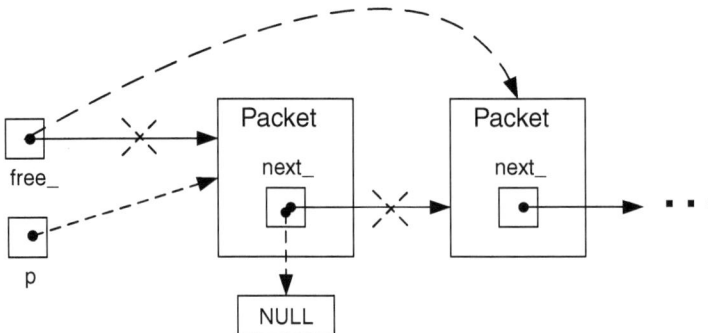

Fig. 8.3. Diagram of packet allocation when the free packet list is non-empty. The dotted lines show the actions caused by function alloc of class Packet.

Program 8.4 Functions `alloc`, `allocdata`, and `copy` of class `Packet`.

```
//~/ns/common/packet.h
1  inline Packet* Packet::alloc(int n)
2  {
3      Packet* p = alloc();
4      if (n > 0)
5          p->allocdata(n);
6      return (p);
7  }

8  inline void Packet::allocdata(int n)
9  {
10     assert(data_ == 0);
11     data_ = new PacketData(n);
12     if (data_ == 0)
13         abort();
14 }

15 inline Packet* Packet::copy() const
16 {
17     Packet* p = alloc();
18     memcpy(p->bits(), bits_, hdrlen_);
19     if (data_)
20         p->data_ = data_->copy();
21     return (p);
22 }
```

the header of packet *p. From Line 5 in Program 8.1, invocation of function `init(p)` executes "`bzero(p-> bits_,hdrlen_)`", which clears `bits_` to zero.[6] Line 19 sets `fflag_` to be `true`, indicating that the packet *p is now in use. Line 23 sets the pointer `p->next_` to be zero. Lines 17, 18, and 20 initialize the common header. We will discuss packet header in greater detail in Section 8.3.2.

Apart from function `alloc()`, other relevant functions include `alloc(n)`, `allocdata(n)`, and `copy()` (See Program 8.4). Function `alloc(n)` allocates a packet (Line 3), and invokes `allocdata(n)` (Line 5). Function `alloc(n)` creates data payload with size "n" bytes (by invoking `new PacketData(n)` in Line 11). We will discuss the details of data payload later in Section 8.4.

Function `copy()` returns a replica of the current `Packet` object. The only difference between the current and the replicated `Packet` objects is the unique ID (`uid_`) field. This function is quite useful, since we often need to create a packet which is the same as or slightly different from an original packet. This

---

[6] Function `bzero` takes two arguments – the first is a pointer to the buffer and the second is the size of the buffer – and sets all values in a buffer to zero.

function first allocates a packet pointer p in Line 17. Then, it copies packet header and data payload to the packet *p in Lines 18 and 20, respectively.

Despite its name, function refcopy() (Line 16 in Program 8.1) does not create a copy of a Packet object. Rather, it returns the pointer to the current Packet object. For example, suppose p is a Packet pointer. Then, x = p and x = p->refcopy() both store p in x. However, function refcopy() also keeps track of the number of objects which share the same Packet object, by using variable ref_count_ (Line 9 in Program 8.1). This variable is initialized to 0 in the constructor of class Packet (Line 13 in Program 8.1). It is incremented by one when function ref_copy() (Line 16 in Program 8.1) is invoked, indicating that a new object starts using the current Packet object. Similarly, it is decremented by one when function free(p) (see Section 8.2.2) is invoked, indicating that an object has stopped using the current Packet object.

8.2.2 Packet Deallocation

When a packet *p is no longer in use, NS2 deallocates the packet by using function free(p). By deallocation, NS2 returns the memory used to store packet header and data payload to the system, sets the pointer data_ to zero, and stores the Packet object in the free packet list. Note that although the value of bits_ is not set to zero, the memory location stored in bits_ is no longer accessible by bits_. It is very important not to use bits_ after packet deallocation. Otherwise, NS2 will encounter a (memory share violation) runtime error.

The details of function free(Packet*) are shown in Program 8.5. Before returning a Packet object to the free packet list, we need to make sure that

(i) The packet is in use (i.e., p->fflag_ = 1 in Line 3), since there is no point in deallocating a packet which has already been deallocated.
(ii) No object is using the packet; the variable ref_count_ is Zero (Line 4), where ref_count_ stores the number of objects which are currently using the packet.
(iii) The packet is no longer on the simulation time line (i.e., p->uid_<=0 in Line 5). Deallocating a packet while it is still on the simulation timeline will cause event mis-sequencing and runtime error. Line 5 asserts that the event unique ID corresponding to the Packet object p (i.e., p->uid_) is non-positive, and therefore is no longer on the simulation timeline.[7]
(iv) The data payload pointer data_ must not point to NULL (p->data_≠0 in Line 6), when returning the memory occupied by data payload to the system.

NS2 allows more than one simulation object to share the same Packet object. To deallocate a packet, NS2 must ensure that the packet is no longer

---

[7] From Fig. 4.2, an event with positive unique ID (e.g, uid_ is 2 or 6) was scheduled but has not been dispatched.

Program 8.5 Function `free` of class `Packet`.

```
     //~/ns/common/packet.h
1    inline void Packet::free(Packet* p)
2    {
3        if (p->fflag_) {
4            if (p->ref_count_ == 0) {
5                assert(p->uid_ <= 0);
6                if (p->data_ != 0) {
7                    delete p->data_;
8                    p->data_ = 0;
9                }
10               init(p);
11               p->next_ = free_;
12               free_ = p;
13               p->fflag_ = FALSE;
14           } else {
15               --p->ref_count_;
16           }
17       }
18   }
```

used by *any* simulation object. Again, NS2 keeps the number of objects sharing a packet in variable `ref_count_`. If `ref_count_>0`–meaning an object is invoking function `free(p)` while other objects are still using the packet `*p`, function `free(p)` will simply reduce `ref_count_` by one, indicating that one object stops using the packet (Line 15).[8] On the other hand, if `ref_count_` is zero–meaning no other object is using the packet, Lines 5–13 will then clear packet header and data payload, and store the `Packet` object in the free packet list.

If all the above four conditions are satisfied, function `free(p)` will execute Lines 6–13 in Program 8.5. The schematic diagram for this part is shown in Fig. 8.4. Line 7 returns the memory used by data payload to the system. Line 8 sets the pointer `data_` to zero. Line 10 returns the memory used by packet header of a packet `*p` to the system by invoking function `init(p)` (see Line 5 of Program 8.1). Function `free(p)` does not set the variable `bit_` to zero. Do not try to access `bit_` after this point, since doing so will cause a runtime error. Lines 11 and 12 place the packet as the first packet on the free packet list. Finally, Line 13 sets `fflag_` to `false`, indicating that the packet is no longer in use.

---

[8] If the `Packet` object is deallocated when `ref_count_ > 0`, simulation objects may later try to access the deallocated `Packet` object and cause a runtime error.

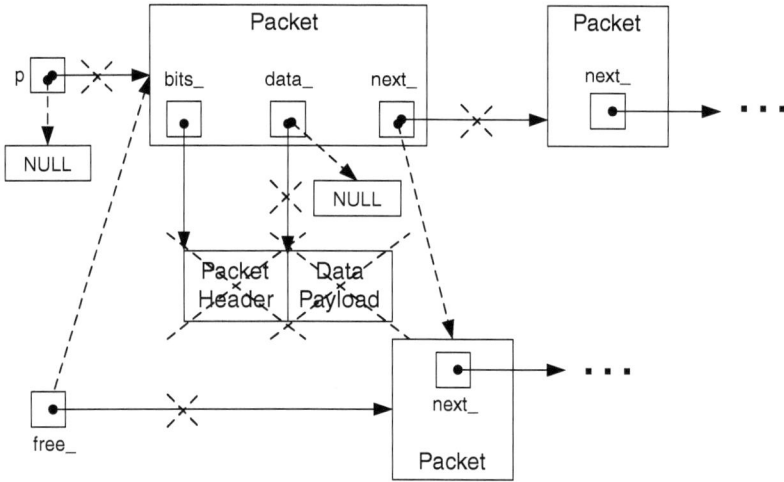

Fig. 8.4. The process of returning a packet to the packet free list. The *dotted lines* show the action caused by function `free` of class `Packet`.

8.3 Packet Header

As a part of a packet, packet header contains packet attributes such as packet unique ID, and IP address. Again, packet header is stored in variable `bits_` of class `Packet` (see Line 3 of Program 8.1). The variable `bits_` is declared as a string (i.e., a *Bag of Bits* (BOB)), and has no structure to store packet attributes. NS2 hence imposes a two-level structure on variable `bits_`, as shown in Fig. 8.5.

The first level divides the entire packet header into protocol specific headers. The location of each protocol specific header on `bits_` is identified by its variable `offset_`. The second level imposes a packet attribute storing structure on each protocol specific header. On this level, packet attributes are stored as members of a C++ `struct` data type.

In practice, a packet contains only relevant protocol specific headers. An NS2 packet on the other hand includes *all* protocol specific headers into a packet header, regardless of packet type. Every packet uses the same amount of memory to store the packet header. The amount of memory is stored in the variable `hdrlen_` of class `Packet` in Line 12 of Program 8.1, and is declared as a static variable. The variable `hdrlen_` has no relationship to simulation packet size. For example, TCP and UPD packets may have different sizes. The values stored in the corresponding variable `hdr_cmn::size_` may be different; however, the values stored in the variable `Packet::hdrlen_` for both TCP and UDP packets are the same.

In the following, we first discuss the first level packet header composition in Section 8.3.1. Sections 8.3.2 and 8.3.3 shows examples of protocol specific headers: common packet header and IP packet header. Section 8.3.4 discusses

Packet

bits_   data_   next_   ...

Packet Header | Data Payload

1st Level

... | TCP Header | IP Header | Common Header | ...

2nd Level

hdr_tcp::offset_

hdr_ip::offset_

hdr_cmn::offset_

uid_
ptype_
offset_
size_

Protocol specific header size is determined during the compilation.

Packet header size is determined during the construction of the simulator

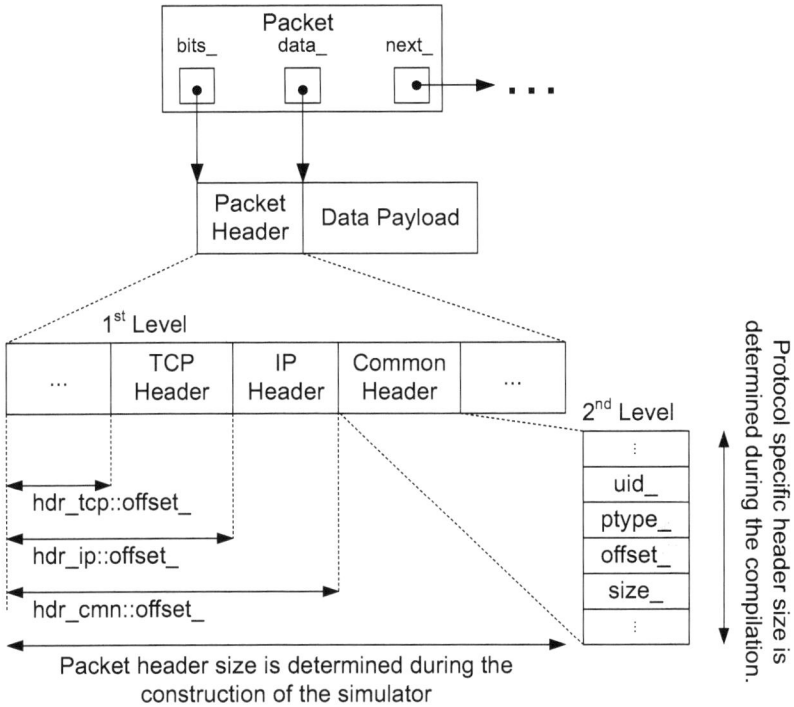

Fig. 8.5. Architecture of packet header.

one of the main packet attributes: packet type. Section 8.3.5 explains the details of protocol specific header (i.e., the second level packet header composition). Section 8.3.6 demonstrates how packet attributes stored in packet header are accessed. Section 8.3.7 discusses one of the main packet header component, a packet header manager, which maintains the active protocol list and sets up the offset value for each protocol. Finally, Section 8.3.8 presents the packet header construction process.

8.3.1 An Overview of First Level Packet Composition: Offseting Protocol Specific Header on the Packet Header

On the first level, NS2 puts together all relevant protocol specific headers (e.g., common header, IP header, TCP header) and composes a packet header (see Fig. 8.5). Conceptually, NS2 allocates a contiguous part on the packet header for a protocol specific header. Each protocol specific header is offset from the beginning of packet header. The distance between the beginning of packet header and that of a protocol specific header is stored in the member variable offset_ of the protocol specific header. For example, hdr_cmn, hdr_ip, and hdr_tcp–which represent common header, IP header, and TCP header–

store their offset values of variables hdr_cmn::offset_, hdr_ip::offset_, and hdr_tcp::offset_, respectively.

8.3.2 Common Packet Header

Common packet header contains packet attributes which are common to all packets. It employs C++ struct data type hdr_cmn to indicate how the packet attributes are stored. Program 8.6 shows a part of hdr_cmn declaration. The main member variables of hdr_cmn are as follows:

Program 8.6 Declaration of C++ hdr_cmn struct data type.

```
    //~/ns/common/packet.h
 1  struct hdr_cmn {
 2      enum dir_t { DOWN= -1, NONE= 0, UP= 1 };
 3      packet_t ptype_;    // packet type
 4      int size_;          // simulated packet size
 5      int uid_;           // unique id
 6      dir_t   direction_; // direction: 0=none, 1=up, -1=down
 7      static int offset_; // offset for this header

 8      inline static hdr_cmn* access(const Packet* p) {
 9          return (hdr_cmn*) p->access(offset_);
10      }
11      inline static int& offset() { return offset_; }
12      inline packet_t& ptype() { return (ptype_); }
13      inline int& size() { return (size_); }
14      inline int& uid() { return (uid_); }
15      inline dir_t& direction() { return (direction_); }
16  };
```

---

ptype_ The packet type (not the type of protocol specific header).

size_ The packet size. Unlike actual packet transmission, the number of bits requires to hold a packet has no relationship to simulation packet size. During simulation, NS2 uses variable hdr_cmn::size_ as the packet size.

uid_ The ID which is unique to every packet.

dir_t The transmitting direction which can be downstream (–1), upstream (1), or not-in-use (0). By default, dir_t is set to downstream (see Line 20 in Program 8.3).

offset_ The memory location relative to the beginning of packet header from which the common header is stored (see Section 8.3.1 and Fig. 8.5).

From Fig. 8.6, most functions of class hdr_cmn act as an interface to access its variables. Apart from these functions, function access(p) in Lines 8–10 is perhaps the most important function of `hdr_cmn`. It is used to access a protocol specific header of the input `Packet` object `*p`. We will discuss the packet header access mechanism in greater detail in Section 8.3.6.

Program 8.7 Declaration of C++ `hdr_ip` struct data type.

```
    //~/ns/common/ip.h
 1  struct hdr_ip {
 2      ns_addr_t    src_;
 3      ns_addr_t    dst_;
 4      int      ttl_;
 5      int      fid_;
 6      int      prio_;
 7      static int offset_;
 8      inline static int& offset() { return offset_; }
 9      inline static hdr_ip* access(const Packet* p) {
10          return (hdr_ip*) p->access(offset_);
11      }
12      ns_addr_t& src() { return (src_); }
13      nsaddr_t& saddr() { return (src_.addr_); }
14      int32_t& sport() { return src_.port_;}
15      ns_addr_t& dst() { return (dst_); }
16      nsaddr_t& daddr() { return (dst_.addr_); }
17      int32_t& dport() { return dst_.port_;}
18      int& ttl() { return (ttl_); }
19      int& flowid() { return (fid_); }
20      int& prio() { return (prio_); }
21  };
```

8.3.3 IP Packet Header

Represented by C++ struct data type `hdr_ip`, IP packet header contains information about source and destination of a packet. Program 8.7 shows a part of `hdr_ip` declaration. IP packet header contains the following five main variables which contain IP-related packet information (see Lines 2–6 in Program 8.7):

`src_`   Source node's address of the packet
`dst_`   Destination node's address of the packet
`ttl_`   Time to live for the packet
`fid_`   Flow ID of the packet
`prio_` Priority level of the packet

NS2 utilizes data type `ns_addr_t` defined in file ~`ns`/config.h to store node address. From Program 8.8, `ns_addr_t` is a struct data type, which contains two members: `addr_` and `port_`. Both members are of type `int32_t`, which is simply an alias for `int` data type (see Line 5 and file ~`ns`/autoconf-win32.h). While `addr_` specifies the node address, `port_` identifies the attached port (if any).

Program 8.8 Declaration of C++ `ns_addr_t` struct data type, and its `int32_t` alias

```
     //~/ns/config.h
1    struct ns_addr_t {
2        int32_t addr_;
3        int32_t port_;
4    };

     //~/ns/autoconf-win32.h
5    typedef int int32_t;
```

The variables `src_` and `dst_` of IP header are of class `ns_addr_t`. Hence, `src_.addr_` and `src_.port_` store the node address and the port of the sending agent, respectively. Similarly, the packet will be sent to a receiving agent attached to port `dst_.port_` of a node with address `dst_.addr_`.

Lines 7–11 in Program 8.7 declare variable `offset_`, function `offset(off)` and function `access(p)`, which are essential to access IP header of a packet. We will discuss the packet access mechanism later in Section 8.3.6. Lines 12–20 in Program 8.7 are functions which return the values of the variables.

8.3.4 Packet Type

Although stored in common header, packet type is attributed to an entire packet, not to a protocol specific header. Each packet corresponds to only one packet type but may contain several protocol specific headers. For example, a packet can be encapsulated by both TCP and IP protocols. However, its type can be either audio *or* TCP packet, *but not both*.

NS2 stores a packet type in a member variable `ptype_` of a common packet header. The type of the variable `ptype_` is `enum packet_t` defined in Program 8.9. Again, members of `enum` are integers which are mapped to strings. From Fig. 8.9, `PT_TCP` (Line 2) and `PT_UDP` (Line 3) are mapped to 0 and 1, respectively. Since `packet_t` declares `PT_NTYPE` (representing undefined packet type) as the last member, the value of `PT_NTYPE` is $N_p - 1$, where N_p is the number of `packet_t` members. NS2 provides 60 built-in packet types, meaning the default value of `PT_NTYPE` is 59.

From Lines 11–30 in Program 8.9, class `p_info` maps each member of `packet_t` to a description string. It has a static associative array variable, `name_` (Line 28). The index and value of `name_` are the packet type, and the corresponding description string, respectively. Class `p_info` also has one

Program 8.9 Declaration of enum `packet_t` type and class `p_info`.

```
    //~/ns/common/packet.h
1   enum packet_t {
2       PT_TCP,
3       PT_UDP,
4       PT_CBR,
5       PT_AUDIO,
6       PT_VIDEO,
7       PT_ACK,
8       ...
9       PT_NTYPE // This MUST be the LAST one
10  }

11  class p_info {
12  public:
13      p_info() {
14          name_[PT_TCP]= "tcp";
15          name_[PT_UDP]= "udp";
16          name_[PT_CBR]= "cbr";
17          name_[PT_AUDIO]= "audio";
18          name_[PT_VIDEO]= "video";
19          name_[PT_ACK]= "ack";
20          ...
21          name_[PT_NTYPE]= "undefined";
22  }
23      const char* name(packet_t p) const {
24          if ( p <= PT_NTYPE ) return name_[p];
25          return 0;
26      }
27  private:
28      static char* name_[PT_NTYPE+1];
29  };
30  extern p_info packet_info; /* map PT_* to string name */
```

important function `name(p)` (Lines 23–26), which translates a `packet_t` variable to a description string.

At the declaration, NS2 declares a global variable `packet_info` (using `extern`), which is of class `p_info` (Line 30). Accessible at the global scope, the variable `packet_info` provides an access to function `name(p)` of class `p_info`. To obtain a description string of a `packet_t` object p, one may invoke

```
packet_info.name(ptype)
```

Example 8.1. Class **Agent** is responsible for creating and destroying network layer packets (see Chapter 11). It is the base class of TCP and UDP transport layer protocol modules. Class **Agent** provides a function `allocpkt()`, which is responsible for allocating (i.e., creating) a packet.

To print out the type of every allocated packet on the screen, we modify function `allocpkt()` of class `Agent` in file ~*ns*/common/agent.cc as follows:

```
//~/ns/common/agent.h
1   Packet* Agent::allocpkt() const
2   {
3       Packet* p = Packet::alloc();
4       initpkt(p);
5       //------- Begin Additional Codes -------------
6       hdr_cmn* ch = hdr_cmn::access(p);
7       packet_t pt = ch->ptype();
8       printf("Example Test: Class Agent allocates a packet
                        with type %s\n", packet_info.name(pt));
9       getchar();
10      //------- End Additional Codes ---------
11      return (p);
12  }
```

where Lines 5–10 are added to the original codes. Line 6 retrieves the reference "ch" to the common packet header (see Section 8.3.6). Line 7 obtains the packet type stored in the common header by using function `ptype()`, and assigns the packet type to variable `pt`. Note that, variable `packet_info` is a global variable of class `p_info`. When the variable `pt` is fed as an input argument, function `packet_info.name(pt)` returns the description string corresponding to the `packet_t` object "pt" (Line 8).

After re-compiling the code, the simulation should show the type of every allocated packet on the screen. For example, when running the Tcl simulation script in Programs 2.1–2.2 provided in Chapter 2, the following result should appear on the screen:

```
>> ns myfirst_ns.tcl
Example Test: Class Agent allocates a packet with type cbr
Example Test: Class Agent allocates a packet with type cbr
Example Test: Class Agent allocates a packet with type cbr
 .
 .
 .
 .
```

8.3.5 Protocol Specific Headers

A protocol specific header stores packet attributes relevant to the underlying protocol only. For example, common packet header holds basic packet attributes such as packet unique ID, packet size, packet type, and so on. IP packet header contains IP packet attributes such as source and destination IP addresses and port numbers. There are 48 classifications of packet headers.

The complete list of protocol specific headers with their descriptions is given in [15].

Each protocol specific header involves three classes discussed below.

A Protocol Specific Header C++ Class

In C++, NS2 uses a **struct** data type to represent a protocol specific header. It stores packet attributes and its offset value in members of a **struct** data type. It also provides a function **access(p)** which returns the reference to the protocol specific header of a packet ***p**. Representing a protocol specific header, each **struct** data type is named using format **hdr_<XXX>**, where XXX is an arbitrary string representing the type of a protocol specific header. For example, the C++ class name for common packet header is **hdr_cmn**.

In the C++ domain, protocol specific headers are declared but not instantiated. Therefore, NS2 uses a **struct** data type (rather than a class) to represent protocol specific headers, and no constructor is required for a protocol specific header. Hereafter, we will refer to **struct** and **class** interchangeably.

A Protocol Specific Header OTcl Class

NS2 defines a shadow OTcl class for each C++ protocol specific header class. An OTcl class acts as an interface to the OTcl domain. It is named with the format **PacketHeader/<XXX>**, where XXX is an arbitrary string representing a protocol specific header. For example, the OTcl class name for common packet header is **PacketHeader/Common**.

A Protocol Specific Header Mapping Class

A mapping class is responsible for binding OTcl and C++ class names together. All the packet header mapping classes derive from class **PacketHeaderClass** which is a child class of class **TclClass**. A mapping class is named with format **<XXX>HeaderClass**, where XXX is an arbitrary string representing a protocol specific header. For example, the mapping class name for common packet header is **CommonHeaderClass**.

Program 8.10 shows the declaration of class **PacketHeaderClass**, which has two key variables: **hrdlen_** in Line 8 and **offset** in Line 9. The variable **hdrlen_** represents the length of the protocol specific header.[9] It is the system memory needed to store a protocol specific header C++ class. Variable **offset_** indicates the location on packet header where the protocol specific header is used.

---

[9] While variable **hdrlen_** in class **PacketHeaderClass** represents the length of a protocol specific header, variable **hdrlen_** in class **Packet** represents total length of packet header.

Program 8.10 Declaration of class `PacketHeaderClass`.

```
//~/ns/common/packet.h
1   class PacketHeaderClass : public TclClass {
2   protected:
3       PacketHeaderClass(const char* classname, int hdrlen) :
4           TclClass(classname), hdrlen_(hdrlen), offset_(0);{};
5       virtual int method(int argc, const char*const* argv);
6       inline void bind_offset(int* off) { offset_ = off; };
7       inline void offset(int* off) {offset_= off;};
8       int hdrlen_;        // # of bytes for this header
9       int* offset_;       // offset for this header
10  public:
11      TclObject* create(int argc,const char*const* argv){return 0;};
12      virtual void bind(){
13          TclClass::bind();
14          Tcl& tcl = Tcl::instance();
15          tcl.evalf("%s set hdrlen_ %d", classname_, hdrlen_);
16          add_method("offset");
17      };
18  };
```

The constructor of class `PacketHeaderClass` in Lines 3–4 takes two input arguments. The first input argument `classname` is the name of the corresponding OTcl class name (e.g., `PacketHeader/Common`). The second one, `hdrlen`, is the length of the protocol specific header C++ class. In Lines 3–4, the constructor feeds `classname` to the constructor of class `TclClass`, stores `hdrlen` in the member variable `hdrlen_`, and resets `offset_` to zero.

Function `method(argc,argv)` in Line 5 is an approach to take a C++ action from the OTcl domain. Functions `bind_offset(off)` in Line 6 and `offset(off)` in Line 7 are used to configure and retrieve, respectively, the value of variable `offset_`. Function `create(argc,argv)` in Line 11 does nothing, since no protocol specific header C++ object is ever. It will be overridden by the derived classes of class `PacketHeaderClass`. Function `bind()` in Lines 12–17 glues the C++ class to the OTcl class. Line 13 first invokes function `bind()` of class `TclClass`, which performs the basic binding actions. Line 15 then exports variable `hdrlen_` to the OTcl domain. Line 16 registers the *OTcl method* `offset` using function `add_method("offset")`.

Apart from the commands discussed in Section 3.4.4, an *OTcl method* is another way to invoke C++ functions from the OTcl domain. It is implemented in C++ via the following two steps. The first step is to define a function `method(ac,av)`. As can be seen from Program 8.11, the structure of function `method` is very similar to that of function `command`. A *method* "offset" sets the value of `*offset_` to be what specified in the input argument (Line 7 in Program 8.11). The second step in method implementation is to register the name of the *method* by using a function "`add_method(str)`", which

Program 8.11 Function method of class `PacketHeaderClass`.

```
//~/ns/common/packet.cc
1   int PacketHeaderClass::method(int ac, const char*const* av)
2   {
3       Tcl& tcl = Tcl::instance();
4       ...
5       if (strcmp(argv[1], "offset") == 0) {
6           if (offset_) {
7               *offset_ = atoi(argv[2]);
8               return TCL_OK;
9           }
10          tcl.resultf("Warning: cannot set
                          offset_ for %s",classname_);
11          return TCL_OK;
12      }
13      ...
14      return TclClass::method(ac, av);
15  }
```

takes the method name as an input argument. For class `PacketHeaderClass`, the *method* `offset` is registered from within function `bind(...)` (Line 16 of Program 8.10).

A protocol specific header is implemented using a `struct` data type, and hence does not derive function `command(...)` from class `TclObject`[10]. It resorts to OTcl *method*s defined in the mapping class to take C++ actions from the OTcl domain. We will show an example use of the method `offset` later in Section 8.3.8, when we discuss packet construction mechanism.

Program 8.12 Declaration of class `CommonHeaderClass`.

```
//~/ns/common/packet.cc
1   class CommonHeaderClass : public PacketHeaderClass {
2   public:
3       CommonHeaderClass() : PacketHeaderClass("PacketHeader/Common",
                          sizeof(hdr_cmn)) {
4           bind_offset(&hdr_cmn::offset_);
5       }
6   } class_cmnhdr;
```

Consider, for example, a common packet header. Its C++, OTcl, and mapping classes are `hdr_cmn`, `PacketHeader/Common`, and `CommonPacketHeader Class`, respectively (see Table 8.1). Program 8.12 shows the declaration of

---

[10] Since NS2 does not instantiate a protocol specific header object, it models a protocol specific header using `struct` data type.

Table 8.1. Classes and objects related to common packet header

Class/Object	Name
C++ class	`hdr_cmn`
OTcl class	`PacketHeader/Common`
Mapping class	`CommonHeaderClass`
Mapping variable	`class_cmnhdr`

class `CommonPacketHeaderClass`. As a child class of `TclClass`, a class mapping variable `class_cmnhdr` is instantiated at the declaration. Line 3 of the constructor invokes the constructor of its parent class `PacketHeaderClass`, which takes the OTcl class name (i.e., `PacketHeader/Common`) and the amount of memory needed to hold the C++ class (i.e., `hdr_cmn`) as input arguments. Here, "`sizeof (hdr_cmn)`" computes such the required amount of memory, which is fed as the second input argument. In Line 6 of Program 8.10, function `bind_offset(&hdr_cmn::offset_)` sets its variable `offset_` to share the address with the input argument. Therefore, a change in `hdr_cmn::offset_` will result in an automatic change in variable `*offset_` of class `CommonHeader-Class`, and vice versa.

8.3.6 Packet Header Access Mechanism

This section demonstrates how packet attributes stored in packet header can be retrieved and modified. NS2 employs a two-level packet header structure to store packet attributes. On the first level, protocol specific headers are stored within a packet header. On the second level, each protocol specific header employs a C++ `struct` data type to store packet attributes. The header access mechanism consists of two major steps: (1) Retrieve a reference to a protocol specific header, and (2) Follow the structure of the protocol specific header to retrieve or modify packet attributes. In this section, we will explain the access mechanism through common packet header (see the corresponding class names in Table 8.1).

Retrieving a Reference to Protocol Specific Header

NS2 obtains a reference to a protocol specific header by of a packet *p using a function `access(p)` in the C++ class. A reference to the common header of a `Packet` object *p can be obtained by executing `hdr_cmn::access(p)` (see Example 8.2 below).

Example 8.2. Consider function `allocpkt()` of class `Agent` shown in Program 8.13, which shows the details of functions `allocpkt()` and `initpkt(p)`. Function `allocpkt()` in Lines 1–6 creates a `Packet` object and returns a pointer to the created object to the caller. Function `allocpkt()` first invokes

Program 8.13 Functions `allocpkt` and `initpkt` of class `Agent`.

```
   //~/ns/common/agent.cc
1  Packet* Agent::allocpkt() const
2  {
3       Packet* p = Packet::alloc();
4       initpkt(p);
5       return (p);
6  }

7  Packet* Agent::initpkt(Packet* p) const
8  {
9       hdr_cmn* ch = hdr_cmn::access(p);
10      ch->uid() = uidcnt_++;
11      ch->ptype() = type_;
12      ch->size() = size_;
13      ...
14      hdr_ip* iph = hdr_ip::access(p);
15      iph->saddr() = here_.addr_;
16      iph->sport() = here_.port_;
17      iph->daddr() = dst_.addr_;
18      iph->dport() = dst_.port_;
19      ...
20 }
```

function `alloc()` of class `Packet` in Line 3 (see the details in Section 8.2.1). Then, Line 4 initializes the allocated packet, by invoking function `initpkt(p)`. Finally, Line 5 returns the pointer p to the initialized `Packet` object to the caller.

Function `initpkt(p)` follows the structure defined in the protocol specific header C++ classes to set packet attributes to the default values. Lines 9 and 14 in Program 8.13 execute the first step in the access mechanism: retrieve references to common packet header `ch` and IP header `iph`, respectively.

After obtaining pointers `ch` and `iph`, Lines 10–12 and Lines 15–18 carry out the second step in the access mechanism: access packet attributes through the structure defined in the protocol specific headers. In this step, the relevant packet attributes such as unique packet ID, packet type, packet size, source IP address and port, destination IP address and port, are configured through pointers `ch` and `iph`. Note that `uidcnt` (i.e., uid count) is a static member variable of class `Agent` which represents the total number of generated packets. We will discuss the details of class `Agent` later in Chapter 9.

Figure 8.6 shows an internal mechanism of function `hdr_cmn::access(p)` where p is a `Packet` pointer. When `hdr_cmn::access(p)` is executed Line 9 in Program 8.6 executes `p->access(offset_)`, where `offset_` is the member variable of class `hdr_cmn`, specifying the location on the packet header allocated to the common header (see also Fig. 8.5). On the right hand side of

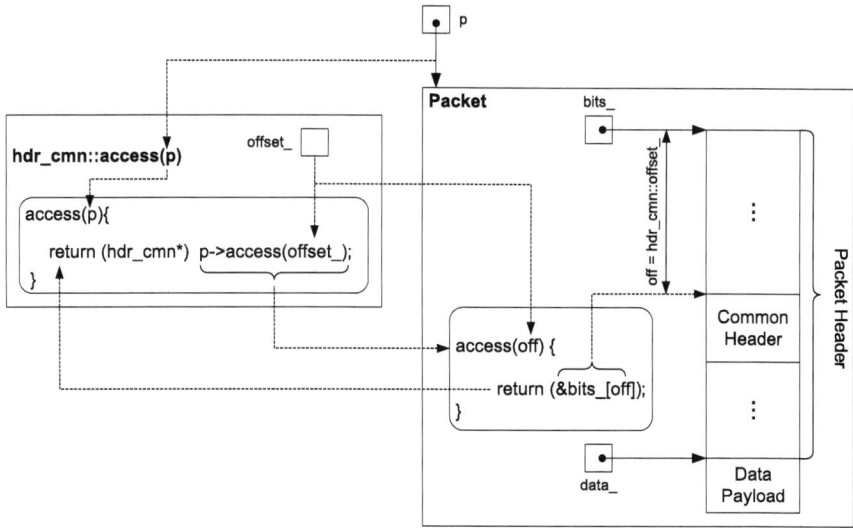

Fig. 8.6. The internal mechanism of function `access(p)` of the `hdr_cmn` struct data type, where p is a pointer to a `Packet` object.

Fig. 8.6, function `access(off)` simply returns `&bits_[off]`, where `bits_` is the member variable of class `Packet` storing the packet header. Since `hdr_cmn` feeds its variable `offset_` as the input argument, function `access(offset_)` essentially returns `&bits_[hdr_cmn::offset_]`, which is the reference to the common header stored in the `Packet` object *p. This reference is returned as an `unsigned char*` variable. Then, class `hdr_cmn` casts the returned reference to type `hdr_cmn*`, and returns it to the caller.

Accessing Packet Attributes in a Protocol Specific Header

After obtaining a reference to a protocol specific header, the second step is to access the packet attributes according to the structure specified in the protocol specific header C++ class. Since NS2 declares a protocol specific header as a `struct` data type, it is fairly straightforward to access packet attributes once the reference to the protocol specific header is obtained (see Example 8.2).

8.3.7 Packet Header Manager

A packet header manager is responsible for keeping the list of active protocols and setting the offset values of all the active protocols. It is implemented using a C++ class `PacketHeaderManager` which is bound to an OTcl class with the same name. Program 8.14 and Fig. 8.7 show the declaration of the C++ class `PacketHeaderManager` as well as the corresponding binding class, and the diagram of the OTcl class `PacketHeaderManager`, respectively.

Program 8.14 Declarations of C++ class `PacketHeaderManager` and mapping class `PacketHeaderManagerClass`.

```
    //~/ns/common/packet.cc
1   class PacketHeaderManager : public TclObject {
2   public:
3       PacketHeaderManager() {bind("hdrlen_", &Packet::hdrlen_);}
4   };

5   static class PacketHeaderManagerClass : public TclClass {
6   public:
7       PacketHeaderManagerClass() : TclClass("PacketHeaderManager") {}
8       TclObject* create(int, const char*const*) {
9           return (new PacketHeaderManager);
10      }
11  } class_packethdr_mgr;
```

The C++ class `PacketHeaderManager` has only one constructor (Line 3) and has neither variables nor functions. The constructor binds the instvar `hdrlen_` of OTcl class `PacketHeaderManager` to variable `hdrlen_` of class `Packet` (see also Fig. 8.1). The OTcl class `PacketHeaderManager` has two main instvars: `hdrlen_` and `tab_`. Instvar `hdrlen_` stores the length of packet header. It is initialized to zero in Line 1 of Program 8.15, and is incremented as protocol specific headers are added to the packet header. Representing the active protocol list, instvar `tab_` (Line 2 in Program 8.16) is an associative array whose indexes are protocol specific header OTcl class names and values

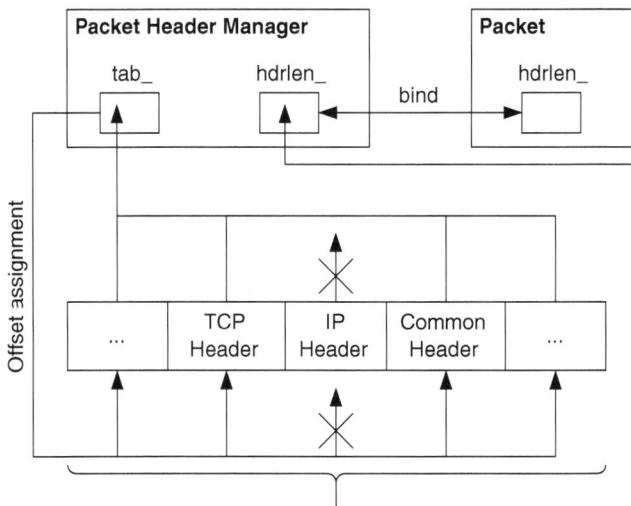

Fig. 8.7. Architecture of an OTel `PacketHeaderManager` object.

Program 8.15 Initialization of a `PacketHeaderManager` object.

```
//~/tcl/ns-packet.tcl
1  PacketHeaderManager set hdrlen_ 0

2  foreach prot {
3      Common
4      Flags
5      IP
6      ...
7  } {
8      add-packet-header $prot
9  }

10 proc add-packet-header args {
11     foreach cl $args {
12         PacketHeaderManager set tab_(PacketHeader/$cl) 1
13     }
14 }
```

Program 8.16 Function `create_packetformat` of class `Simulator` and function `allochdr` of class `PacketHeaderManager`.

```
//~/tcl/ns-packet.tcl
1  Simulator instproc create_packetformat { } {
2      PacketHeaderManager instvar tab_
3      set pm [new PacketHeaderManager]
4      foreach cl [PacketHeader info subclass] {
5          if [info exists tab_($cl)] {
6              set off [$pm allochdr $cl]
7              $cl offset $off
8          }
9      }
10     $self set packetManager_ $pm
11 }

12 PacketHeaderManager instproc allochdr cl {
13     set size [$cl set hdrlen_]
14     $self instvar hdrlen_
15     set NS_ALIGN 8
16     set incr [expr ($size + ($NS_ALIGN-1)) & ~($NS_ALIGN-1)]
17     set base $hdrlen_
18     incr hdrlen_ $incr
19     return $base
20 }
```

are 1 if the protocol specific header is active (see Line 12 in Program 8.5). If the protocol specific header is inactive, the corresponding value of `tab_` will not be available (i.e., NS2 `unsets` all entries corresponding to inactive protocol specific headers; see Line 7 in Program 8.20).

8.3.8 Protocol Specific Header Composition and Packet Header Construction

Packet header is constructed through the following three-step process:

Step 1: At the Compilation Time

During the compilation, NS2 translates all C++ codes into an executable file. It sets up all necessary variables (including the length of all protocol specific headers) for all built-in protocol specific headers, and includes all built-in protocol specific headers into the active protocol list. There are three main tasks in this step.

Task 1: Construct all mapping variables, configure the variable `hdrlen_`, and register the OTcl class name, and binds the offset value

Since all mapping variables are instantiated at the declaration, they are constructed during the compilation using their constructors. As an example, consider the common packet header[11] whose construction process shown in Fig. 8.8 proceeds as follows:

1. Store the corresponding OTcl class name (e.g., `PacketHeader/Common`) in the variable `classname_` of class `TclClass`.
2. Determine the size (using function `sizeof (...)`) of the protocol specific header, and store it in the variable `hdrlen_` of class `PacketHeaderClass`.
3. Bind the variable `PacketHeader:: offset_` to that of the C++ class `hdr_cmn`.

Task 2: Invocation of function `bind()` of class `TclClass` which exports the variable `hdrlen_`

The main NS2 function (i.e., `main(argc,argv)`) invokes function `init(...)` of class `Tcl`, which in turn invokes function `bind()` of class `TclClass` of all mapping variables. Function `bind()` registers and binds an OTcl class name to the C++ domain (see file ~*tclcl*/Tcl.cc). This function is overridden by class `PacketHeaderClass`.

---

[11] NS2 repeats the following process for all protocol specific headers. For brevity, we show the construction process through common packet header only.

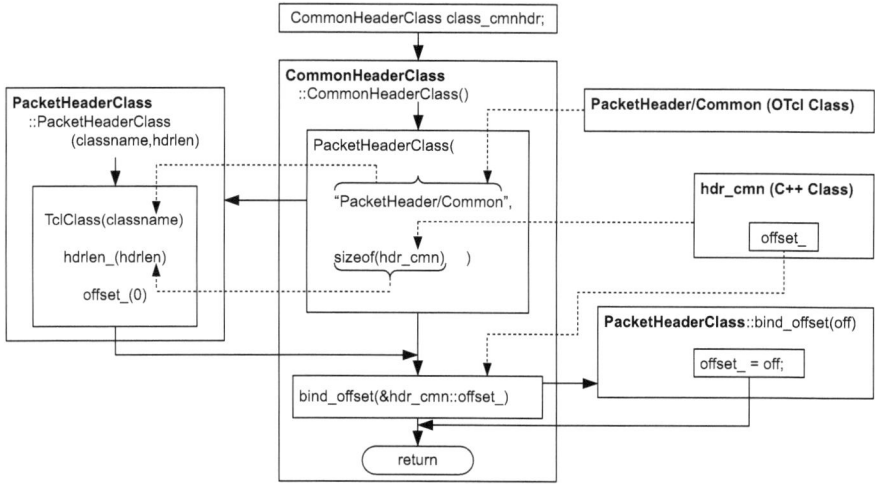

Fig. 8.8. Construction of the static mapping variable `class_cmnhdr`.

As shown in Lines 12–17 of Program 8.10, class `PacketHeaderClass` over-rides function `bind()` of class `TclClass`. Line 13 first invokes the function `bind()` of class `TclClass`. Line 15 exports the variable `hdrlen_` to the OTcl instvar with the same name. Finally, Line 16 registers the OTcl method `offset`.

In case of class `CommonHeaderClass`, `classname_` is `PacketHeader/Common` and `hdrlen_` is 104 bytes. Therefore, Line 15 of Program 8.10 executes the following command in the OTcl domain:

`PacketHeader/Common set hdrlen_ 104`

which sets instvar `hdrlen_` of class `PacketHeader/Common` to be 104. Note that this instvar `hdrlen_` is not bound to the C++ domain.

After Task 1 and Task 2 are completed, the related protocol specific classes, namely, `hdr_cmn`, `PacketHeader/Common`, and `CommonHeaderClass`, would be as shown in Fig. 8.9. The mapping object `class_cmnhdr` is of class `CommonHeaderClass`, which derives from classes `PacketHeaderClass` and `TclClass`, respectively. It inherits variables `classname_`, `hdrlen_`, and `offset_` from its parent classes. After object construction is complete, variable `classname_` will store the name of the OTcl common header class (i.e., `PacketHeader/Common`), `hdr_len_` will store the amount of memory in bytes needed to store common header, and `offset_` will point to `hdr_cmn::offset_`. Here, variable `offset_` of class `CommonHeaderClass` only points to variable `offset_` of class `hdr_cmn`. However, at this moment, the offset value is set to zero. The dashed arrow in Fig. 8.9 indicates that the value of variable `hdr_cmn::offset_` will be later set to store an offset from the beginning of a packet header to the point where the common packet header is stored. Also, after function `Tcl::init()` invokes function `bind()` of class

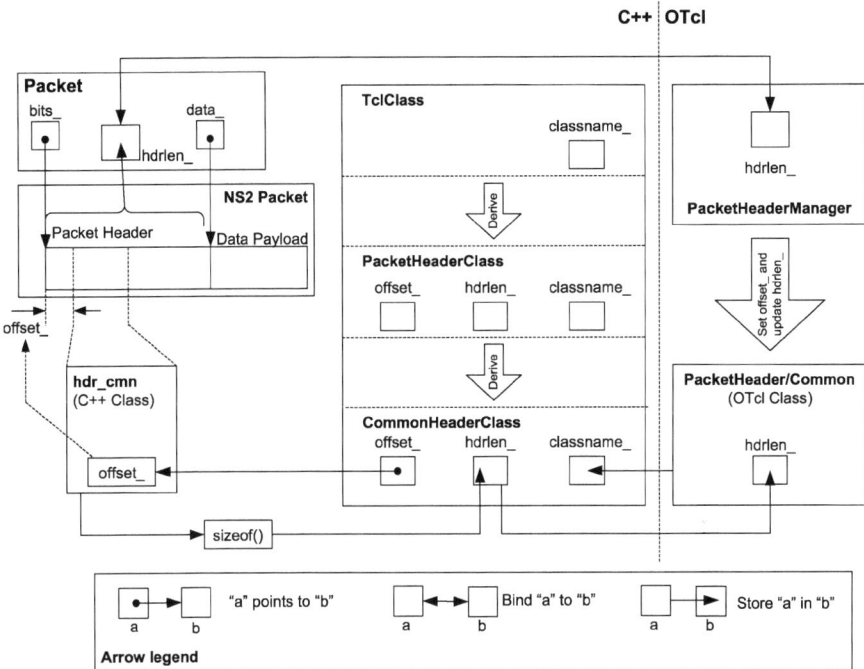

Fig. 8.9. A schematic diagram of a static mapping object `class_cmnhdr`, class `hdr_cmn`, class `PacketHeader/Common`, and class `Packet`.

`PacketHeaderClass`, instvar `hdrlen_` of class `PacketHeader/Common` will store the value of variable `hdrlen_` of class `CommonHeaderClass`. Note that tasks and 2 only set up C++ OTcl class, and mapping class. However, the packet header manager is not configured at this phase.

Task 3: Sourcing the file ~ns/tcl/lib/ns-packet.tcl to setup an active protocol list

As discussed in Section 3.7, NS2 sources all scripting Tcl files during the compilation process. In regards to packet header, Program 8.15 shows a part of the file ~*ns*/tcl/lib/ns-packet. Here, Line 8 invokes procedure `add-packet-header{prot}` for all built-in protocol specific headers indicated in Lines 3–6. In Line 12, this procedure sets the value of the associated array `tab_` whose index is the input protocol specific header name to be 1.

Step 2: During the Network Configuration Phase

In regards to packet header construction, the main task in the Network Configuration Phase is to setup variables `offset_` of all active protocol specific headers and formulate a packet header format. Subsequent packet creation will follow the packet format created in this step.

The offset configuration process takes place during the simulator construction. From Line 2 of Program 4.11, the constructor of the simulator invokes instproc `create_packetformat{}` of class `Simulator` shown in Program 8.16.

Instproc `create_packetformat{}` creates a `PacketHeaderManager` object `pm` (Line 3), computes the offset value of all active protocol specific headers using instproc `allochdr{cl}` (Line 6), and configures the offset values of all protocol specific headers (Line 7). The `foreach` loop in Line 4 runs for all built-in protocol specific headers which are child classes of class `PacketHeader`. Then Line 5 filters out those which are not in the active protocol list (see Section 8.3.7). Lines 6–7 are executed for all active protocol specific headers specified in variable `tab_` (which was configured in Step 1 – Task 3) of the `PacketHeaderManager` object "pm". Line 7 configures offset values by using the OTcl *method* `offset` (see Program 8.11) of protocol specific header mapping classes. The OTcl *method* `offset` stores the input argument in variable `*offset_` of the protocol specific header mapping class (e.g., `CommonHeaderClass`).

Lines 12–19 in Program 8.16 and Fig. 8.10 show the OTcl source codes and the diagram, respectively, of the instproc `allochdr{cl}` of an OTcl class `PacketHeaderManager`. Instproc `allochdr{cl}` takes one input argument "cl" (in Line 12) which is the name of a protocol specific header, and returns the offset value corresponding to the input argument "cl". Line 13 stores header length of a protocol specific header "cl" (e.g., variable `hdrlen_`

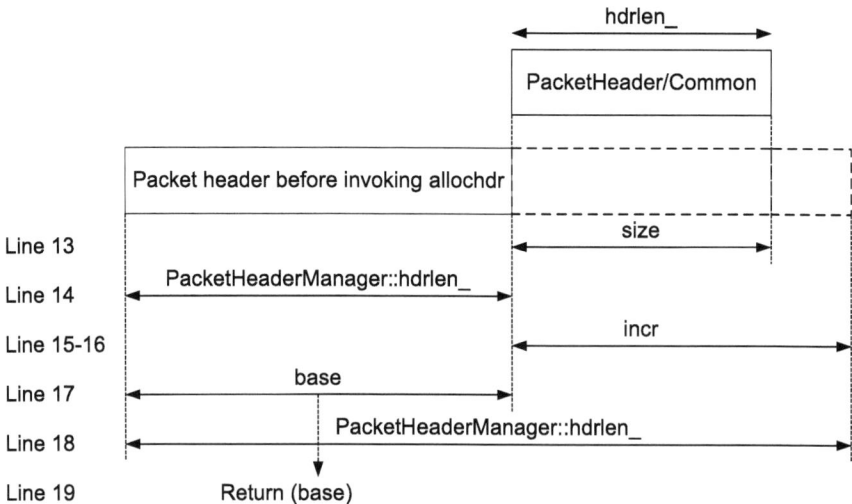

Fig. 8.10. A diagram representing instproc allochdr of class PacketHeaderManager. Line numbers shown on the left correspond to the lines in Program 8.16. The action corresponding to each line is shown on the right.

of class `PacketHeader/Common`) in a local variable `size`.[12] Based on `size`, Lines 15–16 compute the amount of memory "`incr`" needed to store the header based on `size`.[13] Line 17 stores the current packet header length (excluding the input protocol specific header) in a local variable "`base`". Since "`base`" is an offset distance from the beginning of packet header to the input protocol specific header, it is returned to the caller as the offset value in Line 19. Finally, Line 18 increases the header length (i.e., the instvar `hdrlen_` of class `PacketHeaderManager`) by "`incr`".

During the `Simulator` construction, the packet header manager also updates its variable `hdrlen_` (Line 19 in Program 8.16). Note that the instvar `hdrlen_` of class `PacketHeaderManager` was set to zero at the compilation (Line 1 in Program 8.15). As Lines 6–7 in Program 8.16 repeat for every protocol specific header, the offset value is added to the instvar `hdrlen_` of an OTcl class `PacketHeaderManager`. At the end, the instvar `hdrlen_` will represent the total header length, which embraces all protocol specific headers.

Step 3: During the Simulation Phase

During the Simulation Phase, NS2 creates packets based on the format defined in the former two steps. For example, an `Agent` object creates and initializes a packet using its function `allocpkt()`. Here, a packet is created using function `alloc()` of class `Packet`, and initialized using function `initpkt(p)` of class `Agent`. Again, function `alloc()` takes a packet from the free packet list, if it is non-empty. Otherwise, it will create a new packet using `new`. After retrieving a packet, it clears the values stored in the packet header and data payload. Function `initpkt(p)` assigns default values to packet attributes such as packet unique ID, packet type, and packet size (see Program 8.13). The initialization is performed by retrieving a reference (e.g., `ch`) to the relevant protocol specific header and accessing packet attributes using the predefined structure.

8.4 Data Payload

Implementation of data payload in NS2 differs from actual data payload. In practice, user information is transformed into bits, and are stored into data payload. Such a transformation is not necessary in simulation, since NS2 stores the user information in the packet header. NS2 rarely needs to maintain data payload. In Line 11 of Program 7.3, packet transmission time, i.e., the time required to send out a packet, is computed as $\frac{\text{packet size}}{\text{bandwidth}}$. Class `LinkDelay`

---

[12] Variable `hdrlen_` of a protocol specific header OTcl class was configured in Step 1 – Task 2.

[13] Variable "`incr`" could be greater than `size`, depending on the underlying hardware.

determines the size of a packet by `hdr_cmn::size_` (not by counting the number of bits stored in packet header and data payload) to compute packet transmission time. In most cases, users do not need to explicity deal with data payload.

Program 8.17 Declaration of enum `AppDataType` and class `AppData`.

```
    //~/ns/common/ns-process.h
1   enum AppDataType {
2       ...
3       PACKET_DATA,
4       HTTP_DATA,
5       ...
6       ADU_LAST
7
8   };

9   class AppData {
10  private:
11      AppDataType type_;        // ADU type
12  public:
13      AppData(AppDataType type) { type_ = type; }
14      AppData(AppData& d) { type_ = d.type_; }
15      virtual ~AppData() {}
16      AppDataType type() const { return type_; }
17      virtual int size() const { return sizeof(AppData); }
18      virtual AppData* copy() = 0;
19  };
```

NS2 also provides a support to hold data payload. In Line 4 of Program 8.1, class `Packet` provides a pointer `data_` to an `AppData` object.[14] Program 8.17 shows the declaration of an abstract class `AppData`. Class `AppData` has only one member variable `type_` in Line 11. Among its functions, and one is a pure virtual function `copy()` shown in Line 18. Indicating the type of application, variable `type_` is of type enum `AppDataType` defined in Lines 1–8. Function `copy()` duplicates an `AppData` object to a new `AppData` object. It is a pure virtual function, and must be overridden by child instantiable classes of class `AppData`. Function `size()` in Line 17 returns the amount of memory required to store an `AppData` object.

Class `AppData` provides two constructors. One is in Line 13, where the caller feeds an `AppData` type as an input argument. Another is in Line 14, where a reference to a `AppData` object is fed as an input argument. In both the cases, the constructor simply sets the variable `type_` to a value as specified in the input argument.

---

[14] However, no memory is allocated to the `AppData` object unless it is needed.

Program 8.18 Declaration of class `PacketData`.

```
    //~/ns/common/packet.h
1   class PacketData : public AppData {
2   private:
3       unsigned char* data_;
4       int datalen_;
5   public:
6       PacketData(int sz) : AppData(PACKET_DATA) {
7           datalen_ = sz;
8           if (datalen_ > 0)
9               data_ = new unsigned char[datalen_];
10          else
11              data_ = NULL;
12      }
13      PacketData(PacketData& d) : AppData(d) {
14          datalen_ = d.datalen_;
15          if (datalen_ > 0) {
16              data_ = new unsigned char[datalen_];
17              memcpy(data_, d.data_, datalen_);
18          } else
19              data_ = NULL;
20      }
21      virtual ~PacketData() {
22          if (data_ != NULL)
23              delete []data_;
24      }
25      unsigned char* data() { return data_; }
26      virtual int size() const { return datalen_; }
27      virtual AppData* copy() { return new PacketData(*this); }
28  };
```

Program 8.18 shows the declaration of class `PacketData`, a child class of class `AppData`. This class has two new member variables: `data_` (a string variable which stores data payload) in Line 3 and `datalen_` (the length of `data_`) in Line 4. Line 25 defines a function `data()` which simply returns `data_`. Lines 26 and 27 override the virtual functions `size()` and `copy()`, respectively, of class `AppData`. Function `size()` simply returns `datalen_`, while function copy() creates a new `PacketData` object which has the same content as the current `PacketData` object, and returns the pointer to the created object to the caller.

Class `PacketData` has two constructors. One is to construct a new object with size "sz", using the constructor in Lines 6–12. This constructor simply sets the default application data type to be `PACKET_DATA` (Line 6), stores "sz" in `datalen_` (Line 7), and allocates memory of size `datalen_` to `data_` (Line

9). Another construction method[15] is to create a copy of an input `PacketData` object (Lines 13–20). In this case, the constructor feeds an input `PacketData` object "d" to the parent class (Line 13), copies the variable `datalen_` (Line 14), and duplicates its data payload (Line 17).[16]

NS2 creates a `PacketData` object through two functions of class `Packet`: `alloc(n)` and `allocdata(n)`. In Program 8.4, function `alloc(n)` allocates a packet in Line 3, and creates data payload using function `allocdata(n)` in Line 5. Function `allocdata(n)` creates a `PacketData` object of size "n", by executing `new PacketData(n)` in Line 11.

Program 8.19 Functions `accessdata`, `userdata`, `setdata` and `datalen` of class `Packet`.

```
      //~/ns/common/packet.h
1   class Packet : public Event {
2       ...
3   public:
4       inline unsigned char* accessdata() const {
5           if (data_ == 0)
6               return 0;
7           assert(data_->type() == PACKET_DATA);
8           return (((PacketData*)data_)->data());
9       }
10      inline AppData* userdata() const {return data_;}
11      inline void setdata(AppData* d) {
12          if (data_ != NULL)
13              delete data_;
14          data_ = d;
15      }
16      inline int datalen() const { return data_ ? data_->size() : 0; }
17      ...
18 };
```

Program 8.19 shows four functions which can be used to manipulate data payload. Functions `accessdata()` (Lines 4–9) and `userdata()` (Line 10) are both data payload access functions. The difference is that `accessdata()` returns a direct pointer to *a string* `data_` which contains data payload while `userdata()` returns a pointer to *an* `AppData` *object* which contains data payload. Function `setdata(d)` (Lines 11–15) sets the pointer `data_` to point to the input argument "d". Finally, function `datalen()` in Line 16 returns the size of data payload.

---

[15] Function `copy()` in Line 27 employs this constructor to create a copy of a `PacketData` object.

[16] Function `memcpy(dst,src,num))` copies "num" data bytes from the location pointed by "src" to the memory block pointed by "dst".

8.5 Customizing Packets

8.5.1 Creating Your Own Packet

When designing a new protocol, a user may need to change the packet format. This section gives a guideline of how packet header, data payload, or both can be modified. Note that, it is recommended *not to* use data payload in simulation. If possible, include information related to the new protocol in a protocol specific header.

Defining a New Packet Header

Suppose we would like to include a new protocol specific header, namely "My Header", into the packet header. We need to define a C++ class (e.g., hdr_myhdr), an OTcl class (e.g., PacketHeader/MyHeader), and a mapping class (e.g., MyHeaderClass) for My Header, and include the OTcl class into the active protocol list. In particular, we need to perform the following four steps:

- **Step 1:** Define a protocol specific header C++ struct hdr_myhdr (e.g., see Program 8.6).
  - Declare variable offset_.
  - Define function access(p) (see Lines 8–10 in Program 8.6).
  - Include all member variables required to hold new packet attributes.
  - [Optional] Include a new packet type into enum packet_t and class p_info (e.g., see Program 8.9). Again, a new packet type does not need to be added for every new protocol specific header.
- **Step 2:** Define a protocol specific header OTcl class PacketHeader/MyHeader.
- **Step 3:** Derive a mapping class MyHeaderClass from class PacketHeaderClass (e.g, see class CommonHeaderClass in Program 8.12).
  - At the construction, feed the corresponding OTcl class name (i.e., PacketHeader/MyHeader) and the size needed to hold the protocol specific header (i.e., sizeof(hdr_myhdr)) to the constructor of class PacketHeaderClass (e.g., see Line 3 in Program 8.12).
  - From within the constructor, invoke function bind_offset(...) feeding the address of the variable offset_ of the C++ struct data type as an input argument. (i.e., invoke bind_offset(&hdr_myhdr::offset_)).
  - Instantiate a mapping variable class_myhdr at the declaration.
- **Step 4:** Activate My Header by including class PacketHeader/MyHeader into the active protocol list. The simplest way is to modify Lines 2–9 of Program 8.15 as follows:

```
foreach prot {
    Common
    Flags
```

```
    ...
    MyHeader
} {
    add-packet-header $prot
}
```

where only the suffix of the new protocol specific header (i.e., `MyHeader`) is added to the `foreach` loop.

Defining a New Data Payload

Data payload can be created in four levels:

(i) None: NS2 rarely uses data payload in simulation. To avoid any complicacy, it is suggested not to use data payload in simulation.

(ii) Use class `PacketData`: The simplest form of storing data payload is to use class `PacketData` (see Program 8.18). Class `Packet` has a variable `data_` whose class is `PacketData` and provides functions (in Program 8.19) to manipulate the variable `data_`.

(iii) Derive a class (e.g., class `MyPacketData`) from class `PacketData`: This option is suitable when new functionalities (i.e., functions and variables) in addition to those provided by class `PacketData` are needed. After deriving a new `PacketData` class, a user may also derive a new class (e.g., class `MyPacket`) from class `Packet`, and override the variable `data_` of class `Packet` to be a pointer to a `MyPacketData` object.

(iv) Define a new data payload class: A user can also define a new payload type if needed. This option should be used when the new payload has nothing in common with class `PacketData`. The followings are the main tasks needed to define and use a new payload type `MY_DATA`.

- Include the new payload type (e.g., `MY_DATA`) into `enum AppDataType` data type (see Program 8.17).
- Derive a new payload class `MyData` from class `AppData`.
  – Feed the payload type `MY_DATA` to the constructor of class `AppData`.
  – Include any other necessary functions and variables.
  – Override functions `size()` and `copy()`.
- Derive a new class `MyPacket` from class `Packet`
  – Declare a variable of class `MyData` to store data payload.
  – Include functions to manipulate the above `MyData` variable.

8.5.2 Activate/Deactivate a Protocol Specific Header

By default, NS2 includes *all* built-in protocol specific headers into packet header (see Program 8.15). This inclusion can lead to unnecessary wastage of memory especially in a *packet-intensive* simulation, where numerous packets are created. For example, common, IP, and TCP headers together use only 0.1 kB, while the default packet header consumes as much as 1.9 kB [15].

Again, NS2 does not return the memory allocated to a `Packet` object until the simulation terminates. Selectively including protocol specific header can lead to huge memory saving.

The packet format is defined when the Simulator is created. Therefore, a protocol specific headers must be activated/deactivated prior to the creation of the Simulator. NS2 provides the following OTcl procedures to activate/deactivate protocol specific headers:

- To add a protocol specific header `PacketHeader/MH1`, execute

  `add-packet-header MH1`

  In Lines 10–14 of Program 8.15, the above statement includes `PacketHeader/MH1` to the variable `tab_` of class `PacketHeaderManager`.
- To remove a protocol specific header `PacketHeader/MH1` from the active list, execute

  `remove-packet-header MH1`

  The details of procedure `remove-packet-header{args}` are shown in Lines 1–9 of Program 8.20. From Line 7, the above statement removes the entries with index `PacketHeader/MH1` from the variable `tab_` of class `PacketHeaderManager`.

Program   8.20   Procedures   remove-packet-header,   and remove-all-packet-header.

```
//~/tcl/ns-packet.tcl
1  proc remove-packet-header args {
2      foreach cl $args {
3          if { $cl == "Common" } {
4              warn "Cannot exclude common packet header."
5              continue
6          }
7          PacketHeaderManager unset tab_(PacketHeader/$cl)
8      }
9  }

10 proc remove-all-packet-headers {} {
11     PacketHeaderManager instvar tab_
12     foreach cl [PacketHeader info subclass] {
13         if { $cl != "PacketHeader/Common" } {
14             if [info exists tab_($cl)] {
15                 PacketHeaderManager unset tab_($cl)
16             }
17         }
18     }
19 }
```

- To remove all protocol specific headers, execute

  `remove-all-packet-header`

  In Lines 10–19 of Program 8.20, the procedure `remove-all-packet-header{}` uses `foreach` to remove all protocol specific headers (except for common header) from the active protocol list (i.e., the instvar `tab_`).

8.6 Chapter Summary

Consisting of packet header and data payload, a packet is represented by a C++ class `Packet`. Class `Packet` consists of pointers `bits_` to its packet header and `data_` to its data payload. It employs a pointer `next_` to form a linked list of packets. It also has a pointer `free_` which points to the first `Packet` object on the free packet list. When a `Packet` object is no longer in use, NS2 stores the `Packet` object in the free packet list for future reuse. Again, `Packet` objects are not destroyed until the simulation terminates. When allocating a packet, NS2 first tries to take a `Packet` object from the free packet list. Only when the free packet list is empty, will NS2 create a new `Packet` object.

During simulation, NS2 usually stores relevant user information (e.g., packet size) in packet header, and rarely uses data payload. It is recommended not to use data payload if possible, since storing all information in packet header greatly simplifies the simulation yet yields the same simulation results.

Packet header consists of several protocol specific headers. Each protocol specific header occupies a contiguous part in packet header, and identifies the occupied location by using its variable `offset_`. NS2 employs a packet header manager (represented by an OTcl class `PacketHeaderManager`) to maintain a list of active protocols, and define packet header format using the list when the Simulator is created. The packet header construction process proceeds in the three following steps:

(i) *At the Compilation*: NS2 defines the following three classes for each of protocol specific headers:
   - *A C++ class*: NS2 uses C++ `struct` data type to define how packet attributes are stored in a protocol specific header. One of the important member variables is `offset_`, which indicates the location of the protocol specific header on the packet header.
   - *An OTcl class*: During the Network Configuration Phase, the packet header manager configures packet header from the OTcl domain. It accesses a protocol specific header from the OTcl class which acts as an interface from the OTcl to the C++ domains.
   - *A mapping class*: A mapping class binds the OTcl and C++ class together. It declares a *method* `offset`, which is invoked by a packet header manager from the OTcl domain to configure the value of variable `offset_` of the C++ class `PacketHeaderClass`.

In this step, NS2 also stores all built-in protocol specific headers in instvar `tab_` of class `PacketHeaderManager`, which represents the active protocol list.

(ii) *At the Network Configuration Phase*: A user may add/remove protocol specific headers to/from the active protocol list. When the Simulator is created, the packet header manager computes and assigns appropriate offset values to all protocol specific headers specified in the active list.

(iii) *At the Simulation Phase*: NS2 follows the above packet header definitions when allocating a packet.

9

Transport Control Protocols Part 1 – An Overview and User Datagram Protocol implementation

A typical communication system consists of applications, transport layer agents, and a low level network. An application models user demand to transmit data. Taking user demand as an input, a sending transport layer agent creates packets and forwards them to the associated receiving transport layer agent through a low-level network. Having discussed the details of low level networks in Chapters 5–7, the details of transport layer agents are presented here in Chapters 9–10. Also, the details of applications will be presented in Chapter 11.

This chapter provides an overview of transport layer agents, and shows NS2 implementation of User Datagram Protocol (UDP) agents. In particular, Section 9.1 introduces two most widely used transport control protocols: Transmission Control Protocol (TCP) and User Datagram Protocol (UDP). Section 9.2 explains NS2 implementation of basic agents. Section 9.3 shows the implementation of UDP agents and Null agents. Finally, the chapter summary is given in Section 9.4.

9.1 UDP and TCP Basics

9.1.1 UDP Basics

Defined in [18] and [19], User Datagram Protocol (UDP) is a connectionless transport-layer protocol, where no connection setup is needed prior to data transfer. UDP offers minimal transport layer functionalities – non-guaranteed data delivery – and gives applications a direct access to the network layer. Aside from the multiplexing/demultiplexing functions and some light error checking, it adds nothing to IP packets. In fact, if the application developer employs UDP as a transport layer protocol, then the application is communicating almost directly with the network layer.

T. Issariyakul, E. Hossain, *Introduction to Network Simulator NS2*,
DOI: 10.1007/978-0-387-71760-9_9, © Springer Science+Business Media, LLC 2009

UDP takes messages from an application process, attaches source and destination port for the multiplexing/demultiplexing service, adds two other fields of error checking and length information, and passes the resulting packet to the network layer [19]. The network layer encapsulates the UDP packet into a network layer packet and then delivers the encapsulated packet at the receiving host. When a UDP packet arrives at the receiving host, it is delivered to the receiving UDP agent identified by the destination port field in the packet header.

9.1.2 TCP Basics

As shown in Fig. 9.1, Transmission Control Protocol (TCP) [20] is a connection-oriented reliable transport protocol consisting of three phases of operations: connection setup, data transfer, and connection termination. In the *connection setup* phase, a TCP sender initiates a three-way handshake (i.e., sending `SYN`, `SYN-ACK`, and `ACK` messages). After a connection is established, TCP enters the *data transfer* phase where a TCP sender transfer data to a TCP receiver. Finally, after the data transfer is complete, TCP tears down the connection in the *connection termination* phase by using a four-way handshake (i.e., sending two pairs of `FIN-ACK` messages.)

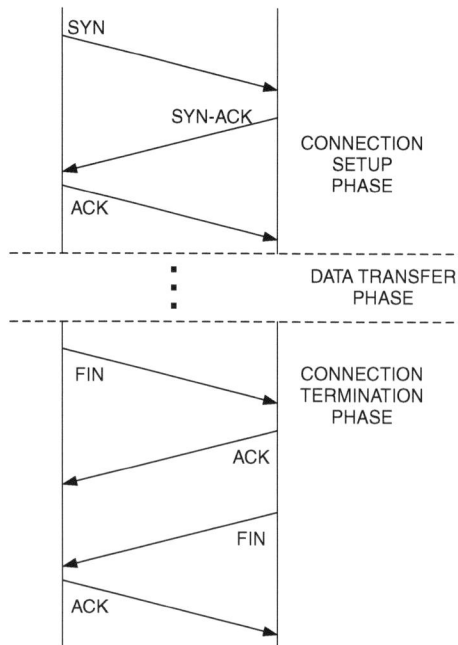

Fig. 9.1. Main phases of TCP operation: Connection setup, data transfer, and connection termination.

The main operation of TCP lies in the data transfer phase, which implements two following mechanisms: (1) Error control using basic acknowledgement and timeout, and (2) Congestion control using a window-based mechanism.

Error Control Using Basic Acknowledgement and Timeout

As a reliable transport layer protocol, TCP provides connection reliability by means of acknowledgement (ACK). For every received packet, a TCP receiver returns an ACK packet to the sender. If an ACK packet is not received within a given *timeout* value, the TCP sender will assume that the packet is lost, and will retransmit the lost packet. Note that in the literature, a timeout period is also referred to as Retransmission TimeOut (RTO). Hereafter, we will refer to these two terms interchangably.

TCP employs a *cumulative* acknowledgement mechanism. With this mechanism, a TCP receiver always acknowledges to the sender with the highest sequence number up to which all packets have been successfully received. For example, in Fig. 9.2, packet 3 is lost. Therefore, the TCP receiver returns ACK for packet 2 (A2) even when packets 4, 5, and 6 have been received. These ACK packets (e.g., A2), which acknowledge the same TCP packet (e.g., packet 2), are referred to as the *duplicated acknowledgement packets*. From Fig. 9.2, the TCP sender does not receive an ACK packet which acknowledges packet 3. After a period of RTO, the sender will assume that packet 3 was lost and will retransmit packet 3.

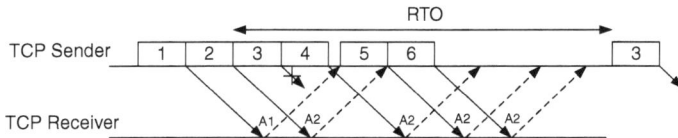

Fig. 9.2. An example of TCP error control using acknowledgement: A TCP sender realizes the loss of TCP packet number 3 after transmitting the packet number 3 for a period of RTO (ie., timeout).

The RTO value is optimized according to the following tradeoff: a small RTO value leads to unnecessary packet retransmission while a large RTO value results in high latency of packet loss detection. In general, an RTO value should be a function of network Round-Trip Time (RTT), which is the time required for a data bit to travel from a source node to the destination node and travel back to the source node. Due to network dynamics, RTT of one packet could be different from that of another. In TCP, smoothed (i.e., average) RTT ($\bar{t}$) and RTT variation (σ_t) are computed based on the collected RTT samples, and are used to compute the RTO value.

According to [21], instantaneous smoothed RTT, RTT variation, and instantaneous RTO are computed as follows. Let $t(k)$ be the kth RTT sample collected upon ACK reception. Also, let $\bar{t}(k)$, $\sigma_t(k)$, and $RTO(k)$ be the values of $\bar{t}$, σ_t, and RTO, respectively, when the kth RTT sample is determined. Then,

$$\bar{t}(k+1) = \alpha \times \bar{t}(k) + (1-\alpha) \times t(k+1), \tag{9.1}$$

$$\sigma_t(k+1) = \beta \times \sigma_t(k) + (1-\beta) \times |t(k+1) - \bar{t}(k+1)|, \tag{9.2}$$

$$RTO(k+1) = \min\{ub, \max\{lb, \gamma \times [\bar{t}(k+1) + 4 \times \sigma_t(k+1)]\}\} \tag{9.3}$$

where ub and lb are fixed upper and lower bounds on the RTO value. The constants $\alpha \in (0,1)$ and $\beta \in (0,1)$ are usually set to $7/8$ and $3/4$, respectively. The variable γ is a *binary exponential backoff* (BEB) factor. It is initialized to 1, and doubled for every timeout event, and is reset to 1 when a new ACK packet arrives.

Window-Based Congestion Control

A transport layer protocol is also responsible for network congestion. It limits the transmission rate of a data flow to help control network congestion. As a window-based congestion control protocol, TCP limits the transmission rate by adjusting the *congestion window* (i.e., transmission window) which basically refers to the amount of data that a sender can transmit without waiting for acknowledgement. For example, the congestion window size of the TCP connection in Fig. 9.2 is initialized to 4. Therefore, the TCP sender pauses after sending packets 1–4. After receiving ACK corresponding to packet 1 (i.e., A1), the number of unacknowledged packets becomes 3 and TCP is able to send out packet 5.

Congestion window refers to a range of sequence numbers of TCP packets which can be transmitted at a moment. For example, the congestion window at the beginning of Fig. 9.2 is $\{1,2,3,4\}$ and the congestion window size is 4. When A1 is received, the congestion window becomes $\{2,3,4,5\}$. In this case, we say that the congestion window *slides* to the right. Suppose that the congestion window changes to $\{2,3,4,5,6\}$ (the size is 5). In this case, we say that the congestion window is *opened* by one unit. On the contrary, if the window becomes $\{2,3,4\}$, we say the congestion window is *closed* by one unit. Again, a larger window size allows the sender to transmit more data in a given interval implying a higher transmission rate at the transport layer. TCP increases and decreases its transmission rate by opening and closing its congestion window.

Window Increasing Mechanism

One of the key features of TCP is network-based rate adaptability. TCP slowly opens its congestion window to fill up the underlying network, when the network is underutilized. When the network is overutilized, TCP rapidly closes the congestion window to help relieve the congestion. TCP window opening mechanism consists of two phases, each of which is identified by the current congestion window size (w) and a *slow-start* threshold (w_{th}):

(i) **Slow-start phase:** If $w < w_{th}$, TCP increases w by one for every received ACK packet.

(ii) **Congestion avoidance phase:** If $w \geq w_{th}$, TCP increases w by $\frac{1}{w(t)}$ for every received ACK packet.

Note that, TCP receiver may advertise its maximum window size $(w_{\max})$ which does not fill its buffer too rapidly. This $w_{\max}$ acts as an upper-bound for the above window increasing mechanism. In NS2, congestion window (ω) evolves according to the above two phases, regardless of ω_{max}. However, TCP uses the minimum of ω and ω_{max} to determine amount of data it can transmit.

Packet Loss Detection Mechanism

In the literature, various TCP variants use different combinations of the following packet loss detection mechanisms:

- **Timeout**: As discussed earlier, TCP starts its retransmission timer for every transmitted packet, and assumes a packet loss upon timer expiration.
- **Fast Retransmit**: By default, an RTO has granularity of 0.5 seconds, which could lead to large latency in packet loss detection. Fast Retransmit expedites the packet loss detection by means of duplicated acknowledgement detection. Upon detection of the kth (which is equal to 3 by default) duplicated acknowledgement (excluding the first one which is a new acknowledgement), the TCP sender stops waiting for the timeout, assumes a packet loss, and retransmits the lost packet. From Fig. 9.2, if the fast retransmit mechanism is used, the TCP sender will assume that packet 3 is lost and retransmits packet 3 upon receiving the 4th A2 packet (i.e., the 3rd duplicated acknowledgement). Note that based on the cumulative acknowledgement principle, upon receiving the retransmitted packet 3, TCP receiver sends A6 back to the sender, since packets 4, 5, and 6 have been successfully received earlier.

Window Decreasing Mechanism

Originally conceived to combat congestion in a wired network, TCP assumes that all packet losses occurs due to congestion (i.e., buffer overflow at the routers in the network). It reacts to every packet loss by reducing its transmission rate (or window size) to lessen the congestion. The following approaches are among the most popular window decreasing mechanisms for a TCP variant used in the literature.

- **Reset to 1**: Conventionally, TCP reacts to packet loss by resetting the window size to 1, and setting the slow-start threshold to half of the current congestion window size. However, this option is usually deemed too radical and could lead to TCP throughput degradation in presence of random packet loss.

- **Fast Recovery**: Upon detection of a packet loss, the fast recovery mechanism sets both current window size and slow-start threshold to half of the current congestion window size. Then, it increases the window size by one for each incoming duplicated acknowledgement. At this moment, the sender may transmit a new packet if the congestion window allows. Upon receiving a new acknowledgement, the sender exits Fast Recovery and sets the window size to the slow-start threshold value, after which TCP operates normally in a congestion avoidance phase.

TCP Variants

There are numerous TCP variants in the literature. This section discusses only de facto TCP variants namely Old Tahoe, Tahoe, Reno, and new Reno. These TCP variants utilize the same window increasing mechanism (i.e., slow start and congestion avoidance). However, they differ in how they detect a packet loss and decrease the window size. Table 9.1 shows the differences in window size adjustment mechanism, when packet loss is detected through timeout and Fast Retransmit (i.e., duplicated ACKs).

Table 9.1. Differences among basic TCP variants: Different window closing mechanisms upon detection of a packet loss.

TCP Variant	Loss Detection	
	Timeout	Fast Retransmit
Old-Tahoe	Reset w to 1	N/A
Tahoe	Reset w to 1	Reset w to 1
Reno	Reset w to 1	Fast Recovery (single packet)
New Reno	Reset w to 1	Fast Recovery (all packets)

The very first TCP variant, Old-Tahoe, detects packet loss through timeout only. When packet loss is detected it always resets congestion window size to 1. Developed from Old-Tahoe, TCP Tahoe uses the Fast Retransmit mechanism to expedite packet loss detection rather than waiting for the timeout. It always sets the window size to 1 upon detection of a packet loss. Both TCP Reno and New-Reno reset the window size to 1, when a packet loss is detected through timeout. However, they will employ Fast Recovery if packet loss is detected through Fast Retransmit. The difference between TCP Reno and TCP New- Reno is that TCP Reno exits the fast recovery process as soon as the lost packet which triggered Fast Retransmit is acknowledged. If there are multiple packet losses within a congestion window, Fast Recovery could be invoked for several times, and the window size will decrease significantly. To avoid the multiple window closures, TCP New-Reno stays in the Fast Recovery phase until all packets in the loss window are acknowledged or until timeout occurs.

9.2 Basic Agents

An agent is an NsObject which is responsible for creating and destroying packets. There are two main types of NS2 agents: routing agents and transport-layer agents. A routing agent creates and receives routing control packets, and commands routing protocols to act accordingly. Connecting an application to a low level network, a transport-layer agent controls the congestion and reliability of a data flow based on an underlying transport layer protocol (e.g., UDP or TCP). This book focuses on transport layer agents only

---

Program 9.1 Class `AgentClass` which binds OTcl and C++ class `Agent`.

```
  //~/ns/common/agent.cc
1 static class AgentClass : public TclClass {
2 public:
3     AgentClass() : TclClass("Agent") {}
4     TclObject* create(int, const char*const*) {
5         return (new Agent(PT_NTYPE));
6     }
7 } class_agent;
```

NS2 implements agents in a C++ class `Agent`, which is bound to an OTcl class with the same name (see Program 9.1). In the following, we first discuss the relationship among a transport-layer agent, an application, and a low-level network in Section 9.2.1. Agent configuration and internal mechanisms are discussed in Sections 9.2.2 and 9.2.3, respectively. Finally, Section 9.2.4 provides guidelines to define a new transport-layer agent.

9.2.1 Applications, Agents, and a Low-level Network

An agent acts as a bridge which connects an application and a low-level network. Based on the user demand provided by an application, a sending agent constructs packets and transmits them to a receiving agent through a low-level network. Figure 9.3 shows an example of such a connection.

Consider Fig. 9.3. On the top level, a CBR (constant bit rate) application, which models a user demand to periodically transmit data, is used as an application. The demand is passed to a UDP sending agent, which in turn creates UDP packets. Here, the UDP agent stores source and destination IP addresses and transport layer ports in the packet header, and forwards the packet to the attached node (e.g., Node 1 in Figure 9.3). Using the pre-calculated routing table, the low-level network delivers the packet to the destination node (e.g., Node 3 in Fig. 9.3) specified in the packet header. The destination node employs its demultiplexer to forward the packet to the agent attached to the port specified in the packet header. Finally, a Null receiving agent simply destroys the received packet.

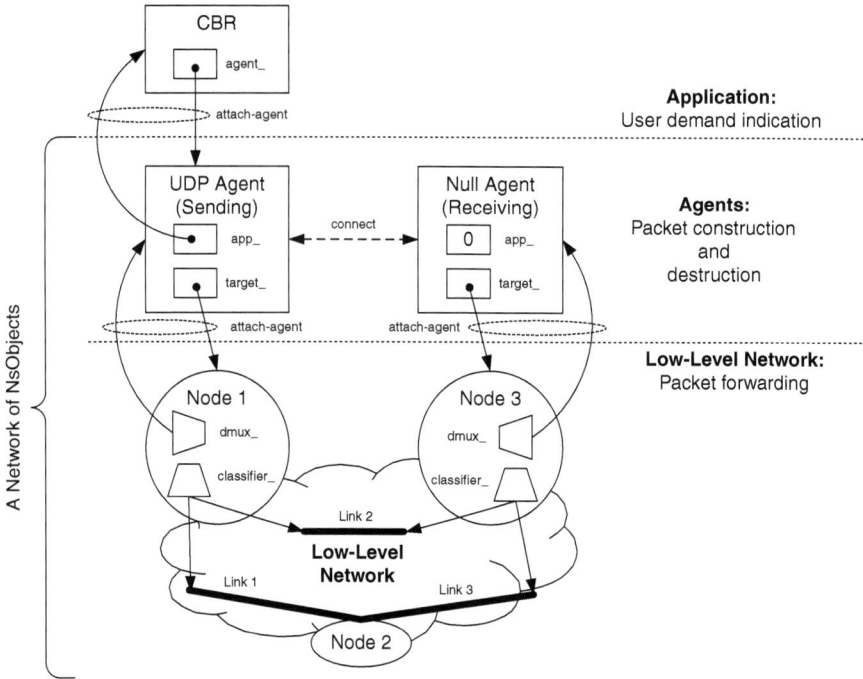

Fig. 9.3. A CBR application over UDP configuration.

From the above discussion, an agent can be used as a sending agent (e.g., a UDP agent) or a receiving agent (e.g., a Null agent). A sending agent has connections to both an application and a low-level network, while a receiving agent may not have a connection to an application (because it does not need any). An application (e.g., CBR) uses its variable `agent_` as a reference to an agent (e.g., UDP and Null agents), while an agent uses its variable `app_` as a reference to an application. It is mandatory to configure the variables `agent_` and `app_` (i.e., create the connection) for a sending agent, while it is optional for a receiving agent. This is mainly because the application needs to inform the agent of user demand (i.e., by invoking function `sendmsg(...)`), and the sending agent needs to inform the application of the completion of data transmission (i.e., by invoking function `resume()`). Since a receiving agent simply destroys the received packet, it does not need a connection to an application.

Both sending and receiving agents connect to a low-level network in the same manner. They use a pointer `target_`, to point to the attached node. The Node, on the other hand, *install*s the agent slot number "port" of its demultiplexer `dmux_` (which is of class `PortClassifier`), where "port" is the corresponding port number of the agent (see Section 6.6.3).

Table 9.2. Key differences between a sending and a receiving agent.

	Sending agent	Receiving agent
Upstream		
- object	Application	Node
- packet forwarding function	sendmsg	recv
Downstream object		
- object	Node	N/A
- packet forwarding function	recv	N/A

Table 9.2 shows the key differences between a sending agent and a receiving agent. The upstream object of a sending agent is an application, which informs a sending agent of incoming user demand through function sendmsg(...) of the sending agent. The upstream object of a receiving agent, on the other hand, is a Node object, which forwards packets to the receiving agent via function recv(p,h). The downstream object of a sending agent is a Node object. The sending agent passes a packet *p to a Node object by executing target_->recv(p,h). A receiving agent usually has no downstream object, since it simply destroys the received packets.

9.2.2 Agent Configuration

From Fig. 9.3, agent configuration consists of four main steps:

(i) Create a sending agent, a receiving agent, and an application using "new{...}".
(ii) Attach agents to the application using OTcl Command attach-agent-{agent} of class Application.
(iii) Attach agents to the a low-level network using instproc attach-agent-{node agent} of class Simulator.
(iv) Associate the sending agent with the receiving agent using instproc connect{src dst} of class Simulator.

Example 9.1 (A Network Construction Example). The example network in Fig. 9.3 employs CBR, a UDP agent, and a Null agent as an application, a sending agent, and a receiving agent, respectively. To setup the example network, we may use the Tcl simulation script in Program 9.2.

While Lines 1–7 create a low-level network (see the details in Chapters 6 and 7), Lines 8–14 set up a CBR application, a UDP agent, and a Null agent on top of the low-level network. Again, there are 4 major steps to create connections among agent, an application, a low-level network:

(i) Create agent and application objects (Lines 8–10).
(ii) Use command attach-agent of class Application to create a connection between an application and a sending agent (Line 11).

Program 9.2 A simulation script which creates the network in Fig. 9.3.

```
1   set ns [new Simulator]
2   set n1 [$ns node]
3   set n2 [$ns node]
4   set n3 [$ns node]
5   $ns duplex-link $n1 $n2 5Mb 2ms DropTail
6   $ns duplex-link $n2 $n3 5Mb 2ms DropTail
7   $ns duplex-link $n1 $n3 5Mb 2ms DropTail

    #=== UDP-Null peering starts here ===
8   set udp [new Agent/UDP]
9   set null [new Agent/Null]
10  set cbr [new Application/Traffic/CBR]
11  $cbr attach-agent $udp
12  $ns attach-agent $n1 $udp
13  $ns attach-agent $n3 $null
14  $ns connect $udp $null
```

(iii) Use instproc attach-agent{node agent} of class Simulator to create a connection between each agent and a node entry (Lines 12 and 13).

(iv) Use instproc connect{src dst} of class Simulator to associate a sending agent with a receiving agent (Line 14).

9.2.3 Internal Mechanism for Agents

The internal mechanisms for agents are defined in the C++ domain as follows:

- *A sending agent*: Receive user demand by having the associated application invoke its function sendmsg(...). From within sendmsg(...), create packets using function allocpkt() and forward the created packets to the low-level network by executing target_->recv(p,h).
- *A receiving agent*: Receive packets by having a low-level network demultiplexer invoke its function recv(p,h). Destroy received packets by invoking function free(p) of class Packet.

In this section, we will discuss the detail of variables and functions required to perform the above mechanisms.

Related C++ and OTcl Variables

The main variables of C++ class Agent and their bound OTcl instvars are shown in Table 9.3. Of type ns_addr_t (see Section 8.3.3), variables here_ and dst_ contain addresses and ports of the Node attached to the agent and the peering agent, respectively. An IPv6 priority level is stored in variable prio_. Variable app_ acts as a reference to an Application object. Since

Table 9.3. The list of C++ and OTcl variables of class `Agent`.

C++ Type	C++ variable	OTcl instvar	Description
ns_addr_t	here_		
	here_.addr_	agent_addr_	Address of the attached node
	here_.port_	agent_port_	Port where the agent is attached
ns_addr_t	dst_		
	dst_.addr_	dst_addr_	Address of the node attaching to a peering agent
	dst_.port_	dst_port_	Port where the peering agent is attached
int	fid_	fid_	Flow ID
int	prio_	prio_	IPv6 priority field (e.g., 0 = unspecified, 1 = background traffic)
int	flags_	flags_	Flags
int	defttl_	ttl_	Default time to live value
int	size_	N/A	Packet size
packet_t	type_	N/A	Packet type
int	seqno_	N/A	Current sequence number
Application*	app_	N/A	A pointer to an application
int	uidcnt_	N/A	Total number of packets generated by all agents

class `Agent` is responsible for packet generation, it must assign a unique ID to every packet. Therefore, it maintains a static variable `uidcnt_`, which counts the total number of generated packets. When a packet is created, an `Agent` object sets the unique ID of the packet to be `uidcnt_`, and increases `uidcnt_` by one (see function `initpkt(p)` in Line 10 of Program 9.3).

Key C++ Functions

A list of key C++ functions with their descriptions is given below (see the declaration of class `Agent` in file ~`ns`/common/agent.cc,h). Since class `Agent` is a template for transport layer agents, it provides no implementation for some of its functions. The child classes of class `Agent` are responsible for implementing these functions.

recv(p,h)	Receive a packet "*p".
send(p,h)	Send a packet "*p".
send(nbytes)	Send a message with "nbytes" bytes.
sendmsg(nbytes)	Send a message with "nbytes" bytes (no implementation).
timeout(tno)	Action to be performed at timeout (No implementation)

connect(dst) Connect to a dynamic destination dst (no implementation).

close() Close a connection-oriented session (no implementation).

listen() Wait for a connection-oriented session (no implementation).

attachApp(app) Store app in the variable app_.

allocpkt() Create a packet.

initpkt(p) Initialize the input packet "*p".

recvBytes(bytes) Send data of "bytes" bytes to the attached application.

idle() Tell the application that the agent has nothing to transmit.

The Constructor

Class Agent has no default constructor. Its only constructor takes a packet_t (see Section 8.3.4 and Program 8.9) object as an input argument (see Line 1 of Program 9.3). The constructor sets the variable type_ to be as specified in the input argument, and resets other variables to zero. This packet type setting implies that one agent is able to transmit packets of one type only. We need several agents to transmit packets of several types.

Functions allocpkt() and initpkt(p)

Shown in Program 9.3, function allocpkt() is the main packet construction function. It creates a packet by invoking function alloc() of class Packet in Line 4, and initializes the packet by invoking function initpkt(p) in Line 5. After initialization, function allocpkt() returns the constructed packet pointer p to the caller.

The details of function initpkt(p) are shown in Lines 8–20 of Program 9.3. Function initpkt(p) sets the initial values in the packet header of the input packet "*p" to the default values. The uniqueness of the unique ID field uid_ in the common header is assured by setting uid_ to be the total number of packets allocated so far. Class Agent stores the total number of allocated packet in its static variable unicnt_. Since the variable unicnt_ is distinct and unique to all agents, assigning this variable to the field uid_ of the common header (Line 11) assures the uniqueness of packet unique ID.

Other initialization includes setting up the packet type in the common header to be as specified in the variable type_ (Line 12). Also, Lines 14–18 configure source and destination IP addresses and port numbers in the variables here_ and dst_.

Function attachApp(app)

Lines 1–4 in Program 9.4 show the details of function attachApp(app). To bind an application to an agent, function attachApp(app) stores the input

Program 9.3 Constructor, function `allocpkt`, and function `initpkt` of class Agent.

```
//~/ns/common/agent.cc
1  Agent::Agent(packet_t pkttype):size_(0),type_(pkttype),app_(0){}

2  Packet* Agent::allocpkt() const
3  {
4      Packet* p = Packet::alloc();
5      initpkt(p);
6      return (p);
   }

7  void Agent::initpkt(Packet* p)
   {
8      hdr_cmn* ch = hdr_cmn::access(p);
9      ch->uid() = uidcnt_++;
10     ch->ptype() = type_;
11     ...
12     hdr_ip* iph = hdr_ip::access(p);
13     iph->saddr() = here_.addr_;
14     iph->sport() = here_.port_;
15     iph->daddr() = dst_.addr_;
16     iph->dport() = dst_.port_;
17     ...
18 }
```

pointer "app" in its pointer to a Application object, app_. After this point, the agent may invoke public functions of the attached application through the pointer app_.

Program 9.4 Functions `attachApp` and `recv` of class Agent.

```
//~/ns/common/agent.cc
1  void Agent::attachApp(Application *app)
2  {
3      app_ = app;
4  }

5  void Agent::recv(Packet* p, Handler*)
6  {
7      if (app_)
8          app_->recv(hdr_cmn::access(p)->size());
9      Packet::free(p);
10 }
```

Functions recv(p,h), send(p,h), and sendmsg(nbytes)

These functions are used by sending and receiving agents in the packet forwarding process. On the sender side, an application informs a sending agent of user demand by invoking functions send(nbytes), and sendmsg(...) of class Agent. As an NsObject, the sending agent forwards an incoming packet *p to a downstream NsObject by executing target_->recv(p,h). As discussed earlier, these functions send(nbytes) and sendmsg(...) have no implementation in the scope of class Agent, and must be implemented by the child classes of class Agent.

On the receiver side, an NsObject forwards packets to a receiving agent by invoking its function recv(p,h). Shown in Lines 5–10 of Program 9.4, function recv(p,h) deallocates the received packet (Line 9) and may inform the attached application (if it exists) of packet reception by invoking function recv(size) of the attached Application object (Lines 7–8), where size is the size of packet *p.

9.2.4 Guidelines to Define a New Transport Layer Agent

Class Agent provides the basic functionalities necessary for most agents. A new agent can be created based on these functionalities, following the guidelines below:

  (i) Define an inheritance structure: Select a base class and derive a new agent class from the selected base class. Bind the C++ and OTcl agent class names together.

  (ii) Define necessary C++ variables and OTcl instvars.

  (iii) Implement the constructors of both C++ and OTcl classes. Bind the variables and the instvars if necessary.

  (iv) Implement the necessary functions including functions send(nbyte), send- msg(...), recv(p,h), and timeout(tno). Also define OTcl inst-procs if necessary.

  (v) Define necessary OTcl commands as interfaces to the C++ domain from the OTcl domain.

  (vi) [Optional]Define a timer (see Section 12.1).

9.3 UDP (User Datagram Protocol) and Null Agents

UDP (User Datagram Protocol) is a connectionless transport layer protocol, which provides neither congestion control nor error control. In NS2, a UDP agent is used as a sending agent. It is usually peered with a Null (receiving) agent, which is responsible for packet destruction. Figure 9.3 shows a network configuration example where a CBR (Constant Bit Rate) traffic source employs a UDP agent and a Null agent as its transport later agents. Here, the CBR asks the UDP agent to transmit a burst of packets for every fixed interval. The UDP agent creates and forwards packets to the low-level network, irrespective of the network condition. On the receiving end, the Null agent simply destroys the packets received from the low-level network. In the following, we will discuss the details of UDP and Null agents.

9.3.1 Null (Receiving) Agents

A Null agent is the simplest but one of the most widely-used receiving agents. The main responsibility of a Null agent is to deallocate packets, through function `free(p)` of class `Packet` (see Line 9 in Program 9.4). A Null agent is represented by an OTcl class `Agent/Null` which is derived directly from an OTcl class `Agent` (see file ~ns/tcl/lib/ns-agent.tcl). Due to its simplicity, Null agents have no implementation in the C++ domain.

9.3.2 UDP (Sending) Agent

A UDP agent is perhaps the simplest form of sending agents. It receives user demand to transmit data by having the attached application invoke its function (e.g., `sendmsg(...)`), creates packets based on the demand, and forwards the created packet to a low-level network. An application may use three following ways to tell a UDP agent to send out packets: via a C++ function `sendmsg(...)` of class `UdpAgent`, via an OTcl command `send{...}` of OTcl class `Agent/UDP`, or via an OTcl command `sendmsg{...}` of OTcl class `Agent/UDP`.

Again, NS2 defines a UDP sending agent based on the guideline in Section 9.2.4. Since a UDP agent implements no acknowledgement mechanism and needs no timer, we can skip the last step in the guideline.

Step 1: Define Inheritance Structure

A UDP agent is represented by a C++ class `UdpAgent` and an OTcl class `Agent/UDP`. These two classes derive from class `Agent` in their domains, and are bound by using a mapping class `UdpAgentClass` (see Program 9.5).

Step 2: Define C++ Variables and OTcl Instvars

The key variable of class `UdpAgent` is `seqno_` (Line 12 in Program 9.6), which counts the number of packets generated by a `UdpAgent` object. Note that every packet has a unique ID `uid_`. Also, every packet generated by the *same agent* has a unique sequence number `seqno_`. However, two packets generated by different agents may have the *same* sequence number `seqno_` but they must have different unique ID `uid_`.

Program 9.5 Mapping class `UdpAgentClass` which binds a C++ class `UdpAgent` to an OTcl class `Agent/UDP`.

```
   //~/ns/apps/udp.cc
1  static class UdpAgentClass : public TclClass {
2  public:
3      UdpAgentClass() : TclClass("Agent/UDP") {}
4      TclObject* create(int, const char*const*) {
5          return (new UdpAgent());
6      }
7  } class_udp_agent;
```

Step 3: Implement the Constructors in the C++ and OTcl Domains

NS2 implements constructors for a UDP agent in the C++ domain only. From Program 9.6, the default constructor in Lines 14–16 feeds UDP packet type (i.e., `PT_UDP`) to constructor of class `Agent`, essentially storing `PT_UDP` in the variable `type_`. It also sets the sequence number (i.e., `seqno_`) to be –1. By specifying the packet type, the constructor in Lines 17–19 sets the packet type to be as specified in the input argument. The constructor in this case does not set the value of `seqno_` since the packets of specified type may not have sequence number. For both constructors, the C++ variable `size_`, which specifies the packet size, is bound to instvar `packetSize_` in the OTcl domain (Lines 15 and 19). By default, the packet size is set to 1,000 bytes in file ~*ns*/tcl/lib/ns-default.tcl (Line 20).

Step 4: Define the Necessary C++ Functions

As a sending agent, a UDP agent needs to define a function `sendmsg(...)` to receive a user demand from the application. Program 9.7 shows the details of function `sendmsg(nbytes,data,flags)`, which takes three input arguments: `nbytes`, `data`, and `flags`. Function `sendmsg(...)` divides data payload with size `nbytes` bytes into "n" (see Line 4) or "n+1" parts (depending on `nbytes`), stores each part into a UDP packet (which contains a payload of `size_` bytes), and transmits all ("n" or "n+1") packets to a low-level network.

Since NS2 rarely sends actual payload along with a packet (see Section 8.4), Line 8 only sets the size of packet created in Line 6 to be `size_`. Line 11 sends out the created packet, by executing `target_->recv(p)`.[1] Lines 6–11 are repeated for "n" times to transmit all "n" packets.

After transmitting the first "n" packets, the entire application payload is left with `nbytes % size_`, where % is a modulus operator. If the remainder is

---

[1] Variable `target_` is configured to point to a node entry during the network configuration phase (see Section 9.2.2).

Program 9.6 Declaration and the constructors of class `UdpAgent` as well as the default value of the instvar `packetSize_` of class `Agent/UDP`.

```
    //~/ns/apps/udp.h
1   class UdpAgent : public Agent {
2   public:
3       UdpAgent();
4       UdpAgent(packet_t);
5       virtual void sendmsg(int nbytes, const char *flags = 0){
6           sendmsg(nbytes, NULL, flags);
7       }
8       virtual void sendmsg(int nbytes, AppData* data, ...
                                           const char *flags = 0);
9       virtual void recv(Packet* pkt, Handler*);
10      virtual int command(int argc, const char*const* argv);
11  protected:
12      int seqno_;
13  };

    //~/ns/apps/udp.cc
14  UdpAgent::UdpAgent() : Agent(PT_UDP), seqno_(-1){
15      bind("packetSize_", &size_);
16  }

17  UdpAgent::UdpAgent(packet_t type) : Agent(type){
18      bind("packetSize_", &size_);
19  }

    //~/ns/tcl/lib/ns-default.tcl
20  Agent/UDP set packetSize_ 1000
```

nonzero, Lines 15–20 will transmit the remaining application payload in another packet. Finally, Line 22 invokes function `idle()` to inform the attached application that the UDP agent has finished data transmission. From Line 24, function `idle()` does so by invoking function `resume()` of the attached application (if any).

There are two important notes for UDP agents. First, since a UDP agent is a sending agent its function `recv(p,h)` is generally not to be used. Secondly, in Program 9.7, function `sendmsg(...)` transmits packets, irrespective of network condition.

Step 5: Define OTcl Commands and Instprocs

Class `Agent/UDP` defines the two following OTcl commands defined in Program 9.8:

Program 9.7 Function `sendmsg` of class `UdpAgent` and function `idle` of class `Agent`.

```
//~/ns/apps/udp.cc
1  void UdpAgent::sendmsg(int nbytes,AppData* data,const char* flags)
2  {
3      Packet *p;
4      int n = nbytes / size_;
5      while (n-- > 0) {
6          p = allocpkt();
7          /* packet header configuration */
8          hdr_cmn::access(p)->size() = size_;
9          ...
10         /* -------------------------- */
11         target_->recv(p);
12     }

13     n = nbytes % size_;
14     if (n > 0) {
15         p = allocpkt();
16         /* packet header configuration */
17         hdr_cmn::access(p)->size() = n;
18         ...
19         /* -------------------------- */
20         target_->recv(p);
21     }
22     idle();
23 }

   //~/ns/common/agent.cc
24 void Agent::idle() { if (app_) app_->resume(); }
25 }
```

- `send{nbytes str}`: Send a payload of size "nbytes" containing a message "str".
- `sendmsg{nbytes str flags}`: Similar to the command `send` but also passes the input flag "flags" when sending a packet.

Lines 5–8 in Program 9.8 show the details of the OTcl command `send{...}`. Line 5 creates a `PacketData` object. Line 6 stores the input message `str` in the created `PacketData` object. Line 7 sends out the application payload by invoking function `sendmsg(...)`. Note that the size of application payload does not depend on the size of the message in the `PacketData` object (i.e., `argv[3]` or `str`). Rather, it is specified in the first input argument (i.e., `argv[2]` or `nbytes`). The implementation of the OTcl command `sendmsg(...)` is similar to that of an OTcl command `send{...}`. However, it also feeds a flag "flags" as an input argument of function `sendmsg(...)` (see Line 14).

Program 9.8 OTcl Commands send and sendmsg of class Agent/UDP.

```
    //~/ns/apps/udp.cc
1   int UdpAgent::command(int argc, const char*const* argv)
2   {
3       if (argc == 4) {
4           if (strcmp(argv[1], "send") == 0) {
5               PacketData* data = new PacketData(1 + strlen(argv[3]));
6               strcpy((char*)data->data(), argv[3]);
7               sendmsg(atoi(argv[2]), data);
8               return (TCL_OK);
9           }
10      } else if (argc == 5) {
11          if (strcmp(argv[1], "sendmsg") == 0) {
12              PacketData* data = new PacketData(1 + strlen(argv[3]));
13              strcpy((char*)data->data(), argv[3]);
14              sendmsg(atoi(argv[2]), data, argv[4]);
15              return (TCL_OK);
16          }
17      }
18      return (Agent::command(argc, argv));
19  }
```

9.3.3 Setting Up a UDP Connection

A UDP connection can be created by the network configuration method provided in Section 9.2.2. An example connection where a UDP agent, a Null agent, and a CBR traffic source are used as a sending agent, a receiving agent, and an application is shown in Example 9.1.

9.4 Chapter Summary

An agent is a connector which bridges an application to a low-level network. Its main responsibilities are to create packets based on user demand received from an application, to forward packets to a low-level network, and to destroy packets received from a low-level network. From this point of view, an agent can be used to model transport layer protocols and routing protocols. This chapter focuses on transport layer (protocol) agents only.

Class Agent is a base class, which represents both sending and receiving agents. It connects to an application and a low-level network using pointers app_ and target_. An application also has a pointer agent_ to an agent, while a low-level network uses a pointer target_ as a reference to an agent. Class Agent provides basic functionalities for creating, forwarding, and destroying packets. Its functions send(...) and sendmsg(...) are invoked by an attached application to pass on user demand. An agent creates packets based on the demand, and forwards the created packet to a low-level network by executing `target_->recv(p,h)`. A low level network sends a packet to a receiving agent which may destroy the packet by invoking function `recv(p,h)` of the receiving agent.

User Datagram Protocol (UDP) and Transmission Control Protocol (TCP) are among the most widely used transport layer protocols. UDP is a simple transport layer protocol and it can be flexibly used by other protocols. In NS2, UDP is implemented in the C++ class `UdpAgent` which is bound to an OTcl class `Agent/UDP`. A UDP agent is usually peered with a Null agent, which simply destroys a received packet.

TCP is a reliable transport control protocol. Its main features are end-to-end error control and network congestion control. It implements timeout and acknowledgement to provide end-to-end error control, and adopts a window-based rate adjustment to control network congestion. We will discuss the details of TCP implementation in NS2 in the next chapter.

10

Transport Control Protocols
Part 2 – Transmission Control Protocol (TCP)

As a transport control protocol, TCP (Transmission Control Protocol) bridges an application to a low-level network, controls network congestion, and provides reliability to an end-to-end connection. This chapter discusses the details of TCP agents. Section 10.1 gives an overview of TCP agents. Here, we show a TCP network configuration method, a brief overview of TCP internal mechanism, TCP header format, and the main steps in defining TCP senders and TCP receivers. Sections 10.2 and 10.3 discuss the implementation of TCP receivers and senders, respectively. Sections 10.4–10.7 presents the implementation of four main functionalities of a TCP sender. Finally, the chapter summary is provided in Section 10.8.

10.1 An Overview of TCP Agents in NS2

Based on user demand from an application, a TCP sender creates and forwards packets to a low-level network. It controls the congestion by limiting the rate (i.e., by adjusting the congestion window) at which packets are fed to the low-level network. It enforces an acknowledgment mechanism to provide connection reliability. A TCP receiver must acknowledge every received TCP packet. Based on the acknowledgment pattern, a TCP sender determines whether the transmitted packet was lost or not. If so, it will retransmit the packet. A TCP sender is responsible for sending packets as well as controlling the transmission rate, while the role of a TCP receiver is only to return acknowledgments to the associated TCP sender.

10.1.1 Setting Up a TCP Connection

As a transport layer agent, TCP can be incorporated into a network by using the method discussed in Section 9.2.2.

T. Issariyakul, E. Hossain, *Introduction to Network Simulator NS2*,
DOI: 10.1007/978-0-387-71760-9_10, © Springer Science+Business Media, LLC 2009

Example 10.1. Consider Fig. 9.3. Replace the CBR application with FTP (File Transfer Protocol), the UDP agent with a TCP sender, and the Null agent with a TCP receiver. The modified network can be created by using the following Tcl simulation script.

```
1  set ns [new Simulator]
2  set n1 [$ns node]
3  set n2 [$ns node]
4  set n3 [$ns node]
5  $ns duplex-link $n1 $n2 5Mb 2ms DropTail
6  $ns duplex-link $n2 $n3 5Mb 2ms DropTail
7  $ns duplex-link $n1 $n3 5Mb 2ms DropTail

   #=== TCP connection setup starts here ===
8  set tcp [new Agent/TCP]
9  set sink [new Agent/TCPSink]
10 set ftp [new Application/FTP]
11 $ns attach-agent $n1 $tcp
12 $ns attach-agent $n3 $sink
13 $ftp attach-agent $tcp
14 $ns connect $tcp $sink
15 $ns at 0.0 "$ftp start"
```

Similar to those in Example 9.1, Lines 8–14 above create a TCP connection on top of a low-level network.

10.1.2 Packet Transmission and Acknowledgment Mechanism

TCP provides connection reliability by means of acknowledgment and packet retransmission. Figure 10.1 shows a diagram for TCP packet transmission and acknowledgment mechanisms. The process starts when an application (e.g., FTP) informs a TCP sender (e.g., `TcpAgent`) of user demand by invoking function `sendmsg(nbytes)` of the `TcpAgent` object through its variable `agent_`. The TCP sender creates TCP packets, and forwards them to its downstream object by executing `target_->recv(p,h)`. The low-level network delivers the packets to the destination node attached to the TCP receiver (i.e., `TcpSink`). The destination node forwards the packet to the TCP receiver (i.e., a `TcpSink` object) by invoking function `recv(p,h)` of the TCP receiver installed in its demultiplexer (e.g., `dmux_`). Upon receiving a TCP packet, the TCP receiver creates an ACK packet and returns it to the TCP sender by executing `target_->recv(p,h)`. The low-level network delivers the ACK packet to the sending node, which forwards the ACK packet to the TCP sender via its demultiplexer.

If a TCP packet or an ACK packet is lost (or delayed for a long period of time), the TCP sender will assume that the packet is lost. In this case, the

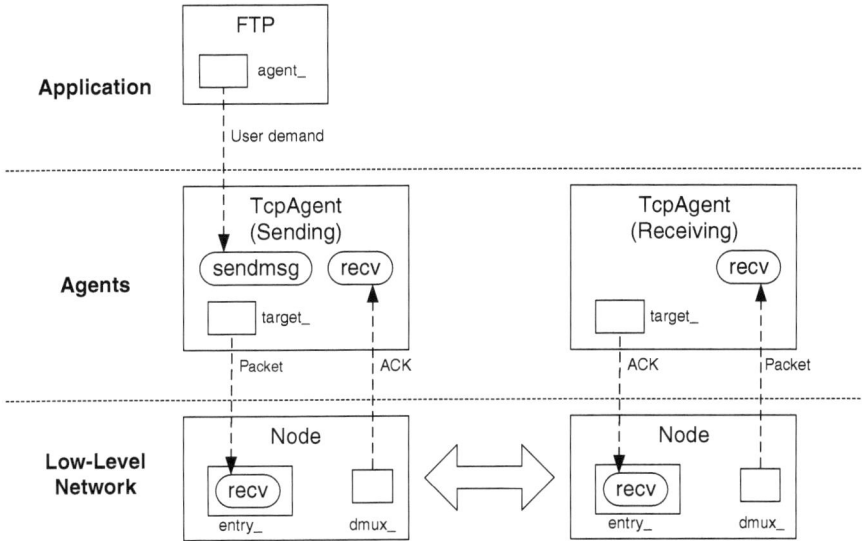

Fig. 10.1. TCP packet transmission and acknowledgment mechanisms.

TCP sender will retransmit the lost TCP packet using `target_->recv(p,h)` (see the description of the retransmission process in Section 9.1.2).

10.1.3 TCP Header

TCP packet header is defined in the "`hdr_tcp`" `struct` data type shown in Program 10.1. The key variables of `hdr_tcp` include

`seqno_`   TCP sequence number
`ts_`   Timestamp: The time when the packet was generated
`ts_echo_`   Timestamp echo: The time when the peering TCP received the packet
`reason_`   Reason for packet transmission (e.g., 0 = normal transmission)

In common with other packet header, `hdr_tcp` contains function `access(p)` (Lines 8–10), which can be used to obtain the reference to a TCP header stored in the input packet `*p`. This reference can then be used to access the attributes of a TCP packet header.

10.1.4 Defining TCP Sender and Receiver

We follow the guidelines provided in Section 9.2.4 to define a TCP sender and a TCP receiver.

Program 10.1 Declaration of `hdr_tcp` `struct` data type.

```
   //~/ns/tcp/tcp.h
1  struct hdr_tcp {
2      double ts_;                /*time packet generated (at source)*/
3      double ts_echo_;           /*the echoed timestamp*/
4      int seqno_;                /*sequence number */
5      int reason_;               /*reason for a retransmit */
6      static int offset_; // offset for this header
7      inline static int& offset() { return offset_; }
8      inline static hdr_tcp* access(Packet* p) {
9          return (hdr_tcp*) p->access(offset_);
10     }
11     int& seqno() { return (seqno_); }
12     ...
13 };
```

Step 1: Define the Inheritance Structure

NS2 defines TCP sender in a C++ class `TcpAgent` which is bound to an OTcl class `Agent/TCP` through a mapping class `TcpClass`, as shown in Lines 1–7 of Program 10.2. Similarly, TCP receiver is defined in a C++ class `TcpSink` which is bound to an OTcl class `Agent/TCPSink` through a mapping class `TcpSinkClass`, as shown in Lines 8–14 of Program 10.2.

Step 2: Define Necessary C++ and OTcl Variables

While class `TcpSink` has only one C++ key variable `acker_` which is of class `Acker`,[1] class `TcpAgent` has several variables. We classify the key C++ variables of class `TcpAgent` into four categories. First, C++ variables, whose values change dynamically during a simulation, are shown in Table 10.1. Secondly, C++ variables, which are usually configured once, are listed in Table 10.2. Thirdly, Table 10.3 shows variables which are related to TCP timer mechanism. Finally, Table 10.4 shows the other non-classified variables of class `TcpAgent`.

Step 3: Implement the Constructor

The constructors of both TCP senders and TCP receivers set their variables to the default values, and bind C++ variables to OTcl instvars as specified in Tables 10.1–10.3. In addition, the constructor of the TCP sender invokes the constructor of its parent class (i.e., `Agent`) with an input argument `PT_TCP`, setting the instantiated `TcpAgent` object to transmit TCP packet only. It also initializes the retransmission timer `rtx_timer_` with the pointer "`this`"

---

[1] We will be discuss the details of class `Acker` later in Section 10.2.1.

Program 10.2 Class `TcpClass` which binds a C++ class `TcpAgent` and an OTcl class `Agent/TCP` together, class `TcpSinkClass`, which binds a C++ class `TcpSink` and an OTcl class `Agent/TCPSink` together, and the constructor of class `TcpSink`.

```
    //~/ns/common/tcp.cc
1   static class TcpClass : public TclClass {
2   public:
3       TcpClass() : TclClass("Agent/TCP") {}
4       TclObject* create(int , const char*const*) {
5           return (new TcpAgent());
6       }
7   } class_tcp;

    //~/ns/common/tcp-sink.cc
8   static class TcpSinkClass : public TclClass {
9   public:
10      TcpSinkClass() : TclClass("Agent/TCPSink") {}
11      TclObject* create(int, const char*const*) {
12          return (new TcpSink(new Acker));
13      }
14  } class_tcpsink;

15  TcpSink::TcpSink(Acker* acker) : Agent(PT_ACK),
            acker_(acker) {...}
```

to itself. The details of `TcpAgent` construction and timers are given in file ~*ns*/tcp/tcp.cc and Section 12.1.

A TCP receiver is somewhat different from a TCP sender, since it does not have a default constructor. From Line 15 of Program 10.2, the constructor takes a pointer to an `Acker` object as an input argument (see Section 10.2.1), and initializes its variable `ack_` with this input pointer. It also initializes its parent constructor with `PT_ACK`, an ACK packet type. Finally, it binds few C++ variables to OTcl instvars (see the detailed construction of class `TcpSink` in file ~*ns*/tcp/tcp-sink.cc).

Steps 3, 4, and 5: Implement Necessary Functions, OTcl Commands, and Instprocs, and Define Timers if Necessary

The detailed implementation of C++ functions of TCP receivers are shown in the next section, while those of TCP senders are given in Sections 10.3–10.7. For brevity, we will not discuss the details of implementation of OTcl command and instproc. The readers are encouraged to study the details of TCP senders and TCP receivers in files ~*ns*/tcp/tcp.cc,h, ~*ns*/tcp/tcp-sink.cc,h, and ~*ns*/tcl/lib/ns-agent.tcl.

Table 10.1. Key operating variables of class TcpAgent.

C++ variable	OTcl variable	Default value	Description
t_seqno_	t_seqno_	0	Current TCP sequence number
curseq_	seqno_	0	Total number of packets need to be transmitted specified by the application. A TCP sender transmits packets as long as its sequence number is less than curseq_.
highest_ack_	ack_	0	Highest ACK number (not frozen during Fast Recovery)
lastack_	N/A	0	Highest ACK number (frozen during Fast Recovery)
cwnd_	cwnd_	0	Congestion window size in packets
ssthresh_	ssthresh_	0	Slow-start threshold
dupacks_	dupacks_	0	Duplicated ACK counter
maxseq_	maxseq_	0	Highest transmitted sequence number
t_rtt_	rtt_	0	RTT sample
t_srtt_	srtt_	0	Smoothed RTT
t_rttvar_	rttvar_	0	RTT deviation
t_backoff_	backoff_	0	Current RTO backoff multiplicative factor
rtt_active_	N/A	0	Status of the RTT collection process
rtt_ts_	N/A	−1	Time at which the packet is transmitted
rtt_seq_	N/A	0	Sequence number of the tagged packet
t_rtxcur_	N/A	0	Current value of unbounded retransmission timeout
ts_peer_	N/A	0	Latest timestamp provided by the peering TCP receiver
rtx_timer_	N/A	N/A	Retransmission timer object

10.2 TCP Receiver

A TCP receiver is responsible for deallocating received TCP packets and returning cumulative ACK packets to the TCP sender. As discussed in Section 9.1.2, a cumulative ACK packet acknowledges a TCP packet with the highest contiguous sequence number. Upon receiving a cumulative ACK packet, the TCP sender assumes that all packets whose sequence numbers are lower than or equal to that of the ACK packet have been successfully received. A cumulative ACK packet has the capability of acknowledging multiple packets. For example, suppose Packet 3 in Fig. 9.2 is not lost but is delayed and that it arrives the receiver right after Packet 6 is received. Upon receiving Packet 3, the receiver acknowledges with A6, since it has received Packets 4–6 earlier.

Table 10.2. Key variables of class `TcpAgent`.

C++ variable	OTcl variable	Default value	Description
wnd_	window_	20	Upper bound on window size
numdupacks_	numdupacks_	3	Number of duplicated ACKs which triggers Fast Retransmit
wnd_init_	windowInit_	2	Initial value of window size
size_	packetSize_	1,000	TCP packet size in bytes
tcpip_base_	tcpip_base_hdr_size_	40	TCP basic header size in bytes
useHeaders_	useHeaders_	true	If `true`, TCP and IP header size will be added to packet size
maxburst_	maxburst_	0	Maximum number of bytes that a TCP sender can transmit in one transmission
maxcwnd_	maxcwnd_	0	Upper bound on `cwnd_`
control_	control_increase_	0	If set to 1, do not open the congestion window when the network is limited (See Section 10.5).
increase_			

Table 10.3. Timer related variables of class `TcpAgent`.

C++ variable	OTcl variable	Default value	Description
srtt_init_	srtt_init_	0	Initial value of `t_srtt_`
rttvar_init_	rttvar_init_	12	Initial value of `t_rttvar_`
rtxcur_init_	rtxcur_init_	3.0	Initial value of `t_rtxcur_`
T_SRTT_BITS	T_SRTT_BITS	3	Multiplicative factor for smoothed RTT
T_RTTVAR_BITS	T_RTTVAR_BITS	2	Multiplicative factor for RTT deviation
rttvar_exp_	rttvar_exp_	2	Multiplicative factor for RTO computation
decrease_num_	decrease_num_	0.5	Window decreasing factor
increase_num_	increase_num_	1.0	Window increasing factor
tcpTick_	tcpTick_	0.01	Timer granularity in seconds
maxrto_	maxrto_	100,000	Upper bound on RTO in seconds
minrto_	minrto_	0.2	Lower bound on RTO in seconds

Table 10.4. Miscellaneous variables of class `TcpAgent`.

C++ variable	Default value	Description
cong_action_	0	true when the congestion has occurred.
sigledup_	1	If set to 1, the TCP sender will transmit new packets upon receiving first few duplicated ACK packets.
prev_highest_ack_	N/A	Sequence number of an ACK packet received prior to the current ACK packet.
last_cwnd_action_	N/A	The latest action on congestion window
recover_	N/A	The highest transmitted sequence number during the previous packet loss event

In NS2, C++ implementation of TCP receivers involves two main classes: `Acker` and `TcpSink`. Class `Acker` is a helper class responsible for generating ACK packets. Class `TcpSink` contains an `Acker` object, and acts as interfaces to a peering TCP sender and the OTcl domain.

10.2.1 Class `Acker`

Program 10.3 Declaration of class `Acker`.

```
     //~/ns/tcp/tcp-sink.h
1    class Acker {
2    public:
3        Acker();
4        virtual ~Acker() { delete[] seen_; }
5        inline int Seqno() const { return (next_ - 1); }
6        inline int Maxseen() const { return (maxseen_); }
7        int update(int seqno, int numBytes);
8    protected:
9        int next_;
10       int maxseen_;
11       int wndmask_;
12       int *seen_;
13       int is_dup_;
14   };
```

Program 10.3 shows the declaration of a C++ class `Acker`.[2] Class `Acker` stores necessary information required to generate cumulative ACK packets in the following variables:

---

[2] Class `Acker` is not implemented in the OTcl domain.

seen_   An array whose index and value are the sequence number and the corresponding packet size, respectively

next_   Expected sequence number

maxseen_ Highest sequence number ever received

wndmask_ Modulus mask, initialized to maximum window size-1 (set to 63 by default; see Section 12.4.1)

is_dup_   True if the latest received TCP packet was received earlier

Figure 10.2 shows an example of information stored in an Acker object. In this case, Packets 1, 2, 3, 5, and 7 are received, but Packets 4 and 6 are missing. Therefore, next_ and maxseen_ are set to 4 and 7, respectively. Also, variable seen_ stores the size in bytes of Packets 1–7 in its respective entries. To determine whether packet n is missing, class Acker checks the value of seen_[n]. The packet is missing if and only if seen_[n] is zero. Suppose a TCP receiver receives a TCP packet number 4 when the status of the Acker object is as in Fig. 10.2. The Acker object will generate an ACK packet with sequence number 5. However, if the sequence number of the received packet is not 4, the Acker object will create an ACK packet with sequence number 3.

As discussed in Section 9.1.2, a TCP connection can have at most w unacknowledged packets in a network, where w is the current congestion window size. Let MWS be the Maximum Window Size in a simulation (see Line 6 in Program 10.4). Then, $w \in \{0, \cdots, \text{MWS}\}$ and there can be at most MWS unacknowledged packets during the entire simulation. An Acker object needs only MWS entries in the array variable seen_ to store information about unacknowledged packets.

Program 10.4 shows the constructor of the C++ class Acker. The constructor resets next_ and maxseen_ to zero in Line 1. Line 3 allocates memory space for array variable seen_ with MWS entries. Line 4 clears the allocated memory to zero. Also, wndmask_ is set to MWM (Maximum Window Mask which is set to 63 in Line 7).

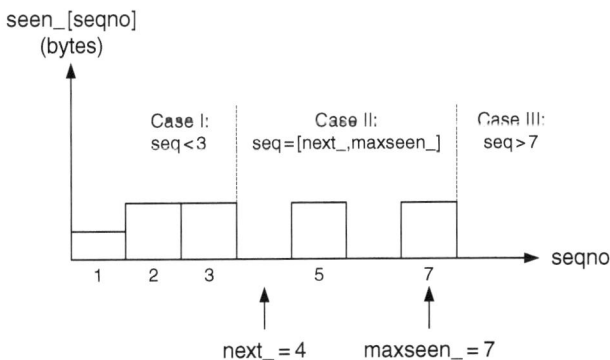

Fig. 10.2. Information necessary to generate a cumulative acknowledgement.

Program 10.4 The constructor of class `Acker`.

```
//~/ns/tcp/tcp-sink.cc
1  Acker::Acker() : next_(0), maxseen_(0), wndmask_(MWM)
2  {
3      seen_ = new int[MWS];
4      memset(seen_, 0, (sizeof(int) * (MWS)));
5  }

   //~/ns/tcp/tcp-sink.cc
6  #define MWS 64
7  #define MWM (MWS-1)
```

The above MWS (set by default to 64 in Line 6 of Program 10.4) entries of `seen_` are reused to store the packet size corresponding to all incoming TCP sequence numbers. Class `Acker` employs a modulus operation to map a sequence number to an array index. Upon receiving a TCP packet with sequence number `seqno`, an `Acker` object stores the packet size in the entry `seqno % MWS` (which is the remainder of `seqno/MWS`), of the array `seen_`, where "%" is a modulus operator. When `seqno` exceeds MWS, `seqno % MWS` will be restarted from the first entry to reuse the memory allocated to `seen_`.

As discussed in Section 12.4.1, a modulus operation can also be implemented by bit masking. In particular, `seqno % MWS` is in fact equivalent to `seqno & wndmask_`, where `wndmask_` is set initially to MWM in the constructor (Line 1 in Program 10.4), and MWM (Maximum Window Mask) is defined as 63 (Lines 6–7 in Program 10.4). To facilitate the understanding, readers may simply regard `seqno & wndmask_` as `seqno % 63`.

Class `Acker` has two key functions: `Seqno()` and `update(seq,numBytes)`. Function `Seqno()` (Line 5 in Program 10.3) returns the highest sequence number of a burst of contiguously received packets. As shown in Program 10.5, function `update(seq,numBytes)` updates its internal variables according to the input arguments.

Function `update(seq,numBytes)` takes two input arguments: `seq` and `numBytes` which are the sequence number and the size of an incoming TCP packet, respectively. It updates variables `next_`, `maxseen_`, `seen_`, and `is_dup_` and returns the number of in-sequence bytes which is ready to be delivered to the application. From Fig. 10.2, "seq" can be (I) less than `next_`, (II) between `next_` and `maxseen_`, and (III) greater than `maxseen_`. Function `update(seq,numBytes)` reacts to these three cases as follows:

(i) If `seq < next_`, function `update(seq,numBytes)` will set `is_dup_` to be true (Line 17). This case implies that this packet was received earlier, and therefore, this packet is a duplicated packet.

(ii) If `seq` lies in between `next_` and `maxseen_`, function `update(seq, numBytes)` will execute Lines 19–26. Line 19 determines whether `seq` was received earlier. This happens to be true under the two following

Program 10.5 Function update of class Acker.

```
//~/ns/tcp/tcp-sink.cc
1   int Acker::update(int seq, int numBytes)
2   {
3       bool just_marked_as_seen = FALSE;
4       is_dup_ = FALSE;
5       int numToDeliver = 0;
6       if (seq > maxseen_) {
7           int i;
8           for (i = maxseen_ + 1; i < seq; ++i)
9               seen_[i & wndmask_] = 0;
10          maxseen_ = seq;
11          seen_[maxseen_ & wndmask_] = numBytes;
12          seen_[(maxseen_ + 1) & wndmask_] = 0;
13          just_marked_as_seen = TRUE;
14      }
15      int next = next_;
16      if (seq < next)
17          is_dup_ = TRUE;
18      if (seq >= next && seq <= maxseen_) {
19          if (seen_[seq & wndmask_] && !just_marked_as_seen)
20              is_dup_ = TRUE;
21          seen_[seq & wndmask_] = numBytes;
22          while (seen_[next & wndmask_]) {
23              numToDeliver += seen_[next & wndmask_];
24              ++next;
25          }
26          next_ = next;
27      }
28      return numToDeliver;
29  }
```

conditions: (1) the corresponding entry of seen_ is nonzero and (2) just_marked_as_seen is false. The latter condition is added since seen_ could have been set by Line 11. In this case, is_dup_ is set to true. Line 21 stores the packet size in seen_[seq & wndmask_].[3] Lines 22–26 update next_ by advancing next_ until seen_[next_ & wndmask_] is empty. Also, Line 23 keeps adding the packet size to numToDeliver, which are returned in Line 28. Essentially, the returned value is the number of bytes which corresponds to next_ advancement.

(iii) If seq > maxseen_, implying a new TCP packet, function update(...) will execute Lines 7–13. Lines 8–9 and 12 clear the seen_[maxseen_+1] through seen_[seq-1]. It updates maxseen_ in Line 10 and stores the

---

[3] Bit masking with wndmask_ has the same impact as a modulus with wndmask_+1 does.

packet size in seen_[seq & wndmask_] in Line 11. Since Line 10 stores seq in maxseen_, the condition in Line 18 is satisfied and Lines 19–26 are to be executed. If Case (iii) is executed, Case (ii) will also be executed. Therefore, Line 13 sets just_marked_as_seen to be true, which simply indicates that the current packet is not a duplicated packet, and Line 20 should be skipped.

10.2.2 Class TcpSink

Representing TCP receivers, class TcpSink reacts to received TCP packets as follows:

(i) Extract the sequence number (seq) from the received TCP packet,
(ii) Inform the Acker object of the sequence number (seq) and the size of the TCP packet (numBytes) through function update(seq,numBytes) of class Acker,
(iii) Create and send an ACK packet to the TCP sender by invoking function ack(p) of class TcpSink. The sequence number in the ACK packet is obtained from function Seqno() of the Acker object (invoked from within function ack(p)).

Program 10.6 shows the declaration of a C++ class Tcpsink, which is bound to an OTcl class Agent/TCPSink. The only key variable of class TcpSink is a pointer to an Acker object, acker_ in Line 8. Two main functions of class TcpSink include recv(p,h) and ack(p).

Program 10.6 Declaration of class TcpSink.

```
    //~/ns/tcp/tcp-sink.cc
1   class TcpSink : public Agent {
2   public:
3       TcpSink(Acker*);
4       void recv(Packet* pkt, Handler*);
5       int command(int argc, const char*const* argv);
6   protected:
7       void ack(Packet*);
8       Acker* acker_;
9   };
```

Shown in Program 10.7, function recv(p,h) is invoked by an upstream object to hand a TCP packet over to a TcpSink object. Lines 4–6 inform the Acker object, acker_, of an incoming TCP packet "pkt". Here, the sequence number (i.e., th->seqno()) and packet size (i.e., numBytes) are passed to acker_ through this function. Again, function update(seq,numBytes) returns the number of in-order bytes which can be delivered to the application. If

this number is nonzero, it will be delivered to the application through function
recvBytes(bytes) in Line 8. Line 9 invokes function ack(pkt) to generate
an ACK packet and send it to the TCP sender. Finally, Line 10 deallocates
the received TCP packet.

Program 10.7 Function recv of class TcpSink.

```
//~/ns/tcp/tcp-sink.cc
1  void TcpSink::recv(Packet* pkt, Handler*)
2  {
3      int numToDeliver;
4      int numBytes = hdr_cmn::access(pkt)->size();
5      hdr_tcp *th = hdr_tcp::access(pkt);
6      numToDeliver = acker_->update(th->seqno(), numBytes);
7      if (numToDeliver)
8          recvBytes(numToDeliver);
9      ack(pkt);
10     Packet::free(pkt);
11 }
```

Program 10.8 shows the details of function ack(p). In this function, variables whose name begins with "o" and "n" are used for an old packet and a new
packet, respectively. Line 6 puts an ACK number in the ACK packet. Lines
7–8 and 9–11 configure timestamp and flow ID of the ACK packet, respectively. Finally, the configured packet is sent out using function send(npkt,0)
of class Agent in Line 12, where a new packet npkt is transmitted along with
a Null handler.

Program 10.8 Function ack of class TcpSink.

```
//~/ns/tcp/tcp-sink.cc
1  void TcpSink::ack(Packet* opkt)
2  {
3      Packet* npkt = allocpkt();
4      hdr_tcp *otcp = hdr_tcp::access(opkt);
5      hdr_tcp *ntcp = hdr_tcp::access(npkt);
6      ntcp->seqno() = acker ->Seqno();
7      double now = Scheduler::instance().clock();
8      ntcp->ts() = now;
9      hdr_ip* oip = hdr_ip::access(opkt);
10     hdr_ip* nip = hdr_ip::access(npkt);
11     nip->flowid() = oip->flowid();
12     send(npkt, 0);
13 }
```

10.3 TCP Sender

A TCP sender has the following four main responsibilities:

- **Packet transmission**: Based on user demand from an application, a TCP sender creates and forwards TCP packets to a TCP receiver.
- **ACK processing**: A TCP sender observes a received ACK pattern and determines whether transmitted packets were lost. If so, it will retransmit the lost packets. From the ACK pattern, it can also estimate the network condition (e.g., end-to-end bandwidth) and adjust the congestion window accordingly.
- **Timer related mechanism**: A retransmission timer is used to provide connection reliability. Unless reset by an ACK packet arrival, the retransmission timer informs the TCP sender of packet loss after the packet has been transmitted for a period of *Retransmission TimeOut (RTO)*.
- **Window adjustment**: Based on the ACK pattern and timeout event, a TCP sender adjusts its congestion window to fully utilize network resource and prevent network congestion.

The details of these four responsibilities will be discussed in the next four sections.

10.4 TCP Packet Transmission Functions

Class `TcpAgent` provides the following four main packet transmission functions:

- `sendmsg(nbytes)`: Send `nbytes` of application payload. If `nbytes=-1`, the payload is assumed to be infinite.

- `sendmuch(force,reason,maxburst)`: out a packet whose sequence number is `t_seqno_`. Keep sending out packets as long as the congestion window allows and the total number of transmitted packets during a function invocation does not exceed `maxburst`.

- `output(seqno,reason)`: Create and send a packet with a sequence number and a transmission reason as specified by `seqno` and `reason`, respectively.

- `send_one()`: Send a TCP packet with a sequence number `t_seqno_`.

Among the above functions, function `sendmsg(nbytes)` is the only `public` function derived from class `Agent`, while the other three functions are internal to class `TcpAgent`. Again, function `sendmsg(nbytes)` is invoked by an application to inform a `TcpAgent` object of user demand. Function `sendmsg(nbytes)` does not directly send out packets. Rather, it computes the number of TCP packets required to hold "nbytes" of data payload, and increases variable curseq_ by the computed value. In NS2, a TcpAgent object keeps transmitting TCP packets as long as the sequence number does not exceed curseq_. Increasing curseq_ is therefore equivalent to feeding data payload to a TcpAgent object.

Another important variable is t_seqno_, which contains the default TCP sequence number. Unless otherwise specified, a TCP sender always transmits a TCP packet with the sequence number stored in t_seqno_. Both functions sendmuch(force,reason,maxburst) and send_one() use function output(t_seqno_,reason) to send out a TCP packet whose sequence number is t_seqno_.

Function send_much(...) acts as a foundation for TCP packet transmission. In most cases, TCP agent first stores the sequence number of packet to be transmitted in t_seqno_. Then, it invokes function send_much(...) to send TCP packets starting with that with sequence number t_seqno_ as long as the transmission window permit. As we shall see in Program 10.10, each packet transmission is carried out using function output(t_seqno_, reason).

10.4.1 Function sendmsg(nbytes)

Function send_msg(nbytes) is the main data transmission interface function derived from class Agent. A user (e.g., application) informs a TCP sender of transmission demand through this function. Function sendmsg(nbytes) usually takes one input argument, nbytes, which is the amount of application payload in bytes that a user needs to send to the TCP receiver. When the user has infinite demand, nbytes is specified as −1.

Program 10.9 shows the details of function sendmsg(nbytes). Lines 4–7 transform the input user demand to the number of TCP packets to be

---

Program 10.9 Function sendmsg of class TcpAgent.

```
    //~/ns/tcp/tcp.h
1 #define TCP_MAXSEQ 1073741824

    //~/ns/tcp/tcp.cc
2 void TcpAgent::sendmsg(int nbytes, const char* /*flags*/)
3 {
4     if (nbytes == -1 && curseq_ <= TCP_MAXSEQ)
5         curseq_ = TCP_MAXSEQ;
6     else
7         curseq_ += (nbytes/size_ + (nbytes%size_ ? 1 : 0));
8     send_much(0, 0, maxburst_);
9 }
```

---

transmitted (i.e., `curseq_`). Line 8 starts data transmission by invoking function `send_much(0,0,maxburst_)`. Note that Line 1 specifies the limit (i.e., `TCP_MAXSEQ`) on the number of TCP sequence numbers which can be created by a certain TCP sender. Again, if `nbytes = -1`, the TCP sender will be backlogged until "TCP_MAXSEQ" TCP packets are transmitted. If `nbytes` is greater than –1, Line 7 will compute the number of TCP packets (each with size "`size_`" bytes) which can accommodate `nbytes` of application payload.

10.4.2 Function `send_much(force,reason,maxburst)`

There are three important points in regards to function `send_much(force, reason,maxburst)`. First, it creates and sends out as many packets as the current transmission window allows, but not greater than `maxburst` packets. Secondly, every packet is transmitted by executing `output(t_seqno_,reason)`. Finally, function `send_much(...)` always sends out TCP packets with a sequence number `t_seqno_`.

Function `send_much(force,reason,maxburst)` takes three following input arguments, where a typical invocation of this function is `send_much(0,0, maxburst_)`:

- `force`: This value is usually set to zero. When `force = 1`, TCP sender will try to transmit data packets even if some conditions are not met.[4]
- `reason`: This value specifies the reason for data transmission. For a normal transmission, `reason` is set to 0. Other possible values of `reason` are shown in Lines 1–4 in Program 10.10. This input argument is later placed in the field `reason_` of TCP packet header (i.e., `hdr_tcp::reason_`) and will be used for various purposes in simulation.
- `maxburst`: The maximum number of packets which can be transmitted for each invocation of function `send_much(force,reason,maxburst)`.

Program 10.10 shows the details of function `send_much(force, reason, maxburst)`. Function `send_much(force,reason,maxburst)` first stores the current congestion window[5] in a variable `win` and sets the variable `npackets` to zero in Line 7. Then, Line 8 checks whether the TCP sender is allowed

---

[4] For example, variable `overhead_` adds a certain delay time specified by a `DelSndTimer` object before data transmission. By default, TCP sender does not transmit when `overhead_` is nonzero. However, it can transmit packets immediately when `force = 1`. Note that we do not discuss the details of class `DelSndTimer` here. The readers may find the details of class `DelSndTimer` in files `~ns/tcp/tcp.cc,h`.

[5] From Lines 19–22 of Program 10.10, function `window()` returns the minimum of `window_` (the maximum window size) and `cwnd_` (the current congestion window size) as the current bounded congestion window.

Program 10.10 Function send_much of class TcpAgent.

```
        //~/ns/tcp/tcp.h
1   #define TCP_REASON_TIMEOUT      0x01 //Timeout
2   #define TCP_REASON_DUPACK       0x02 //Duplicated ACK
3   #define TCP_REASON_RBP          0x03 //Rate Based Pacing
4   #define TCP_REASON_PARTIALACK   0x04 //Partial ACK

        //~/ns/tcp/tcp.cc
5   void TcpAgent::send_much(int force, int reason, int maxburst)
6   {
7       int win = window(), npackets = 0;
8       while (t_seqno_ <= highest_ack_ + win && t_seqno_ < curseq_) {
9           if (overhead_ == 0 || force ) {
10              output(t_seqno_, reason);
11              npackets++;
12              t_seqno_++;
13          }
14          win = window();
15          if (maxburst && npackets == maxburst)
16              break;
17      }
18  }

19  int TcpAgent::window()
20  {
21      return (cwnd_ < wnd_ ? (int)cwnd_ : (int)wnd_);
22  }
```

to send a TCP packet with sequence number t_seqno_. If so, Line 10 will invoke function output(t_seqno_,reason) to send out a TCP packet. Again, a TCP sender is allowed to transmit a packet if the following three conditions are satisfied:

(i) Congestion window allows packet transmission: Function window() in Line 7 returns the minimum of the current congestion window and the maximum window size. This minimum value is stored in the variable win in Line 7. Since the latest received ACK number is highest_ack_, TCP sender can transmit TCP packets with sequence numbers t_seqno_ through highest_ack_+win.

(ii) TCP sender still has data to transmit: The sender will send TCP packets unit the sequence number reaches curseq_. specified by user demand, curseq_ is the highest TCP sequence number that the sender needs to transmit.

After a packet transmission, the default sequence number "t_seqno_" (Line 12) and the congestion window size "win" (Line 14) are updated. Lines 15–17 stops the transmission, if TCP sender has sent out maxburst packets. The above process repeats until the condition in Line 8 becomes false.

10.4.3 Function output(seqno,reason)

Taking two input arguments, function output(seqno,reason) creates a packet, sets the sequence number and the reason field of TCP header to the input arguments seqno and reason, respectively, and forwards the packet to the low-level network by using function send(p,h) of class Agent.

Programs 10.11 and 10.12 show the details of function output(seqno, reason), which consists of five main parts. First, Line 5 creates a packet "p" by using function allocpkt() of class Agent. Secondly, Lines 6–26 configure common, TCP, and flag headers of the created packets. For the common packet header, function output(...) configures packet size (Lines 18–26) to be 1,000 bytes (by default). If useHeaders_ is true, tcpip_base_hdr_size_ (40 bytes by default) will be added to the packet size align. Since a SYN

Program 10.11 Function output of class TcpAgent.

```
    //~/ns/tcp/tcp.cc
1   void TcpAgent::output(int seqno, int reason)
2   {
3       int force_set_rtx_timer = 0;
4       int is_retransmit = (seqno < maxseq_);
5       Packet* p = allocpkt();
6       hdr_tcp *tcph = hdr_tcp::access(p);
7       tcph->seqno() = seqno;
8       tcph->ts() = Scheduler::instance().clock();
9       tcph->ts_echo() = ts_peer_;
10      tcph->reason() = reason;
11      tcph->last_rtt() = int(int(t_rtt_)*tcp_tick_*1000);
12      int databytes = hdr_cmn::access(p)->size();
13      if (cong_action_ && !is_retransmit) {
14          hdr_flags* hf = hdr_flags::access(p);
15          hf->cong_action() = TRUE;
16          cong_action_ = FALSE;
17      }
18      if (seqno == 0) {
19          if (syn_) {
20              databytes = 0;
21              curseq_ += 1;
22              hdr_cmn::access(p)->size() = tcpip_base_hdr_size_;
23          }
24      } else if (useHeaders_ == true) {
25          hdr_cmn::access(p)->size() += headersize();
26      }
```

Program 10.12 Function output of class TcpAgent (Cont.).

```
27      send(p, 0);
28      ++ndatapack_;
29      ndatabytes_ += databytes;
30      if (seqno == curseq_ && seqno > maxseq_)
31          idle();
32      if (seqno > maxseq_) {
33          maxseq_ = seqno;
34          if (!rtt_active_) {
35              rtt_active_ = 1;
36              if (seqno > rtt_seq_) {
37                  rtt_seq_ = seqno;
38                  rtt_ts_ = Scheduler::instance().clock();
39              }
40          }
41      } else {
42          ++nrexmitpack_;
43          nrexmitbytes_ += databytes;
44      }
45      if (highest_ack_ == maxseq_)
46          force_set_rtx_timer = 1;
47      if (!(rtx_timer_.status() == TIMER_PENDING)
                                    || force_set_rtx_timer)
48          set_rtx_timer();
49  }
```

packet (with seqno_=0 and syn_=1) contains no pay-load, its size is set to be tcpip_base_hdr_size_ bytes (Line 22). The following TCP header fields are configured in Lines 6–12: sequence number, timestamp, timestamp echo, transmitting reason, and latest observed round trip time (RTT). Finally, function output(...) configures only the congestion flag in the flag header (Lines 13–16). This congestion flag is set to be true if both of the following conditions are true (i.e., Line 13 is true):

(i) Congestion has occurred: During network congestion, TCP sender closes the congestion window by invoking function slowdown(how), within which the variable cong_action_ is set to true. If variable cong_action_ is true, Lines 13–17 will presume that congestion has occurred.

(ii) TCP sender is transmitting a new packet (is_retx = false): This flag set to true, when a *regular* packet (not a *retransmitted* packets) is experiencing congestion.[6]

---

[6] For example, a router in the network may drop packets marked with a *congestion action* flag to help relieve network congestion. However, dropping a retransmitted packet may lead to TCP connection reset. Therefore, a TCP sender does not mark retransmitted packets with congestion action.

The third part of function output(seqno,reason) is used to send out the configured packet using function send(p,h) of class Agent in Line 27. The fourth part updates the relevant variables of the TcpAgent object in Lines 28–48. If the condition in Line 30 is true, TCP sender will no longer have data to transmit. In this case, Line 31 informs the application so by invoking function idle() of class Agent. Relevant variables to be updated are ndatapack_, ndatabytes_, nrexmitpack_, nremitbytes_, in Lines 28, 29, 42, and 43, respectively. The former two variables denote the data transmitted by the TcpAgent object in packets and bytes, while the latter two are those corresponding to the retransmitted packets only. Lines 33-39 update the related variables when seqno > maxseq_. These variables include maxseq_ and other RTT estimation variables. We will discuss about the RTT estimation later in Section 10.6

The final part is to start the retransmission timer by invoking function set_rtx_timer() in Line 48. Note that each TCP sender has only one retransmission timer. Under a normal situation, the timer is started only when it is idle (i.e., its status is not TIMER_PENDING). However, it is also started when highest_ack_ == maxseq_, regardless of the timer's status (see Line 47).

Program 10.13 Function send_one of class TcpAgent.

```
    //~/ns/tcp/tcp.cc
1 void TcpAgent::send_one()
2 {
3     if (t_seqno_ <= highest_ack_ + wnd_ && t_seqno_ < curseq_ &&
                t_seqno_ <= highest_ack_ + cwnd_ + dupacks_ ) {
4         output(t_seqno_, 0);
5         t_seqno_ ++ ;
6     }
7 }
```

10.4.4 Function send_one()

Figure 10.3 shows the details of function send_one(). Function send_one() is very similar to function send_much(...). It prepares sequence numbers starting at t_seqno_ and passes them to function output(t_seqno_, 0) for packet creation and transmission. The main difference is that while function send_much(...) may send out several packets, function send_one(...) sends out only one packet. Function send_one(...) is designed to send a packet during a fast retransmit phase to indicate whether the TCP sender should send out a new packet for every received duplicated ACK packet (see Section 10.5).

In this case, Line 3 inflates the congestion windows with the number of dupli-cated ACK (`dupacks_`). As will be discussed in Section 10.5, this function is invoked if the option `singledup_` is set to 1 during the reception of first and second duplicated ACK packets.

10.5 ACK Processing Functions

The second responsibility of a TCP sender is to process the ACK packets. An ACK packet could be a new ACK packet or a duplicated ACK packet. A new ACK packet slides the congestion window to the right, and opens the congestion window to allow the TCP sender to transmit more packets. A duplicated ACK packet on the other hand indicates out-of-order packet delivery or packet loss (see Fig. 9.2 for example). Again, TCP Tahoe assumes that packet loss upon detecting the $numdupacks\_^{th}$ (3rd by default) duplicated ACK packet. It sets the slow-start threshold to half of the current congestion window, sets the congestion window size to `wnd_init_` (which is usually set to 1), and retransmits the lost packet. During a fast retransmit phase, the TCP sender transmits a new packet for every received duplicated ACK packet (due to inflated congestion window). When a new ACK packet is received, the TCP sender sets its congestion window to e the same as slow start threshold, and returns to its normal operation.

Class `TcpAgent` provides the four following key ACK Processing functions:

- `recv(p,h)`: This is the main ACK reception function. It determines whether the received packet (`*p`) is a new ACK packet or a duplicated ACK packet, and acts accordingly.

- `recv_newack_helper(p)`: This function is invoked from within function `recv(p,h)` when a new ACK packet is received. It invokes function `newack(p)` to update relevant variables, and opens the congestion window if necessary.

- `newack(p)`: Invoked from within function `recv_newack_helper(p)`, this function updates variables related to sequence number, ACK number, and RTT estimation process, and restarts the retransmission timer.

- `dupack_action()`: This function is invoked from within function `recv(p,h)` when a duplicated ACK packet is received and Fast Retransmit process is launched. It cuts down the congestion window, prepares the sequence number of the lost packet for retransmission, and resets the retransmission timer.

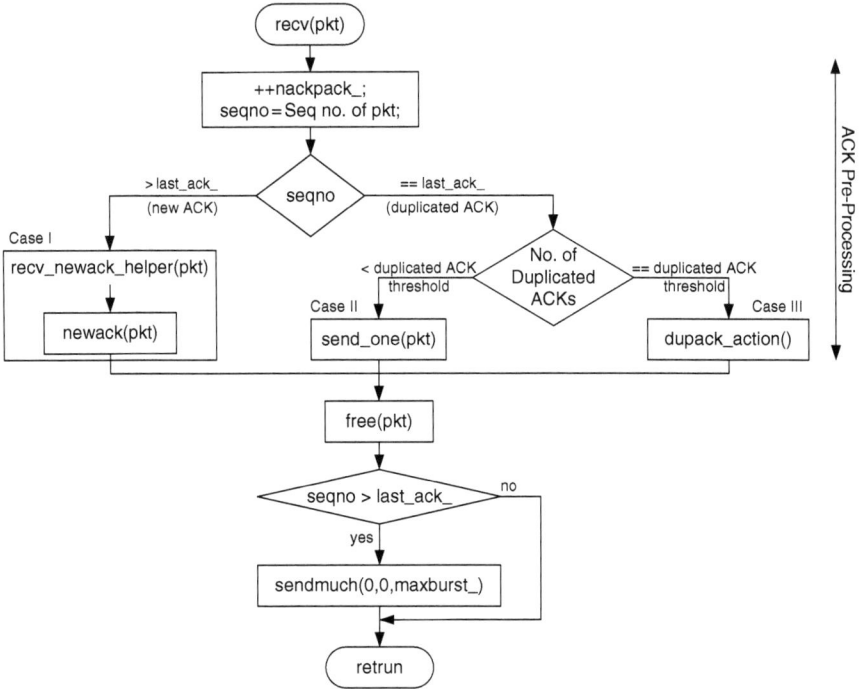

Fig. 10.3. Function `recv(p,h)` of class `TcpAgent`.

10.5.1 Function `recv(p,h)`

Figure 10.3 and Program 10.14 show the diagram and implementation, respectively, for function `recv(p,h)`. Function `recv(p,h)` pre-processes the received ACK packets in Lines 6–14, where `t_seqno_` and `cwnd_` are adjusted. Depending of the received ACK type (i.e., new or duplicated), Lines 6–14 (ACK pre-processing) process an ACK packet according to the following three cases:

- **Case I (New ACK)**: If a new ACK packet is received (i.e., Line 6 returns `true`), Line 7 will invoke function `recv_newack_helper(p)` to adjust congestion window (`cwnd_`) and prepare a new sequence number (`t_seqno_`) for packet transmission.
- **Case II (Duplicated ACK)**: In this case, a duplicated ACK packet is received (i.e., Line 6 returns `false`) but the number of duplicated ACK packets received so far has not reached `numdupacks_` (i.e., Line 9 returns `false`). Line 12 will invoke function `send_one()` to transmit new TCP packets under the congestion window inflated by the number of received duplicated ACK packets. Note that variable `sigledup_` is an NS2 option for congestion window inflation. The above actions are executed when `singledup_` is `true` only. If `singledup_` is `false`, the TCP sender will not send a new packet for every received ACK packet.

- **Case III (Fast retransmit)**: If the received ACK is the last (i.e., numdupacks_th) duplicated ACK packet, the TCP sender will enter the Fast Retransmit phase, by invoking function dupack_action() (Line 10). Note that an option flag noFastRetrans_ is an NS option for a fast retransmit phase. The TCP sender will not enter a Fast Retransmit phase, if noFastRetrans_ is true.

After executing one of the above three cases, Line 17 deallocates the ACK packet *pkt by invoking function free(pkt). If the received ACK is valid (i.e., valid_ack_=1), Line 19 will create and transmit TCP packets using function send_much(0,0,maxburst_). Here a received ACK packet is said to be valid if it is a new ACK packet (i.e., tcph->seqno() > last_ack_) or a duplicated ACK (i.e., tcph->seqno() = last_ack_). If an ACK packet is invalid, a TCP sender will only destroy the ACK packet, but will not create and forward new packets.

Program 10.14 Function recv of class TcpAgent.

```
    //~/ns/tcp/tcp.cc
1   void TcpAgent::recv(Packet *pkt, Handler*)
2   {
3       hdr_tcp *tcph = hdr_tcp::access(pkt);
4       int valid_ack = 0;
5       ++nackpack_;
6       if (tcph->seqno() > last_ack_) {
7           recv_newack_helper(pkt);
8       } else if (tcph->seqno() == last_ack_) {
9           if (++dupacks_ == numdupacks_ && !noFastRetrans_) {
10              dupack_action();
11          } else if (dupacks_ < numdupacks_ && singledup_ ) {
12              send_one();
13          }
14      }
15      if (tcph->seqno() >= last_ack_)
16          valid_ack = 1;
17      Packet::free(pkt);
18      if (valid_ack)
19          send_much(0, 0, maxburst_);
20  }
```

10.5.2 Function recv_newack_helper(pkt)

Function recv_newack_helper(pkt) is a helper function invoked when a new ACK packet is received. As shown in Program 10.15, the function recv_newack_helper(pkt) first invokes function newack(pkt) in Line 2 to update relevant variables and to process the retransmission timer. When Explicit Congestion Notification (ECN) is not enabled (i.e., by default ECT

(ECN Capable Transport System) is set to zero), Line 5 will open the congestion window (by invoking function opencwnd()) when at least one of the following conditions is true (Line 4):

- control_increase_ = 0: Variable control_increase_, when set to 1, suppresses the congestion window opening. When control_increase_ is zero, a TCP sender can freely increase the congestion window.

Program 10.15 Function recv_newack_helper of class TcpAgent.

```
     //~/ns/tcp/tcp.cc
1    void TcpAgent::recv_newack_helper(Packet *pkt) {
2        newack(pkt);
3        if (!ect_) {
4            if (!control_increase_ ||
                 (control_increase_ && (network_limited() == 1)))
5                    opencwnd();
6        }
7        if ((highest_ack_ >= curseq_-1) && !closed_) {
8            closed_ = 1;
9            finish();
10       }
11 }
```

- control_increase_ ≠ 0 and network is limited: When control_increase_ is 1, the TCP sender is allowed to open the congestion window only when the previous congestion window is not sufficient to transmit the current packet (i.e., the network is limited). In NS2, a network is said to be limited when t_seqno_ is less than prev_highest_ack_ + win, where prev_highest_ack_ is the ACK number prior to the reception of the current ACK packet and win is the current congestion window (see Program 10.16). In this case, it is necessary to open the congestion window, even if control_increase_ is enabled. Note that if the TCP sender stops transmission due to any reason other than the reason that the network is limited, function recv_newack_helper(pkt) will not open the congestion window.

Program 10.16 Function network_limited of class TcpAgent.

```
     //~/ns/tcp/tcp.cc
1 int TcpAgent::network_limited() {
2     int win = window () ;
3     if (t_seqno_ > (prev_highest_ack_ + win))
4         return 1;
5     else
6         return 0;
7 }
```

Finally, if the TCP sender no longer has data to transmit, Line 8 in Program 10.15 will close the connection by setting `closed_` to 1, and Line 9 will invoke function `finish()`.

10.5.3 Function `newack(pkt)`

Program 10.17 shows the details of function `newack(pkt)`. Lines 5–10 update variables `dupack_`, `last_ack_`, `prev_highest_ack_`, `highest_ack_`, and `t_seqno_`. Lines 12–19 update RTT estimation variables and timeout backoff value. Finally, Line 20 starts a retransmission timer for the transmitting packet. Again, we will discuss the details of RTT estimation and retransmission timer later in Section 10.6.

Program 10.17 Function `newack` of class `TcpAgent`.

```
    //~/ns/tcp/tcp.cc
1   void TcpAgent::newack(Packet* pkt)
2   {
3       double now = Scheduler::instance().clock();
4       hdr_tcp *tcph = hdr_tcp::access(pkt);
5       dupacks_ = 0;
6       last_ack_ = tcph->seqno();
7       prev_highest_ack_ = highest_ack_ ;
8       highest_ack_ = last_ack_;
9       if (t_seqno_ < last_ack_ + 1)
10          t_seqno_ = last_ack_ + 1;
11      hdr_flags *fh = hdr_flags::access(pkt);
12      if (rtt_active_ && tcph->seqno() >= rtt_seq_) {
13          if (!ect_) {
14              t_backoff_ = 1;
15              ecn_backoff_ = 0;
16          }
17          rtt_active_ = 0;
18          rtt_update(now - rtt_ts_);
19      }
20      newtimer(pkt);
21  }
```

Function `dupack_action()`

The main responsibilities of function `dupack_action()` are to: (1) decrease congestion window size, (2) set `t_seqno_` to the sequence number of the lost TCP packet, and (3) restart retransmission timer. Program 10.18 shows the details of function `dupack_action()`. Line 5 registers fast retransmission event (i.e., `FAST_RETX`) for tracing. Line 6 records `CWND_ACTION_DUPACK` as the latest window adjustment action (i.e., `last_cwnd_action_`). Line 7 closes the congestion window by invoking function `slowdown( CLOSE_SSTHRESH_HALF | CLOSE_CWND_ONE)`, feeding how the slow start threshold and congestion window are to be configured as an input argument. Finally, Line 8 invokes function `reset_rtx_timer(0,0)` to set `t_seqno_` to `highest_ack_+1`, and restarts the retransmission timer. The details of functions `reset_rtx_timer(...)` and `slowdown(...)` will be discussed in Sections 10.6 and 10.7, respectively.

Program 10.18 Function dupack_action of class `TcpAgent`.

```
     //~/ns/tcp/tcp.cc
  1  void TcpAgent::dupack_action()
  2  {
  3      if (highest_ack_ > recover_) {
  4      recover_ = maxseq_;
  5      trace_event("FAST_RETX");
  6      last_cwnd_action_ = CWND_ACTION_DUPACK;
  7      slowdown(CLOSE_SSTHRESH_HALF|CLOSE_CWND_ONE);
  8      reset_rtx_timer(0,0);
  9      return;
 10  }
```

TCP Tahoe reacts to a duplicated ACK packet differently. Lines 4–9 in Program 10.18 are executed only when all the packets transmitted during the previous packet loss have been acknowledged. Here, variable `recover_` records the highest TCP sequence number (i.e., `maxseq_`) transmitted during the previous packet loss event. Line 4 sets `recover_` to be `maxseq_` so that it can be used in the next packet loss event. The condition in Line 3, `highest_ack_ > recover_`, implies that the TCP packet with highest sequence number transmitted during the previous loss must be acknowledged. If this condition is not satisfied, the TCP sender will wait for timeout and retransmit the lost packet.

10.6 Timer Related Functions

Another responsibility of a TCP sender is to use a retransmission timer to provide connection reliability. The main components of this part include estimation of smoothed RTT (round trip time) and RTT variation, computation of RTO (retransmission timeout), implementation of BEB (binary exponential backoff), utilization of a retransmission timer, and defining actions to be performed at timeout.

10.6.1 RTT Sample Collection

A TCP sender needs to collect RTT samples in order to estimate smoothed RTT and RTT variation, and to compute retransmission timeout (RTO) value.

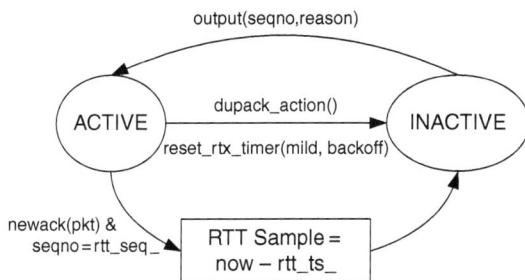

Fig. 10.4. The RTT sampling process.

An RTT sample is measured as the time difference between the point where a packet is transmitted and the point where the associated ACK packet arrives the sender.

In NS2, each TCP sender has only one set of variables including variables `rtt_active_`,`rtt_ts_`, and `rtt_seq_` (see Table 10.1) to track RTT samples. It can collect only one RTT sample at a time – meaning not all the packets are used to collect RTT samples.

Figure 10.4 shows the diagram of the RTT collection process. The process starts in the inactive state where `rtt_active_=0`. The collection is activated (i.e., the process enters the active state) when a TCP sender sends out a new packet using function `output(seqno,reason)`. From Program 10.2, Line 35 sets `rtt_active_` to be 1.[7] Lines 37 and 38 record the TCP sequence number and the current time in the variables `rtt_seq_` and `rtt_ts_`, respectively.

An RTT sample is collected when the associated ACK packet returns (see Lines 12–19 of function `newack(pkt)` in Program 10.17). Given that the collection process is active (i.e., `rtt_active_=1`), Line 12 determines whether the incoming ACK packet belongs to the same collecting sample. It is so if the sequence number in the received ACK packet is the same as that stored in `rtt_seq_` (set at the beginning of the collecting process). Note that the logical relation here is ">=" rather than "==", since some TCP variants may not generate an ACK packet for every received TCP packet. At the end of the collection process, Line 17 sets `rtt_active` to zero indicating that the collecting process has completed (i.e., the process moves back to inactive state), and Line 18 takes an RTT sample by invoking `rtt_update(now-rtt_ts_)` (defined in Program 10.22).

The above RTT collection process operates fairly well under normal situations. However, a packet loss may inflate an RTT sample, and affect the collecting accuracy. In this case, the measured RTT would be the RTT value plus the time used to retransmit the lost packets. To avoid complication, NS2 simply cancels the RTT collection process, when a packet loss

---

[7] If the `rtt_active_` is nonzero, TCP sender will skip the collection process.

occurs. In particular, functions `dupack_action()` (Line 8 in Program 10.18) and `timeout(tno)` (Lines 14 and 16 in Program 10.26) invoke function `reset_rtx_timer(...)` to set `rtt_active_` to zero, essentially cancelling the RTT collecting process.

10.6.2 RTT Estimation

After collecting an RTT sample, a TCP sender feeds a sample "tao" to function `rtt_update(tao)` to estimate smoothed RTT (`t_srtt_`), RTT variation (`t_rttvar_`), and unbounded RTO (`t_rtxcur_`)[8] based on Eqs. (9.1)–(9.3), where $\alpha = 7/8$, $\beta = 3/4$ and $\gamma = 1$. Instead of directly computing these three variables, NS2 manipulates Eqs. (9.1)–(9.3) such that each term in these equations is multiplied with 2^n, where n is an integer. As discussed in Section 12.4.2, multiplication and division by 2^n can be implemented in C++ by shifting the binary value to the left and right, respectively, by n bits. This bit shifting technique is used in function `rtt_update(tao)` to compute `t_srtt_`, `t_rttvar_`, and `t_rtxcur_`.

At time k, let $t(k)$ be the RTT sample, $\bar{t}(k)$ be the smoothed RTT value, $\sigma_t(k)$ be the RTT variation, and Δ refer to $t(k+1) - \bar{t}(k)$. From (9.1)–(9.3),

$$\bar{t}(k+1) = \frac{1}{8}\left(7\bar{t}(k) + t(k+1)\right)$$

$$= \frac{1}{8}\left(7\bar{t}(k) + \bar{t}(k) + t(k+1) - \bar{t}(k)\right)$$

$$= \frac{1}{8}\left(8\bar{t}(k) + \Delta\right) \tag{10.1}$$

$$\sigma_t(k+1) = \frac{1}{4}\left(3\sigma_t(k) + |\Delta|\right)$$

$$= \frac{1}{4}\left(3\sigma_t(k) - 4\sigma_t(k) + 4\sigma_t(k) + |\Delta|\right)$$

$$= \frac{1}{4}\left(-\sigma_t(k) + 4\sigma_t(k) + |\Delta|\right) \tag{10.2}$$

$$RTO_u(k+1) = \gamma \times [t(k+1) + 4\sigma_t(k+1)] \tag{10.3}$$

where $RTO_u(k+1)$ is an unbounded RTO. Equations (10.1)–(10.3) are now arranged in the multiple of $2^n, n = \{0, 2, 3\}$ (i.e., the multiple of 1, 2, and 4). NS2 uses bit shifting operation in place of multiplication to implement Eqs. (10.1)–(10.3).

10.6.3 Overview of State Variables

State variables contain the current status of a TCP agent. Related timer state variables are shown in Tables 10.1 and 10.3. Most the variables are well

---

[8] An actual value of RTO must be bounded by a minimum and a maximum value.

explained by their description. We now discuss a few points related to these variables.

First, C++ timer variables are initialized in function `rtt_init()` (Lines 1–8 in Program 10.19). OTcl timer instvars, on the other hand, are initialized in file ~ns/tcl/lib/ns-default.tcl shown in Lines 9–19 of Program 10.19.

Program 10.19 Default values for the timer-related variables.

```
    //~/ns/tcp/tcp.cc
1   void TcpAgent::rtt_init()
2   {
3       t_rtt_ = 0;
4       t_srtt_ = int(srtt_init_ / tcp_tick_) << T_SRTT_BITS;
5       t_rttvar_ = int(rttvar_init_ / tcp_tick_) << T_RTTVAR_BITS;
6       t_rtxcur_ = rtxcur_init_;
7       t_backoff_ = 1;
8   }

    //~/ns/tcl/lib/ns-default.tcl
9   Agent/TCP set T_SRTT_BITS 3      #in bits
10  Agent/TCP set T_RTTVAR_BITS 2    #in bits
11  Agent/TCP set srtt_init_ 0       #in seconds
12  Agent/TCP set rttvar_init_ 12    #in seconds
13  Agent/TCP set rtxcur_init_ 3.0   #in seconds
14  Agent/TCP set T_SRTT_BITS 3      #in bits
15  Agent/TCP set T_RTTVAR_BITS 2    #in bits
16  Agent/TCP set rttvar_exp_ 2      #in bits
17  Agent/TCP set tcp_tick_ 0.1      #in seconds
18  Agent/TCP set maxrto_ 100000     #in seconds
19  Agent/TCP set minrto_ 0.2        #in seconds
```

Secondly, `tcp_tick` is a simulation time unit (i.e., granularity) in seconds. Hereafter, we will refer to a simulation time unit as a "tick". The default value of `tcp_tick_` is 100 ms. In other words, one "tick" is set by default to 0.01 (see Line 17 in Program 10.23).

Thirdly, `t_back_off_` is used as a binary exponential backoff factor (i.e., γ in Eq. (10.3)). A TCP sender doubles its retransmission timer for every timeout event. In NS2, a TCP sender doubles `t_backoff_` for every timeout event, and computes the unbounded RTO as `t_rtxcur_ * t_backoff_` (see Line 7 in Program 10.23).

Finally, there are two main points related to variables `t_srtt_` and `t_rttvar_`. One is that these variables are stored in "ticks", rather than seconds. However, their initial values are in seconds. Lines 4–5 in Program 10.19 divides the initial values of smoothed RTT and RTT variation by `tcp_tick` to obtain the time in "ticks" (rather than in seconds). Another point is the division operation (by 8 and 4, respectively). To avoid round-off error during a division, these two variables are multiplied by 8 and 4, respectively, at the initialization. Again, Lines 4 and 5 in Program 10.19 shift `t_srtt_` and `t_rttvar_` to the left by `T_SRTT_BITS=3` bits and `T_RTTVAR_BITS = 2` bits, respectively. This bit shifting is equivalent to multiplying 8 and 4 to `t_srtt_` and `t_rttvar_`, respectively.

10.6.4 Retransmission Timer

A TCP sender employs a retransmission timer to provide end-to-end reliability. When transmitting a packet, it starts a retransmission timer. Upon the timer expiration, the timer informs the TCP sender of a packet timeout. Here the TCP sender assumes that the packet is lost and retransmits the lost packet. If an ACK packet is received prior to the timeout, the timer will be stopped (i.e., cancelled). The details of NS2 timer implementation is given in Section 12.1.

Program 10.20 Class `RtxTimer` and related components.

```
     //~/ns/tcp/tcp.h
 1   class RtxTimer : public TimerHandler {
 2   public:
 3       RtxTimer(TcpAgent *a) : TimerHandler() { a_ = a; }
 4   protected:
 5       virtual void expire(Event *e);
 6       TcpAgent *a_;
 7   };

     //~/ns/tcp/tcp.cc
 8   void RtxTimer::expire(Event*)
 9   {
10       a_->timeout(TCP_TIMER_RTX);
11   }

12   void TcpAgent::set_rtx_timer()
13   {
14       rtx_timer_.resched(rtt_timeout());
15   }
```

NS2 models retransmission timers using a C++ class `RtxTimer` shown in Program 10.20. Derived from class `TimerHandler`, class `RtxTimer` has one variable `a_` which is a pointer to a `TcpAgent` object. It derives three main functions – `sched(delay)`, `resched(delay)`, and `cancel()` – and overrides one function `expire(e)` of class `TimerHandler`. Function `sched(delay)` starts the timer and sets the timer to expire at "`delay`" seconds in future. Function cancel() stops the pending the timer. Function resched(delay) restarts the timer and again sets the timer to expire at "delay" seconds in future. Finally, function expire(e) defines a set of actions which are taken at the timer expiration.

Program 10.21 Components of TcpAgent related to TCP retransmission timer.

```
    //~/ns/tcp/tcp.h
1 class TcpAgent : public Agent {
2     ...
3     protected:
4     RtxTimer rtx_timer_;
5     ...
6 }

    //~/ns/tcp/tcp.cc
7 TcpAgent::TcpAgent() : ... rtx_timer_(this), ...
8 {   ...   }
```

NS2 creates a two-way connection between TcpAgent and RtxTimer objects by using the following mechanism. First, class TcpAgent declares an RtxTimer object (rtx_timer_ in Line 4 in Program 10.21) as its member variable. Every TcpAgent object therefore has a direct access to an RtxTimer object. Secondly, on the reverse direction, class RtxTimer declares a pointer "a_" to a TcpAgent object in Line 6 of Program 10.20 as its member variable. Finally, a TcpAgent object is specified as a target of the pointer a_ in the constructor of the RtxTimer object. From Line 7 in Program 10.21, the constructor of class TcpAgent creates a rtx_timer_ by feeding this (i.e., a pointer to itself) as an input argument. From Line 3 in Program 10.20, the constructor of rtx_timer_ stores "this" in its variable "a_", creating a connection from the rtx_timer_ back to the TcpAgent object.

Note that in Line 4 in Program 10.21, a TCP sender has only one retransmission timer. Therefore, the TCP timeout mechanism applies to only one packet at a time. The retransmission timer is started when a new packet is transmitted (by function output(...); see Line 48 in Program 10.12). Here, the timer is not allowed to start if it is in use (i.e., its status is TIMER_PENDING). This is in contrast to the actual TCP implementation where retransmission timers are set for all transmitted packets.

10.6.5 Function Overview

Class TcpAgent provides the following seven key timer-related functions:

- rtt_update(tao): Take an RTT sample "tao" as an input argument, updates smoothed RTT (t_srtt_), RTT variation (t_rttvar_), and unbounded RTO (`t_rtxcur_`) according to Eqs. (10.1), (10.2), and (10.3), respectively.

- `rtt_timeout()`: Computes the bounded RTO value based on `t_rtxcur_`, `minrto_`, and `maxrto_`, as well as TCP binary exponential backoff (BEB) mechanism which make use of the current value of `t_backoff_`.

- `rtt_backoff()`: Double the binary exponential backoff multiplicative factor `t_backoff_`.

- `set_rtx_timer()`: Restart the retransmission timer.

- `reset_rtx_timer(mild,backoff)`: Restart the retransmission counter and cancel the RTT sample collecting process. If `mild` is zero, set `t_seqno_` to `highest_ack_+1`. Also, it invokes function `rtt_backoff()` if `backoff` is nonzero.

- `newtimer(pkt)`: Take an ACK packet "pkt" as an input argument. Start the retransmission timer if TCP connection is active,[9] cancel the timer, otherwise.

- `timeout(tno)`: If the connection is active, close the congestion window, adjust `t_backoff_`, retransmit the lost packet, and restart the retransmission timer. Otherwise, restart the retransmission timer (but does not perform other action).[10]

10.6.6 Function rtt_update(tao)

Function `rtt_update(tao)` updates three main timer variables: smoothed RTT (`t_srtt_`), RTT variable (`t_rttvar_`), and Retransmission TimeOut (RTO; `t_rtxcur_`). Shown in Program 10.22, function `rtt_update(tao)` takes an RTT sample as an input argument. It is invoked from within function `newack(pkt)`, when a new ACK packet is received and a new RTT sample is `now - rtt_ts_` (see Line 18 in Program 10.17).

Function `rtt_update(tao)` aligns the input argument `tao` with `tcp_tick_` and stores the aligned valued in variable `t_rtt_` as the latest RTT sample (Lines 4–6). Before proceeding further, let us define the following variables

---

[9] A TCP connection is said to be active and idle when it has data to transmit and does not have data to transmit, respectively.

[10] As we will see, a retransmission timer does not stop when a TCP connection becomes idle. At the expiration, a TCP sender does nothing but restarts the timer. By keeping the timer running, the timer will be available as soon as the TCP sender becomes active.

Program 10.22 Function `rtt_update` of class `TcpAgent`.

```
    //~/ns/tcp/tcp.cc
1   void TcpAgent::rtt_update(double tao)
2   {
3       double now = Scheduler::instance().clock();
4       double tickoff = fmod(now-tao+boot_time_, tcp_tick_);
5       if ((t_rtt_ = int((tao + tickoff) / tcp_tick_)<1);
6           t_rtt_ = 1;
7       if (t_srtt_ != 0) {
8           register short delta = t_rtt_ - (t_srtt_ >> T_SRTT_BITS);
9           if ((t_srtt_ += delta) <= 0)
10              t_srtt_ = 1;
11          if (delta < 0)
12              delta = -delta;
13          delta -= (t_rttvar_ >> T_RTTVAR_BITS);
14          if ((t_rttvar_ += delta) <= 0)
15              t_rttvar_ = 1;
16      } else {
17          t_srtt_ = t_rtt_ << T_SRTT_BITS;
18          t_rttvar_ = t_rtt_ << (T_RTTVAR_BITS-1);
19      }
20      t_rtxcur_ = (((t_rttvar_ << (rttvar_exp_ + (T_SRTT_BITS -
            T_RTTVAR_BITS)))) + t_srtt_)  >> T_SRTT_BITS ) * tcp_tick_;
21      return;
22  }
```

$$\bar{t} = \frac{\texttt{t\_srtt\_}}{8} = \texttt{t\_srtt\_>>T\_SRTT\_BITS} \qquad (10.4)$$

$$\sigma_t = \frac{\texttt{t\_rttvar\_}}{4} = \texttt{t\_rttvar\_>>T\_RTTVAR\_BITS} \qquad (10.5)$$

$$\Delta = \texttt{t\_rtt\_} - \bar{t} = \texttt{t\_rtt\_} - (\texttt{t\_srtt\_>>T\_SRTT\_BITS}) \qquad (10.6)$$

where `T_SRTT_BITS`, `T_RTTVAR_BITS`, and `rttvar_exp_` are defined in Program 10.19 as 3, 2, and 2, respectively. Again, variables `t_srtt_` and `t_rttvar_` are stored in multiples of 8 and 4 (see Lines 4–5 in Program 10.19). Therefore, their relationship to actual smoothed RTT ($\bar{t}$) and RTT variation (σ_t) is given by Eqs. (10.4) and (10.5), respectively.

Based on the above variables, Lines 8–15 compute the smoothed RTT value. In Eqs. (10.1) and (10.2), we rearrange the variables `t_srtt_` and `t_rttvar_` as follows:

$$\texttt{t\_srtt\_}(k+1) = 8\bar{t}(k+1) = 8\bar{t}(k) + \Delta(k) = \texttt{t\_srtt\_}(k) + \Delta(k) \quad (10.7)$$

$$\texttt{t\_rttvar\_}(k+1) = 4\sigma_t(k+1) = |\Delta| - \sigma_t(k) + 4\sigma_t(k)$$
$$= |\Delta| - [\texttt{t\_rttvar\_>>T\_SRTT\_BITS}](k) + \texttt{t\_rttvar\_}(k). \qquad (10.8)$$

In Program 10.22, Line 8 computes `delta` (i.e., Δ) as indicated in (10.6). Line 9 updates `t_srtt_` according to Eq. (10.7) and Lines 11–12 compute $|\Delta|$. Lines 13–14 update `t_rttvar_` according to Eq. (10.8). From Lines 9–10 and 14–15, both `t_srtt_` and `t_rttvar_` will be set to 1, if their updated values are less than zero. Also, Lines 8–15 are invoked when `t_srtt_` is nonzero only. When `t_srtt_` is zero, `t_srtt_` and `t_rttvar_` are simply set to 8 times (Line 17) and twice of (Line 18) the RTT sample (i.e., `t_rtt_`), respectively.

NS2 computes (using Eq. (10.3)) and stores the unbounded value of RTO in variable `t_rtxcur_` (Line 20). It is computed as $\bar{t} + 4\sigma_t$ shown in Eq. (9.3). The upper-bound and the lower-bound in Eq. (9.3) will be implemented when an unbounded RTO is assigned to the retransmission timer (e.g., in function `rtt_timeout()`). The computation of `t_rtxcur_` in Line 20 consists of 4 steps:

(i) Scale `t_rttvar_`: Variables `t_srtt_` and `t_rttvar_` are stored as multiples of $2^{\text{T\_SRTT\_BITS}} = 8$ and $2^{\text{T\_RTTVAR\_BITS}} = 4$, respectively. Line 20 converts the scale of `t_rtt_var_` into the same scale of `t_srtt_` as follows:

$$\texttt{t\_rttvar\_} \rightarrow \texttt{t\_rttvar\_} \times \frac{8}{4}$$
$$= \texttt{t\_rttvar\_>>T\_RTTVAR\_BITS<<T\_SRTT\_BITS}$$
$$= \texttt{t\_rttvar\_<<(T\_SRTT\_BITS - T\_RTTVAR\_BITS)}$$

(ii) Multiply $2^{\texttt{rttvar\_exp\_}} = 2^2 = 4$ to the value obtained from Step (i). Denote the result from Step (i) as $\texttt{t\_rtt\_var\_}^{(1)}$.

$$\texttt{t\_rttvar\_}^{(1)} \rightarrow 4 \times \texttt{t\_rttvar\_}^{(1)}$$
$$= \texttt{t\_rttvar\_}^{(1)}\texttt{<<rttvar\_exp\_}$$

(iii) Denote the value computed in Step (ii) be $\texttt{t\_rtt\_var\_}^{(2)}$. Add `t_srtt_` to $\texttt{t\_rtt\_var\_}^{(2)}$.

$$\texttt{t\_rttvar\_}^{(2)} \rightarrow \texttt{t\_rttvar\_}^{(2)} + \texttt{t\_srtt\_}$$

(iv) Convert the computed value to seconds: Let $\texttt{t\_rtt\_var\_}^{(3)}$ be the value computed in Step (iii). This value is stored in "ticks" and is in the scale of `t_srtt_` (i.e., multiple of 8). To change the unit of $\texttt{t\_rtt\_var\_}^{(3)}$ to seconds,

$$\texttt{t\_rttvar\_}^{(3)} \rightarrow \texttt{t\_rttvar\_}^{(3)}\texttt{>>T\_SRTT\_BITS*tcp\_tick\_}$$

which is equivalent to Line 20 in Program 10.22.

10.6.7 Function `rtt_timeout()`

Shown in Program 10.23, function `rtt_timeout()` computes the bounded RTO, based on unbounded RTO (`t_rtxcur_`), RTO lower-bound (`minrto_`),

RTO upper-bound (`maxrto_`), and TCP binary exponential backoff (BEB) mechanism. NS2 implements the BEB mechanism by using a multiplicative factor `t_backoff_`. The timeout value which is used by the retransmission timer is a product of `t_rtxcur_` and `t_backoff_` (see Line 7). The lower-bound and the upper-bound are implemented in Lines 3–7 and Lines 8–9, respectively. Note that, while the lower-bound applies to `t_rtxcur_` before applying the BEB mechanism, the upper-bound does so after the BEB. Hence, Lines 11–12 place another lower-bound constraint (i.e., `2.0*tcp_tick_`) for the value after the BEB.

Program 10.23 Functions `rtt_timeout` and `rtt_backoff` of class `TcpAgent`.

```
   //~/ns/tcp/tcp.cc
1  double TcpAgent::rtt_timeout()
2  {
3      double timeout;
4      if (t_rtxcur_ < minrto_)
5          timeout = minrto_ * t_backoff_;
6      else
7          timeout = t_rtxcur_ * t_backoff_;
8      if (timeout > maxrto_)
9          timeout = maxrto_;
10     if (timeout < 2.0 * tcp_tick_)
11         timeout = 2.0 * tcp_tick_;
12     return (timeout);
13 }

14 void TcpAgent::rtt_backoff()
15 {
16     if (t_backoff_ < 64)
17         t_backoff_ <<= 1;
18     if (t_backoff_ > 8) {
19         t_rttvar_ += (t_srtt_ >> T_SRTT_BITS);
20         t_srtt_ = 0;
21     }
22 }
```

10.6.8 Function `rtt_backoff()`

Function `rtt_backoff()` applies TCP binary exponential backoff (BEB) mechanism to a multiplicative factor `t_backoff_`. As discussed in Section 9.1.2, `t_backoff_` is doubled for every timeout and is reset to its initial value when a new ACK packet is received. As we will see, function `rtt_backoff()` is invoked by function `reset_rtx_timer(mild,backoff)` to double `t_backoff_`.

Program 10.23 shows the details of function `rtt_backoff()`. If the current `t_backoff_` is less than 64 (Line 16), it will be doubled (i.e., shifted to the left by one bit in Line 17). Also, a large value of `t_backoff_` (e.g., > 8 in Line 18) implies a long interval between two RTT samples. In this case, smoothed RTT and RTT variation may not well represent the actual network RTT. In this case, RTT should be a function of the most recent RTT sample only. Therefore, Line 20 sets `t_srtt_` to zero. After this point, function `rtt_update(tao)` will invoke Lines 17–18 (rather than Lines 8–15) in Program 10.22 to estimate network RTT.

10.6.9 Function `set_rtx_timer()` and Function `reset_rtx_timer(mild,backoff)`

Programs 10.20 and 10.24 show the details of functions `set_rtx_timer()` and `reset_rtx_timer(mild,backoff)`, respectively. From Line 4 in Program 10.20, function `set_rtx_timer()` simply sets the timer to expire at t seconds in future, where t is the timeout value returned from function `rtt_timeout()` (see also Program 10.23).

Program 10.24 Function `reset_rtx_timer` of class `TcpAgent`.

```
//~/ns/tcp/tcp.cc
1 void TcpAgent::reset_rtx_timer(int mild, int backoff)
2 {
3     if (backoff)
4         rtt_backoff();
5     set_rtx_timer();
6     if (!mild)
7         t_seqno_ = highest_ack_ + 1;
8     rtt_active_ = 0;
9 }
```

From Program 10.24, function `reset_rtx_timer(mild,backoff)` has four main tasks:

(i)   Restart the retransmission timer (Line 5).
(ii) Update the backoff multiplicative factor `t_backoff_`, if the input argument `backoff` is nonzero (Lines 3–4).
(iii) Update the next transmitting sequence number. Store `highest_ack_+1` in `t_seqno_`, if the input argument "mild" is zero (Lines 6–7).
(iv) Cancel the pending RTT sample collection process by setting `rtt_active_` to zero (Line 8).

10.6.10 Function `newtimer(pkt)`

Function `newtimer(pkt)` is invoked from within function `newack(pkt)` when a new ACK packet is received and the TCP sender is about to send out another packet. As shown in Program 10.25, it takes an ACK packet `*pkt` as an input argument. If the TCP sender still has data to transmit (i.e., Line 4 returns `true`), Line 5 will restart the retransmission timer by invoking `set_rtx_timer()`. Otherwise, Line 7 will cancel the timer by invoking `cancel_rtx_timer()`.

Program 10.25 Function `newtimer` of class `TcpAgent`.

```
    //~/ns/tcp/tcp.cc
1 void TcpAgent::newtimer(Packet* pkt)
2 {
3     hdr_tcp *tcph = hdr_tcp::access(pkt);
4     if (t_seqno_ > tcph->seqno() || tcph->seqno() < maxseq_)
5         set_rtx_timer();
6     else
7         cancel_rtx_timer();
8 }
```

10.6.11 Function `timeout(tno)`

Function `timeout(tno)` is invoked when a retransmission timer expires. It adjusts congestion window as well as slow start threshold, and retransmits the lost packet. Again, function `expire(e)` is invoked when the timer expires. From Line 10 in Program 10.20, function `expire(e)` of class `RtxTimer` simply invokes function `timeout(TCP_TIMER_RTX)` of the associated `TcpAgent` object. As shown in Lines 1–19 of Program 10.26, function `timeout(tno)` takes a timer option (`tno`) as an input argument, where the possible values of `tno` are defined in Lines 20–25 of Program 10.26. In this section, we are interested in TCP Tahoe. Therefore, we will discuss the case where only `timeout(TCP_TIMER_RTX)` is invoked.

The basic operation of function `timeout(tno)` is to close the congestion window (in Line 10), restart the retransmission timer (in Lines 14 and 16), and retransmits the lost packet (in Line 18). We will discuss the details of function `slowdown(...)` which closes the congestion window in Section 10.7. The retransmission timer is restarted by using the function `reset_rtx_timer(mild,backoff)` (see Program 10.24). For zero value of "mild" this function sets `t_seqno_` to `highest_ack_+1`. The zero and nonzero values of the second input argument "backoff" inform function `reset_rtx_timer(mild,backoff)` to and not to (respectively) update the binary exponential backoff multiplicative factor (`t_backoff_`). Again, the

Program 10.26 Function `timeout` of class `TcpAgent` and the possible values of its input argument `tno`.

```
    //~/ns/tcp/tcp.cc
1   void TcpAgent::timeout(int tno)
2   {
        ...
3       if (cwnd_ < 1) cwnd_ = 1;
4       if (highest_ack_ == maxseq_ && !slow_start_restart_) {
5       } else {
6           recover_ = maxseq_;
7           if (highest_ack_ < maxseq_) {
8               ++nrexmit_;
9               last_cwnd_action_ = CWND_ACTION_TIMEOUT;
10              slowdown(CLOSE_SSTHRESH_HALF|CLOSE_CWND_RESTART);
11          }
12      }
13      if (highest_ack_ == maxseq_)
14          reset_rtx_timer(0,0);
15      else
16          reset_rtx_timer(0,1);
17      last_cwnd_action_ = CWND_ACTION_TIMEOUT;
18      send_much(0, TCP_REASON_TIMEOUT, maxburst_);
19  }

    //~/ns/tcp/tcp.h
20  #define TCP_TIMER_RTX       0
21  #define TCP_TIMER_DELSND    1
22  #define TCP_TIMER_BURSTSND  2
23  #define TCP_TIMER_DELACK    3
24  #define TCP_TIMER_Q         4
25  #define TCP_TIMER_RESET     5
```

TCP sender assumes that all packets with sequence number lower than `highest_ack_` are successfully transmitted. At a timeout event, it assumes that the first lost packet (i.e., the packet to be retransmitted) is the packet with sequence number `highest_ack_+1`. After preparing `t_seqno_` (i.e., set to `highest_ack_+1`) for retransmission, Line 18 invokes function `send_much(0, TCP_REASON_TIMEOUT, maxburst_)` to transmit the lost packet.

After a TCP sender transmits all the packets provided by an attached application, its variable `t_seqno_` is equal to variable `curseq_`, and variable `maxseq_` stops increasing. After the last packet (with sequence number `maxseq_`) is acknowledged, variable `highest_ack_` is equal to `maxseq_`. At this point, the TCP sender enters an idle state. Its retransmission timer, however, does not stop at this moment. It keeps expiring for every period of RTO. From Line 14 of Program 10.26, function `timeout(tno)` will invoke `reset_rtx_timer(0,0)`, which stores the value of `highest_ack_+1` in variable `t_seqno_` but does not change the multiplicative factor `t_backoff_`. Also, function `send_much(0, TCP_REASON_TIMEOUT, maxburst_)` will not send out any packet since `t_seqno` is not less than `curseq_` (see Program 10.10).

When the application sends more user demand (i.e., data payload) by invoking `sendmsg(nbytes)`, variable `curseq_` is incremented and the TCP connection becomes active. In this case, function `send_much(0,0,maxburst_)` will send out packets, starting with the packet with sequence number `t_seqno_= max_seq_+1 = highest_ack_ + 1`.

There are two important details in function `timeout(tno)`. One is that regardless of whether connection is busy or idle, Line 17 sets the variable `last_cwnd_action_` which records the latest action imposed on the congestion window to be `CWND_ACTION_TIMEOUT`. Another is related to variable `recover_`. Recall that `recover_` contains the highest sequence number among all the transmitted TCP packets at the latest loss event (i.e., either timeout or Fast Retransmit). Line 6 hence records the highest TCP sequence number transmitted so far in the variable `recover_`.

10.7 Window Adjustment Functions

From Section 9.1.2, a TCP sender dynamically adjusts congestion window to fully utilize the network resource. When the network is under utilized, a TCP sender increases transport-level transmission rate by opening the congestion window. In the slow start phase, where the congestion window (`cwnd_`) is less than the slow start threshold (`ssthresh_`), a TCP sender increases the congestion window by one for every received ACK packet. If `cwnd_` is not less than `ssthresh_`, on the other hand, a TCP sender will be in the congestion avoidance phase, and the congestion window is increased by `1/cwnd_` for every received ACK packet.

When the network is congested, a TCP sender closes the congestion window to help relieve network congestion. As discussed in Section 9.1.2, TCP may decrease the window by half or may reset the congestion window size to one, depending on the situation.

Class `TcpAgent` provides two main functions, which can be used to adjust the congestion window:

- `opencwnd()`: Increases the size of the congestion window. The increasing method depends on `cwnd_` and `ssthresh_`.
- `slowdown(how)`: Decreases the size of the congestion window by the method specified in "how".

The possible values of "how" are defined in Program 10.27. All possible values of `how` contain 32 bits, and conform to the following format: 1 of "one" bit and 31 of "zero" bits. The difference among the values defined in Program 10.27 lies in the position of the "one" bit. This format acts as a simple identification of the input method "how" through an "AND" operator. For example, suppose the input argument how is set to CLOSE_CWND_ONE (=2). Let x be a variable which can be any value in Program 10.27. Then, how & x would be nonzero if and only if x=CLOSE_CWND_ONE. This assignment is also able to contain several slowdown methods in one variable using an "OR" operator. For example, let how be CLOSE_CWND_ONE|CLOSE_SSTHRESH_HALF. Then, how & x would be nonzero if and only if x=CLOSE_CWND_ONE or x=CLOSE_SSTHRESH_HALF.

Program 10.27 Possible values of how – the input argument of function slowdown.

```
   //~/ns/tcp/tcp.h
 1 #define CLOSE_SSTHRESH_HALF      0x00000001
 2 #define CLOSE_CWND_HALF          0x00000002
 3 #define CLOSE_CWND_RESTART       0x00000004
 4 #define CLOSE_CWND_INIT          0x00000008
 5 #define CLOSE_CWND_ONE           0x00000010
 6 #define CLOSE_SSTHRESH_HALVE     0x00000020
 7 #define CLOSE_CWND_HALVE         0x00000040
 8 #define THREE_QUARTER_SSTHRESH   0x00000080
 9 #define CLOSE_CWND_HALF_WAY      0x00000100
10 #define CWND_HALF_WITH_MIN       0x00000200
11 #define TCP_IDLE                 0x00000400
12 #define NO_OUTSTANDING_DATA      0x00000800
```

10.7.1 Function opencwnd()

Function opencwnd() is invoked when a new ACK packet is received (see function recv_newack_helper() in Line 5 of Program 10.15). It opens the congestion window, and allows the TCP sender to transmit more packets without waiting for acknowledgement. Program 10.28 shows the details of function opencwnd(). From Line 3, if cwnd_ is less than ssthresh_, the TCP sender will be in the slow start phase and cwnd_ will be increased by 1. Otherwise, the TCP sender must be in a congestion avoidance phase, and cwnd_ will be increased by 1/cwnd_ (Lines 6–7), where increase_num_ is usually set to 1. In both cases, Lines 9–10 bound cwnd_ within maxcwnd_, the predefined maximum congestion window size.

10.7.2 Function slowdown(how)

Function slowdown(how) closes the congestion window based on the method specified in the input argument how. It is invoked from within function dupack_action() and timeout(tno) to decrease transport layer transmission

Program 10.28 Function `opencwnd` of class `TcpAgent`.

```
    //~/ns/tcp/tcp.cc
 1  void TcpAgent::opencwnd()
 2  {
 3      if (cwnd_ < ssthresh_) {
 4          cwnd_ += 1;
 5      } else {
 6          double increment = increase_num_ / cwnd_;
 7          cwnd_ += increment;
 8      }
 9      if (maxcwnd_ && (int(cwnd_) > maxcwnd_))
10          cwnd_ = maxcwnd_;
11  }
```

rate. Function `dupack_action()` invokes function `slowdown(how)` feeding `how` = `CLOSE_SSTHRESH_HALF | CLOSE_CWND_ONE` (Line 7 in Program 10.18) as an input argument. From Program 10.29, this invocation halves the current slow start threshold (Lines 9–13) and resets the congestion window to 1 (Line 26). Function `timeout(tno)` on the other hand invokes function `slowdown(how)` with an input argument `how` = `CLOSE_SSTHRESH_HALF | CLOSE_CWND_RESTART` as an input argument (Line 10 in Program 10.26). From Program 10.29, this invocation halves the current slow start threshold (Lines 9–13) and resets the congestion window to a predifined window-restart value (Line 24). In both cases, NS2 employs an "OR" operator to combine `how` to adjust slow start threshold and `how` to adjust congestion window, and feed it as an input argument to function `slowdown(how)`.

The details of function `slowdown(how)` are shown in Program 10.29. In this function, Lines 4–6 first set a variable `slowstart` to zero and one when TCP is in the slow start phase (i.e., `cwnd_< ssthresh_`) and in the congestion avoidance phase (i.e., `cwnd_>= ssthresh_`), respectively. Line 7 stores half of the window size in a variable `halfwin` and the window size in a variable `win`. Variable `decrease_num_` in Line 8 is set to 0.5 by default. Therefore, the local variable `decreasewin` is half of the current congestion window. The variable `decrease_num_` provides an option for window decrement, where different TCP variants may set the value of `decrease_num_` differently (e.g., 0.3, 0.7). Lines 9–26 show different window closing method, which will be invoked according to the input argument "how". Line 27 ensures that the minimum slow start threshold is 2. Line 29 sets the variable `cong_action_` to be `true` if the window adjustment method, `how`, is either of `CLOSE_CWND_HALF`, `CLOSE_CWND_RESTART`, `CLOSE_CWND_INIT`, or `CLOSE_CWND_ONE`. Again, the variable `cong_action_` is used in function `output(seqno,reason)` to set the congestion flag of the transmitted packet. Finally, Line 32 sets `first_decrease_` to zero, indicating TCP has decreased the congestion window at least once.

Program 10.29 Function `slowdown` of class `TcpAgent`.

```
//~/ns/tcp/tcp.cc
1   void TcpAgent::slowdown(int how)
2   {
3       double win, halfwin, decreasewin;
4       int slowstart = 0;
5       if (cwnd_ < ssthresh_)
6           slowstart = 1;
7       halfwin = windowd() / 2; win = windowd();
8       decreasewin = decrease_num_ * windowd();

9       if (how & CLOSE_SSTHRESH_HALF)
10          if (first_decrease_ == 1||slowstart ||
                  last_cwnd_action_ == CWND_ACTION_TIMEOUT)
11              ssthresh_ = (int) halfwin;
12          else
13              ssthresh_ = (int) decreasewin;
14      else if (how & THREE_QUARTER_SSTHRESH)
15          if (ssthresh_ < 3*cwnd_/4) ssthresh_ = (int)(3*cwnd_/4);

16      if (how & CLOSE_CWND_HALF)
17          if (first_decrease_==1||slowstart||decrease_num_==0.5){
18              cwnd_ = halfwin;
19          } else
20              cwnd_ = decreasewin;
21      else if (how & CWND_HALF_WITH_MIN) {
22          cwnd_ = decreasewin;
23          if (cwnd_ < 1) cwnd_ = 1;
24      } else if (how & CLOSE_CWND_RESTART) cwnd_=int(wnd_restart_);
25      else if (how & CLOSE_CWND_INIT) cwnd_ = int(wnd_init_);
26      else if (how & CLOSE_CWND_ONE) cwnd_ = 1;

27      if (ssthresh_ < 2) ssthresh_ = 2;
28      if (how & (CLOSE_CWND_HALF|CLOSE_CWND_RESTART|
                          CLOSE_CWND_INIT|CLOSE_CWND_ONE))
29          cong_action_ = TRUE;
30      if (first_decrease_ == 1) first_decrease_ = 0;
31  }
```

Lines 9–15 adjust the slow start threshold (`ssthresh_`) based on the value of "how":

- CLOSE_SSTHRESH_HALF (Lines 11 and 13): Sets the slow start threshold `ssthresh_` to the half of the current congestion window size `cwnd_`.
- THREE_QUARTER_SSTHRESH (Line 15): Sets the slow start threshold `ssthresh_` to at least 3/4 of its current value.

Similarly, Lines 16–29 adjust the congestion window (`cwnd_`) based on the value of "`how`":

- `CLOSE_CWND_HALF` (Lines 17–20): Decreases the current congestion window size (i.e., `cwnd_`) by half (i.e., either `halfwin` or `decreasewin`).
- `CWND_HALF_WITH_MIN` (Lines 22–23): Sets the current congestion window size to `decreasewin` but not less than 1.
- `CLOSE_CWND_RESTART` (Line 24): Sets the current congestion window size to the predifined window-restart value `wnd_restart_`.
- `CLOSE_CWND_INIT` (Line 25): Sets the current congestion window size to `wnd_init_` (i.e., initial value of congestion window size).
- `CLOSE_CWND_ONE` (Line 26): Sets the current congestion window size to 1.

10.8 Chapter Summary

TCP is a reliable connection-oriented transport layer protocol. It provides a connection with end-to-end error control and congestion control. NS2 implements TCP senders and TCP receivers using C++ classes `TcpAgent` and `TcpSink`, which are bound to OTcl classes `Agent/TCP` and `Agent/TCPSink`, respectively. A TCP sender has four main responsibilities. First, based on user demand, it creates and forwards packets to a TCP receiver. Secondly, it provides an end-to-end connection with reliability by means of packet retransmission. Thirdly, it implements timer-related components to estimate round trip time (RTT) and retransmission timeout (RTO), used to determine whether a packet is lost. Finally, it dynamically adjusts transport-level transmission rate to fully utilize the network resource without causing network congestion. ATCP receiver is responsible for creating (cumulative) ACK packets and forwards them back to the TCP sender.

11

Application: User Demand Indicator

Operating on top of a transport layer agent, an application models user demand for data transmission. A user is assumed to create bursts of data payload or application packets. These payload bursts are transformed into transport layer packets which are then forwarded to a transport layer receiving agent. Applications can be classified into *traffic generator* and *simulated application*. A traffic generator creates user demand based on a predefined schedule. A simulated application, on the other hand, creates the demand as if the application is running.

In the followings, we first discuss the relationship between an application and a transport layer agent in Section 11.2. Sections 11.3 and 11.4 discuss the detailed implementation of traffic generators and simulated applications, respectively. Finally, the chapter summary is given in Section 11.5.

11.1 Relationship Between an Application and a Transport Layer Agent

From time to time, an application needs to exchange user demand information with a transport layer agent. An application declares a pointer `agent_` to an attached agent. Similarly, an agent defines a pointer `app_` to an attached application. The user demand information is exchanged between an application and an agent through these two pointers. Section 9.2.2 gives a four-step agent configuration method, which binds an application and a transport layer agent together. The details of these four steps are given below:

Step 1: Create a Sending Agent, a Receiving Agent, and an Application

An agent and an application can be created by using instproc `new{..}` as follows:

T. Issariyakul, E. Hossain, *Introduction to Network Simulator NS2*,
DOI: 10.1007/978-0-387-71760-9_11, © Springer Science+Business Media, LLC 2009

```
set agent [new Agent/<agent_type>]
set app [new Application/<app_type>]
```

where <agent_type> and <app_type> denote the type of an agent (e.g., TCP or UDP) and an application (e.g. Traffic/CBR or FTP), respectively.

Step 2: Connect an Agent to an Application

A two way connection between an application and an agent can be created by using of class Application whose syntax is shown below:

```
$app attach-agent $agent
```

where $app and $agent are Application and Agent objects. The details of instproc attach-app{s_type} are shown in Program 11.1. Line 7 stores an input Agent object in the variable agent_. Line 12 invokes function attachApp(this) of class Agent, while Lines 19–22 create a connection from the Agent object to the Application object. From Line 21, function

---

Program 11.1 An OTcl command attach-agent of class Application and function attachApp of class Agent.

```
    //~/ns/apps/app.cc
1   int Application::command(int argc, const char*const* argv)
2   {
3       Tcl& tcl = Tcl::instance();
4       ...
5       if (argc == 3) {
6           if (strcmp(argv[1], "attach-agent") == 0) {
7               agent_ = (Agent*) TclObject::lookup(argv[2]);
8               if (agent_ == 0) {
9                   tcl.resultf("no such agent %s", argv[2]);
10                  return(TCL_ERROR);
11              }
12              agent_->attachApp(this);
13              return(TCL_OK);
14          }
15          ...
16      }
17      return (Process::command(argc, argv));
18  }

    //~/ns/common/agent.cc
19  void Agent::attachApp(Application *app)
20  {
21      app_ = app;
22  }
```

`attachApp(app)` stores an input `Application` object `app` in the variable `app_` of the `Agent` object. Since Line 12 feeds the pointer `this` to function `attachApp(...)` of the `Application` object, which simply sets the pointer `agent_->app_` to point to itself.

Step 3: Attaching an Agent to a Low-Level Network

Here, we consider the case where an agent is connected to a node in a low-level network. As discussed in Section 6.6.3, an agent can be attached to a node by using instproc `attach-agent{node agent}` of class `Simulator`, where `node` and `agent` are the `Node`, and `Agent` objects, respectively. This instproc creates a two-way connection between a `Node` object `node` and an `Agent` object `agent`. It sets variable `agent::target_` to point to `node` and installs `agent` in the demultiplexer (i.e., instvar `dmux_`) of `node`.

The process of attaching an agent to a node involves three OTcl classes: `Simulator`, `Node`, and `RtModule`. Figure 11.1 shows the main operation when "`$ns attach-agent $node $agent`" is invoked:

(i) Instproc `attach-agent{node agent}` of class `Simulator` invokes `$node attach $agent`.
(ii) Instproc `attach{agent}` of class `Node` allocates a port for an input agent `$agent`, configures instvar `agent_addr_` and `agent_port_` of the input

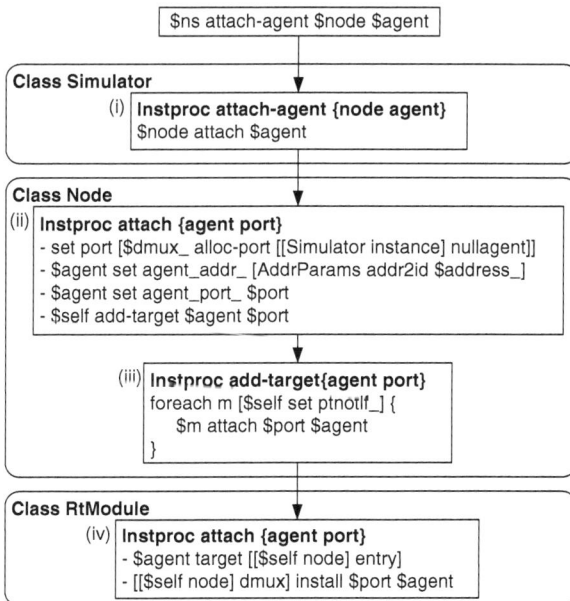

Fig. 11.1. Internal mechanism of instproc `attach-agent{node agent}` of class `Simulator`.

agent $agent, and invokes instproc add-target{agent port} to inform every routing module stored in the instvar ptnotif_.

(iii) Instproc add-target{agent port} of class Node invokes instproc attach{ agent port} of each routing module (of class RtModule) stored in the instvar ptnotif_.

(iv) Instproc attach{agent port} of class RtModule creates a connection between a node and an agent. Here, $agent sets its $target_ to point to the entry of $node, while $node installs $agent in the slot "port" of its demultiplexer "dmux_". This connection is created for both sending and receiving agents.

Step 4: Associating a Sending Agent with a Receiving Agent

To associate a sending agent with a receiving agent, NS2 employs instproc connect of class Simulator, whose syntax is shown below:

    $ns connect $s_agent $r_agent

where $ns, $s_agent, and $r_agent are Simulator, sending Agent, and receiving Agent objects, respectively.

Program 11.2 shows the details of instproc connect{src dst}. Lines 3 and 4 invoke instproc simplex-connect{src dst}, which set up a connection from src to dst_[1], and simplex-connect{dst src} which creates a connection from dst back to src.

Program 11.2 Instprocs connect and simplex-connect of class Simulator.

```
//~/ns/tcl/lib/ns-lib.tcl
1  Simulator instproc connect {src dst} {
2      ...
3      $self simplex-connect $src $dst
4      $self simplex-connect $dst $src
5      ...
6  }

7  Simulator instproc simplex-connect { src dst } {
8      ...
9      $src set dst_addr_ [$dst set agent_addr_]
10     $src set dst_port_ [$dst set agent_port_]
11     ...
12 }
```

---

[1] From Table 9.3, instvars dst_addr_ and dst_port_ are bound to the C++ variables dst_::addr_ and dst_::port_, respectively, in the C++ domain.

Instvars `dst_addr_` and `dst_port_` are configured in Lines 9–10. When an agent creates a packet, it stores values in variables `dst_.addr_` and `dst_.port_` in the packet header. During a packet forwarding process, a low-level network delivers packets to the agent corresponding to whose address and port are specified in the packet header.

11.2 Details of Class Application

An application is defined in a C++ class `Application` as shown in Program 11.3. Class `Application` has only one key variable `agent_` which is a pointer to class `Agent`. Other two variables, `enableRecv_` and `enableResume_`, are flag variables, which indicate whether an `Application` object should react to functions `recv(nbytes)` and `resume()`, respectively. These two flag variables are set to zero by default.

Program 11.3 Declaration of class `Application`.

```
      //~/ns/apps/app.h
 1    class Application : public Process {
 2    public:
 3        Application();
 4        virtual void send(int nbytes);
 5        virtual void recv(int nbytes);
 6        virtual void resume();
 7    protected:
 8        virtual int command(int argc, const char*const* argv);
 9        virtual void start();
10        virtual void stop();
11        Agent *agent_;
12        int enableRecv_;
13        int enableResume_;
14    };
```

11.2.1 Functions of Classes `Application` and `Agent`

After their connection is created, an application and an agent may invoke `public` functions of each other through the pointers `agent_` and `app_`, respectively. The key `public` functions of class `Application` include functions `send(nbytes)`, `recv(nbytes)`, and `resume()`, while those of class `Agent` are functions `send(nbytes)`, `sendmsg(nbytes)`, `close()`, `listen()`, and `set_pkttype(pkttype)`.

Apart from these `public` functions, class `Application` also provides `protected` functions `start()` and `stop()` to start and stop an `Application`

object, respectively. Finally, there are five key OTcl commands for class
Application which can be invoked from the OTcl domain: start{}, stop{},
agent{}, send{nbytes}, and attach-agent{agent}.

11.2.2 Public Functions of Class Application

Program 11.4 shows the details of the three following public functions of
class Application:

- send(nbytes): Inform the attached transport layer agent that a user needs
  to send nbytes of data payload. Line 3 sends the demand to the attached
  agent by executing "agent_->sendmsg(nbytes)".

- recv(nbytes): Receive "nbytes" bytes from a receiving transport layer
  agent. A UDP agent specifies nbytes as the number of bytes in a received
  packet. In case of UDP, nbytes is equal to packet size; on the other hand,
  TCP specifies "nbytes" as the number of in-sequence received bytes. Due
  to possibility of out-of-order packet delivery, nbytes can be greater than
  the size of one packet.

- resume(): Invoked by a sending agent, this function indicates that the
  agent has sent out all data corresponding to the user demand. For a TCP

Program 11.4 Implementation of functions send, recv, and resume of class
Application.

```
    //~/ns/apps/app.cc
1   void Application::send(int nbytes)
2   {
3       agent_->sendmsg(nbytes);
4   }

5   void Application::recv(int nbytes)
6   {
7       if (! enableRecv_)
8           return;
9       Tcl& tcl = Tcl::instance();
10      tcl.evalf("%s recv %d", name_, nbytes);
11  }

12  void Application::resume()
13  {
14      if (! enableResume_)
15          return;
16      Tcl& tcl = Tcl::instance();
17      tcl.evalf("%s resume", name_);
18  }
```

sender, this function is invoked when it sends out all the packets regardless of whether the transmitted packets have been acknowledged.

Note that both functions `recv(nbytes)` and `resume()` will do nothing if `enableRecv_ = 0` and `enableResume_ = 0`, respectively. Otherwise, Line 10 and 17 in Program 11.5 will invoke OTcl commands or instprocs `recv{nbytes}` (Line 10) and `resume{}` (Line 18) in the OTcl domain, respectively. By default, both `enableRecv_` and `enableResume_` are set to zero, and functions `recv(nbytes)` and `resume()` simply do nothing.

A user may specify actions to be done upon invocation of functions `recv(nbytes)` and `resume()` by

(i) Setting `enableRecv_` and/or `enableResume_` to one.
(ii) Specifying the actions in
    (a) Functions `recv(nbytes)` and/or `resume()`,
    (b) Instprocs `recv{nbytes}` and/or `resume{}` in the OTcl domain, or
    (c) OTcl commands `recv{nbytes}` and/or `resume{}` in function `command()`.

It is important to perform both the steps above. Failing to perform the second step will result in a run-time error, since commands or instprocs `recv{nbytes}` and `resume{}` are undefined in class `Application`

Exercise 11.1. Modify (1) C++ functions, (2) OTcl commands, and (3) OTcl instprocs. Force an application to print out a message when its functions `recv(nbyte)` and `resume()` are invoked. Show simulation results to verify the modification.

11.2.3 Related Public Functions of Class `Agent`

Class `Application` may invoke the following functions of class `Agent` through variable `agent_`:

- `send(nbytes)`: Send "nbytes" of application payload (i.e., user demand) to a receiving agent. If `nbytes=-1`, the user demand would be infinite.
- `sendmsg(nbytes,flags)`: Similar to function `send(nbytes)`, but also feed `flags` as an input variable.
- `close()`: Ask an agent to close the connection (applicable only to TCP)
- `listen()`: Ask an agent to listen to (i.e., wait for) a new connection (applicable only to Full TCP)
- `set_pkttype(pkttype)`: Set the variable `type_` of the attach agent to be `pkttype`.

11.2.4 OTcl Commands of Class Application

Defined in function command, OTcl commands associated with class Application are as follows:

- start{}: Invoke function start() to start the application.
- stop{}: Invoke function stop() to stop the application.
- agent{}: Return the name of the attached agent.
- send{nbytes}: Send nbytes bytes of user payload to the attached agent by invoking function send(nbytes).
- attach-agent{agent}: Create a two-way connection between itself and the input Agent object (agent).

The details of the above OTcl command can be found in file ~ns/apps/app.cc.

11.3 Traffic Generators

A traffic generator models user behavior which follows a predefined schedule. In particular, it sends a demand to transmit one burst of user payload to an attached agent at a time specified in the schedule, regardless of the state of the agent. In NS2, there are four main traffic generators:

- Constant Bit Rate (CBR): Send a fixed size payload to the attached agent. By default, the interval between two payloads (i.e., the sending rate) is fixed, but it can be optionally randomized.
- Exponential On/Off: Send fixed size payloads for every randomized interval to an attached agent during an ON period. Stop sending during an OFF period. ON and OFF periods are exponentially distributed, and are alternated when one period terminates.
- Pareto On/Off: Similar to the Exponential On/Off traffic generator. However, the durations of ON and OFF periods follow a Pareto distribution.
- Traffic Trace: Generate traffic according to a given trace file, which contains a series of inter-burst transmission intervals and payload burst sizes.

11.3.1 An Overview of Class TrafficGenerator

NS2 implements traffic generators using class TrafficGenerator. Program 11.5 shows the declaration of the abstract class TrafficGenerator, where function next_interval(size) (Line 4) is pure virtual. Class Traffic Generator consists of the following variables:

timer_ A TrafficTimer object, which determines when a new burst of payload is created.

nextPkttime_ Simulation time that the next payload will be passed to the attached transport layer agent

Program 11.5 Declaration of class `TrafficGenerator`.

```
   //~/ns/tools/trafgen.h
 1 class TrafficGenerator : public Application {
 2 public:
 3     TrafficGenerator();
 4     virtual double next_interval(int &) = 0;
 5     virtual void init() {}
 6     virtual double interval() { return 0; }
 7     virtual int on() { return 0; }
 8     virtual void timeout();
 9     virtual void recv() {}
10     virtual void resume() {}
11 protected:
12     virtual void start();
13     virtual void stop();
14     double nextPkttime_;
15     int size_;
16     int running_;
17     TrafficTimer timer_;
18 };
```

    `size_`   Application payload size

`running_`   true if the `TrafficGenerator` object is running

Class `TrafficGenerator` derives and overrides the following four key functions of class `Application`. It derives functions `recv(nbytes)` and `resume()` (i.e., share the implementation) from class `Application`, and overrides functions `start()`, and `stop()` of class `Application`. Functions `start()` and `stop()` inform the `TrafficGenerator` object to start and stop, respectively, generating user payload. In Program 11.6, function `start()` initializes the `TrafficGenerator` object by invoking function `init()`[2] in Line 3, and sets `running_` to 1 in Line 4. It computes and stores the time until the next payload burst is generated in variable `nextPkttime_` in Line 5. Finally, it sets the `timer_` to expire at `nextPkttime_` seconds in future (Line 6). From Lines 8 to 13 in Program 11.6, function `stop()` does the opposite of function `start()`. It cancels the pending timer (if any) in Line 11, and sets `running_` to 0 in Line 12.

Class `TrafficGenerator` also defines the following three new functions:

`next_interval(size)`   Takes payload size "`size`" as an input argument, and returns the delay time after which a new payload is generated (Line 4). This function is pure virtual and must be implemented by the instantiable derived classes of class `TrafficGenerator`.

---

[2] In Line 5 of Program 11.5, function `init()` simply does nothing.

Program 11.6 Functions start, stop, and timeout of class
TrafficGenerator.

```
      //~/ns/tools/trafgen.cc
1   void TrafficGenerator::start()
2   {
3       init();
4       running_ = 1;
5       nextPkttime_ = next_interval(size_);
6       timer_.resched(nextPkttime_);
7   }

8   void TrafficGenerator::stop()
9   {
10      if (running_)
11          timer_.cancel();
12      running_ = 0;
13  }

14  void TrafficGenerator::timeout()
15  {
16      if (! running_)
17          return;
18      send(size_);
19      nextPkttime_ = next_interval(size_);
20      if (nextPkttime_ > 0)
21          timer_.resched(nextPkttime_);
22      else
23          running_ = 0;
24  }
```

init() Initializes the traffic generator.
timeout() Sends a user payload to the attached application and restart
        timer_. This function is invoked when timer_ expires.

The details of function timeout() are shown in Lines 14–24 of Program 11.6. Function timeout() does nothing if the TrafficGenerator object is not running (Lines 16–17). Otherwise, it will send "size_" bytes of user payload to the attached agent using function send(nbytes) (defined in Program 11.4). Then, Line 19 will compute nextPkttime_. If nextPkttime_ > 0, Line 21 will inform timer_ to expire after a period of nextPkttime_. Otherwise, Line 23 will stop the TrafficGenerator by setting running_ to zero.

11.3.2 Main Mechanism of a Traffic Generator

Figure 11.2 illustrates the main mechanism of a traffic generator, which relies heavily on the variable timer_ whose class is TrafficTimer derived from

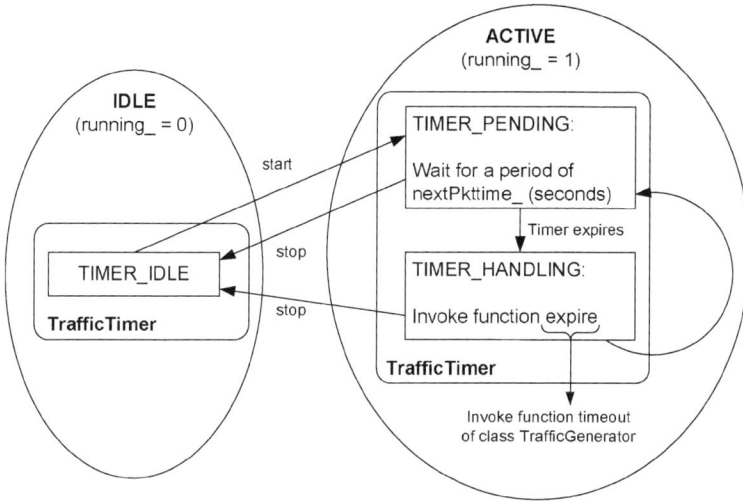

Fig. 11.2. Main mechanism of a traffic generator.

class `TimerHandler`. As discussed in Section 12.1, class `TimeHandler` consists of three states: `TIMER_IDLE`, `TIMER_PENDING`, and `TIMER_HANDLING`. Each of these states corresponds to one of two `TrafficGenerator` states: Idle (i.e., `running_=0`) and Active (i.e., `running_=1`). While state `TIMER_IDLE` corresponds to the idle state of a `TrafficGenerator` object, the other two timer states are within the active state of a `TrafficGenerator` object.

Starting in an idle state, a traffic generator moves to active state when function `start()` is invoked. Here the `timer_` state is set to `TIMER_PENDING`. At the expiration, `timer_` moves to state `TIMER_HANDLING`, and invokes function `timeout()` of class `TrafficGenerator`. After executing function `timeout()`, it reschedule itself, changes the state to `TIMER_PENDING`, reschedules itself, and repeats the above process. When `timer_` state is `TIMER_PENDING` or `TIMER_HANDLING`, the traffic generator can be stopped by invoking function `stop()`.

Program 11.7 shows the declaration of class `TrafficTimer`, which derives from class `TimerHandlor` (see Section 12.1). Class `TrafficTimer` has a key variable `tgen_`, a pointer to a `TrafficGenerator` object (Line 6). At the expiration, NS2 invokes function `expire(e)` of `timer_` (Lines 8–11), which in turn invokes function `timeout()` of the associated `TrafficGenerator` object (i.e., `*tgen_`).

A two-way connection between `TrafficGenerator` and `TrafficTimer` objects is created as follows. Class `TrafficGenerator` declares `timer_` as its pointer to a `TrafficTimer` object (Line 17 in Program 11.5). A `TrafficGenerator` object instantiates `timer_` by feeding a pointer to itself (i.e., `this`)

Program 11.7 Declaration of class `TrafficTimer`, function `expire` of class `TrafficTimer`, and the constructor of class `TrafficGenerator`.

```
   //~/ns/tools/trafgen.h
1  class TrafficTimer : public TimerHandler {
2  public:
3      TrafficTimer(TrafficGenerator* tg) : tgen_(tg) {}
4  protected:
5      void expire(Event*);
6      TrafficGenerator* tgen_;
7  };

   //~/ns/tools/trafgen.cc
8  void TrafficTimer::expire(Event *)
9  {
10     tgen_->timeout();
11 }

12 TrafficGenerator::TrafficGenerator() :
           nextPkttime_(-1), running_(0), timer_(this) {}
```

as an input argument (Line 12 in Program 11.7). The construction of variable `timer_` in turn assigns the input pointer (i.e., `this`) to its pointer to a `TrafficGenerator` object, `tgen_` (Line 3 in Program 11.7), creating a connection back to the `TrafficTimer` object.

11.3.3 Built-in Traffic Generators in NS2

Constant Bit Rate (CBR)

A CBR traffic generator creates a fixed size payload burst for every fixed interval. As shown in Program 11.8, NS2 implements CBR traffic generators by using a C++ class `CBR_Traffic` which is bound to an OTcl class `Application/Traffic/CBR`, whose key instvars with their default values are shown in Table 11.1.

Note that, by default the inter-burst transmission interval, which is the interval between the beginning of two successive payload bursts, can be computed by dividing the payload burst size by the sending rate. By default, the inter-burst transmission interval is $210 \times 8/488.000 \approx 3.44$ ms. The detailed mechanism for class `CBR_Traffic` will be discussed in Section 11.3.4.

Program 11.8 Class `CBRTrafficClass` which binds C++ class `CBR_Traffic` and OTcl class `Application/Traffic/CBR` together.

```
//~/ns/tools/cbr_traffic.cc
1 static class CBRTrafficClass : public TclClass {
2  public:
3     CBRTrafficClass() : TclClass("Application/Traffic/CBR") {}
4     TclObject* create(int, const char*const*) {
5         return (new CBR_Traffic());
6     }
7 } class_cbr_traffic;
```

Table 11.1. Instvars of a CBR traffic generator.

Instvar	Default value	Description
packetSize_	210	Application payload size in bytes
rate_	488×10^3	Sending rate in bps
random_	0 (false)	If **true**, introduce a random time (either positive or negative) to the inter-burst transmission interval.
maxpkts_	16^7	Maximum number of application payload packet that CBR can send

Exponential On/Off

An exponential on/off traffic generator acts as a CBR traffic generator during an ON interval and does not generate any payload during an OFF interval. ON and OFF periods are both exponentially distributed. As shown in Program 11.9, NS2 implements Exponential On/Off traffic generators by using a C++ class `EXPOO_Traffic` which is bound to an OTcl class

Program 11.9 Class `EXPTrafficClass` which binds C++ class `EXPOO_Traffic` and OTcl class `Application/Traffic/Exponential` together.

```
//~/ns/tools/expoo.cc
1 static class EXPTrafficClass : public TclClass {
2  public:
3     EXPTrafficClass() : TclClass("Application/
                                   Traffic/Exponential") {}
4     TclObject* create(int, const char*const*) {
5         return (new EXPOO_Traffic());
6     }
7 } class_expoo_traffic;
```

`Application/Traffic/Exponential`, whose key instvars with their default values are shown in Table 11.2.

Table 11.2. Instvars of an exponential on/off traffic generator.

Instvar	Default value	Description
`packetSize_`	210	Application payload size in bytes
`rate_`	64×10^3	Sending rate in bps during an ON period
`burst_time_`	0.5	Average ON period in seconds
`idle_time_`	0.5	Average OFF period in seconds

Pareto On/Off

A Pareto On/Off traffic generator does the same as an Exponential On/Off generator but the ON and OFF periods conform to a Pareto distribution. As shown in Program 11.10, NS2 implements Pareto On/Off traffic generators by using a C++ class `POO_Traffic` which is bound to an OTcl class `Application/Traffic/Pareto`, whose key instvars with their default values are shown in Table 11.3.

Program   11.10   Class   `POOTrafficClass`   which   binds   C++   class `POO_Traffic` and OTcl class `Application/Traffic/Pareto` together.

```
    //~/ns/tools/pareto.cc
  1 static class POOTrafficClass : public TclClass {
  2  public:
  3      POOTrafficClass() : TclClass("Application/Traffic/Pareto") {}
  4      TclObject* create(int, const char*const*) {
  5          return (new POO_Traffic());
  6      }
  7 } class_poo_traffic;
```

Table 11.3. Instvars of a Pareto/Off traffic generator.

Instvar	Default value	Description
`packetSize_`	210	Application payload in bytes
`rate_`	64×10^3	Sending rate in bps during an ON period
`burst_time_`	0.5	Average ON period in seconds
`idle_time_`	0.5	Average OFF period in seconds
`shape_`	1.5	A "Shape" parameter of a Pareto distribution

Traffic Trace

A traffic trace generates payload bursts according to a given trace file. As shown in Program 11.11, NS2 implements traffic trace by using the C++ class TrafficTrace which is bound to an OTcl class Application/Traffic/Trace. Unlike other traffic generators described before, we need to specify a traffic trace file in the OTcl domain using command attach-tracefile of class Application/Traffic/Trace (see Example 11.2).

Program 11.11 Class TrafficTraceClass which binds C++ class TrafficTrace and OTcl class Application/Traffic/Trace together.

```
//~/ns/trace/traffictrace.cc
1 static class TrafficTraceClass : public TclClass {
2 public:
3     TrafficTraceClass() : TclClass("Application/Traffic/Trace") {}
4     TclObject* create(int, const char*const*) {
5             return(new TrafficTrace());
6     }
7 } class_traffictrace;
```

Example 11.2. A CBR traffic generator in Example 9.1 can be replaced with a traffic trace traffic generator by substituting Lines 10–12 in Program 9.2 with the following lines:

```
set tfile [new Tracefile]
$tfile filename example-trace
set tt [new Applicaiton/Traffic/Trace]
$tt attach-tracefile $tfile
$tt attach-agent $udp
```

A traffic trace file is a pure binary file. A codeword in the binary file consists of two 32-bits fields. The first field indicates inter-burst transmission interval in microseconds, while the second is the payload size in bytes (see file ~ns/tcl/ex/example-trace as an example traffic trace file).

11.3.4 Class CBR_Traffic: An Example Traffic Generator

This section presents a C++ implementation of class CBR_Traffic whose declaration is shown in Program 11.12). Class CBR_Traffic derives from class TrafficGenerator, and has the following main variables:

> rate_ CBR sending rate in bps
> interval_ Packet inter-arrival time in seconds

random_ If **true**, the inter-arrival time will be random
seqno_ CBR sequence number
maxpkts_ Upper bound on the sequence number

Based on the main mechanism discussed in Section 11.3.2, NS2 activates
a traffic generator by invoking function **start()**. When activated, a traffic generator invokes its function **timeout()**, which generates an application
payload, periodically. An interval between two consecutive **timeout()** invocations is determined by the function **next_interval(size)**. The **timeout()**
invocations occur repeatedly until the traffic generator is deactivated (by an
invocation of function **close()**).

As shown in Program 11.12, function **start()** invokes function **init()**
(Line 17) to initialize the traffic generator, sets **running_** to 1 (Line 18), and

---

Program 11.12 Functions **start** and init of class CBR_Traffic.

```
    //~/ns/tools/cbr_traffic.cc
1   class CBR_Traffic : public TrafficGenerator {
2   public:
3       CBR_Traffic();
4       virtual double next_interval(int&);
5       inline double interval() { return (interval_); }
6   protected:
7       virtual void start();
8       void init();
9       double rate_;
10      double interval_;
11      double random_;
12      int seqno_;
13      int maxpkts_;
14  };

15  void CBR_Traffic::start()
16  {
17      init();
18      running_ = 1;
19      timeout();
20  }

21  void CBR_Traffic::init()
22  {
23      interval_ = (double)(size_ << 3)/(double)rate_;
24      if (agent_)
25          if (agent_->get_pkttype() != PT_TCP &&
                    agent_->get_pkttype() != PT_TFRC)
26              agent_->set_pkttype(PT_CBR);
27  }
```

---

invokes function `timeout()` (Line 19). The details of function `init()` are shown in Lines 21–28 of Program 11.12. Line 23 computes the inter-burst transmission interval as transmission rate (`rate_`) divided by payload burst size "`size_<<3`" (in bits).[3] Function `init()` would also set the packet type of the attached agent to be `PT_CBR`, unless the packet type has already been set to `PT_TCP` or `PT_TFRC` (Lines 25–26).

From Program 11.6, function `timeout()`, sends out "`size_`" bytes of application payload (Line 8), recomputes `nextPkttime_` as a value returned from function `next_interval(size_)` (Line 19), and schedules the timer `timer_` to expire at `nextPkttime_` seconds in future (Line 21). Again, function `next_interval(size_)` is pure virtual and must be implemented by instantiable child classes of class `TrafficGenerator`. Class `CBR_Traffic` implements this function (Program 11.13), by returning the packet inter-arrival time converted from payload size "`size_`" and CBR transmission rate "`rate_`" (Lines 3 and 9). Optionally, Line 6 may add or subtract a random value to the computed interval if `random_` is set to `true`. Also, if the application payload are greater than `maxpkts_`, Line 11 will return -1 rather than the computed interval.

Program 11.13 Function `next_interval` of class `CBR_Traffic`.

```
   //~/ns/tools/cbr_traffic.cc
1  double CBR_Traffic::next_interval(int& size)
2  {
3      interval_ = (double)(size_ << 3)/(double)rate_;
4      double t = interval_;
5      if (random_)
6          t += interval_ * Random::uniform(-0.5, 0.5);
7      size = size_;
8      if (++seqno_ < maxpkts_)
9          return(t);
10     else
11         return(-1);
12 }
```

11.4 Simulated Applications

Unlike traffic generators, a simulated application does not have a predefined schedule for payload generation. Rather, it acts as if an actual application is running. NS2 provides two built-in simulated applications: FTP and Telnet.

---

[3] Since the units of the variables `size_` and `rate_` are "bytes" and "bits per second", respectively, Line 9 multiplies `size_` with 8 by shifting `size_` to the left by 3 bits (see Section 12.4.2).

11.4.1 FTP (File Transfer Protocol)

File Transfer Protocol (FTP) is a protocol which divides a given file into small pieces and transfers them to a destination host. Unlike a general FTP in practice, an NS2 FTP module does not need an input file. It simply informs an attached sending transport layer agent of a file size in bytes. Upon receiving user demand (e.g., file size), the agent creates packets which can accommodate the file and forwards them to a connected receiving transport layer agent through a low-level network. Also, an NS2 FTP module is not responsible for specifying a destination host. Destination host identification is responsible by a transport layer agent which identifies the destination by using (through the instproc connect{src dst}; Section 11.1).

Due to its simplicity, an FTP module is implemented in the OTcl domain only. Defined in class Application/FTP which derives class Application, its main OTcl commands and instprocs include

attach-agent{agent}	Register the input "agent" as an attached agent.
start{}	Inform the attached agent of a demand to transmit a file with infinite size by executing "send -1".
stop{}	Stop the pending file transfer session.
send{nbytes}	Send the file with size "nbytes" bytes by invoking function sendmsg(nbytes) of the attached agent.
produce{nbytes}	Inform the attached agent to transmit until its sequence number has reached the minimum of "nbytes" and "maxseq_".
producemore {nbytes}	Inform the attached agent to transmit "nbytes" more packets.

11.4.2 Telnet

Telnet is an interactive client-server text-based application. A Telnet client logs on to a server, and sends text messages to the server. The server in turn executes the received message and returns the result to the client. Clearly, Telnet is not implemented based on a predefined schedule, since its data traffic is created in response to user demand. However, NS2 models a Telnet application in the same way as it does for traffic generators: sending a fixed size packet for every randomized interval.

NS2 defines a Telnet application in C++ class TelnetApp and OTcl class Application/Telnet, which derives from class Application. It uses the value stored in variable size_ of the attached agent as the size of each Telnet packet, and computes the inter-burst transmission interval as follows:

- Case I: If interval_ is nonzero, the inter-burst transmission interval is chosen from an exponential distribution with mean interval_.

- Case II: If `interval_` is zero, the inter-burst transmission interval is chosen from an empirically generated distribution `tcplib` defined in file ˜*ns*/tcp/tcplib-telnet.cc.

Telnet has only one configurable variable `interval_`. In common with other `Application` objects, it can be started and stopped by using command `start{}` and `stop{}`, respectively.

11.5 Chapter Summary

Sitting on top of a transport layer agent, an application informs the attached agent of user demand. Applications can be classified into traffic generators and simulated applications. A traffic generator creates user demand based on a predefined schedule, while a simulated application does so as if the application is running.

Built-in traffic generators in NS2 include CBR (constant bit rate), exponential on/off, Pareto on/off, and Traffic Trace. A CBR traffic generator creates fixed size payloads for every fixed interval. Exponential on/off and Pareto on/off traffic generators create fixed size payloads during an ON period and create no payload during an OFF period. The ON and OFF durations are chosen from an exponential distribution and a Pareto distribution, respectively. Finally, payload size and inter-burst transmission interval for a traffic trace traffic generator are obtained from an input trace file.

NS2 has two built in simulation application: FTP (File Transfer Protocol) and Telnet. FTP informs the attached agent of the file size (in bytes) to be transferred. The attached agent is responsible for creating packets which can accommodate a file, and choosing the destination of the FTP session. In practice, Telnet is a client-server application, whose traffic depends on the interaction between client and server. However, NS2 implements a Telnet as a traffic generator. In particular, it creates a fixed size payload for every random interval, whose distribution is either exponential or `tcplib` defined in ˜*ns*/tcp/tcplib-telnet.cc.

Class `Application` is the base class for all the above applications. It provides few key OTcl commands and instprocs to configure `Application` objects. An instproc `attach-agent{agent}` registers the input "agent" as an attached agent. Instprocs `start{}` and `stop{}` inform an application to start and stop generating data payload. Derived classes of class `Application` reuse these functionalities, and defines their own functionalities for their own purposes.

12

Related Helper Classes

Helper classes generally not a part of a network, but are used in NS2 as an internal mechanism and/or are network components in a special case. This chapter discusses the details of three main NS2 helper classes. In Section 12.1, we first discuss class `Timer`, which is widely used by other NS2 modules such as TCP to implement conditional time-based actions. In Section 12.2, we demonstrate a random number generation process in NS2. In Section 12.3, we explain the details of class `ErrorModel`, which can be used to simulate packet error. Section 12.4 discusses bit masking and bit shifting operations used in various places in NS2. Finally, the chapter summary is given in Section 12.5.

12.1 Timers

Timer is a component which can be used to delay actions. Unless cancelled or restarted, a timer takes actions after it has been started for a given period of time (i.e., at the expiration). For example, a sender starts a retransmission timer as soon as it transmits a packet. Unless cancelled by a reception of an acknowledgement packet, the timer assumes packet loss and asks the sender to retransmit the lost packet at the timer expiration.

12.1.1 Implementation Concept of Timer in NS2

As shown in Fig. 12.1, a timer consists of four following states: IDLE, SET WAITING TIME, WAITING, and EXPIRED. A transition from one state to another occurs immediately when the operation in the current state is complete (i.e., by default), or when the timer receives a start message, a restart message, or a cancel message.

When a timer is created, it sets the state to be IDLE. Upon receiving a start message, the timer moves to state SET WAITING TIME, where it sets its waiting time to be `delay` seconds and moves to state WAITING. The timer

T. Issariyakul, E. Hossain, *Introduction to Network Simulator NS2*,
DOI: 10.1007/978-0-387-71760-9_12, © Springer Science+Business Media, LLC 2009

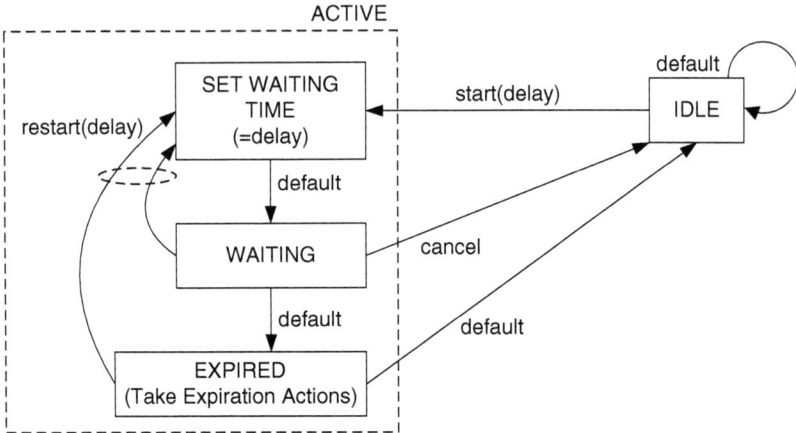

Fig. 12.1. Timer life cycle.

stays in state WAITING for delay seconds, and moves to state EXPIRED. At this point, the timer takes predefined expiration actions and moves back to state IDLE. Hereafter, we will say that the timer *expires* as soon as it enters state EXPIRED. Also, we shall refer to the actions taken in state EXPIRED as *expiration actions*.

The above timer life cycle occurs by default when message start is received. When a cancel messages is received, the timer will stop waiting and move back to state IDLE. If a restart message is received, the timer will restart the waiting process in state SET WAITING TIME.

Implementation of timer in NS2 is a very good example of the *inheritance* concept in OOP. Each timer needs to implement the three following mechanisms: (1) waiting mechanism, (2) interface functions to start, restart, and cancel the waiting process, and (3) expiration actions. The first two mechanisms are common to all timers; however, the last mechanism (i.e., expiration actions) is what differentiates one timer from another. From an OOP point of view, the timer base class must define the waiting mechanism and message receiving interfaces, and leave the implementation of the expiration actions to the derived classes.

In NS2, timers are implemented in both C++ and OTcl. However, both C++ and OTcl timer classes are standalone (i.e., not bound together by TclClass). Relevant functions and variables in both domains are shown in Table 12.1. In both domains, NS2 implements the waiting process by utilizing the Scheduler. Upon entering the state SET WAITING TIME, NS2 places a timer expiration event on the simulation timeline. When the Scheduler fires the expiration event, the timer enters state EXPIRED and executes the expiration actions.

Table 12.1. Timer implementation in C++ and OTcl domains.

Components of a timer	C++ components	OTcl components
State IDLE	status_=TIMER_IDLE	id_ unset
State SET WAITING TIME	status_=TIMER_PENDING	id_ set
State WAITING	status_=TIMER_PENDING	id_ set
State EXPIRATION	status_=TIMER_HANDLING	id_ set
Message start	Function sched	Instprocs sched and resched
Message restart	Function resched	Instprocs sched and resched
Message cancel	Function cancel	Instprocs cancel and destroy
Action at the expiration	Function expire	Instproc timeout

12.1.2 OTcl Implementation

In the OTcl domain, NS2 implements timers using an OTcl class Timer. The implementation of class timer consists of three parts. First, the waiting mechanism is implemented by placing a timer expiration event on the simulation timeline using instproc Simulator::at{...} (See Lines 9 and 15 in Program 12.1). Secondly, the interface of class Timer is defined in instprocs sched{delay}, resched{delay}, cancel{}, and destroy{}. Finally, the expiration actions are specified in instproc timeout{}, which is implemented in child classes of class Timer (see class ConnTimer in file ~ns/tcl/webcache/webtraf.tcl, for example).

Program 12.1 shows the details of various instprocs of OTcl class Timer. Class Timer has two key instvars: ns_ in Line 6 and id_ in Line 7. Instvar ns_ is a reference to the Simulator. It is configured at the construction of a Timer object (see Lines 2–4). The constructor of class Timer takes the Simulator as its input argument and stores the input instance in its instvar ns_. Instvar id_ (Line 7) indicates the state of the timer. If the timer is idle, id_ will not exist (since it is unset). If the timer is active, id_ will contain the unique ID of the timer expiration event on the simulation timeline.

Instprocs sched{delay} (Lines 5–10) and resched{delay} (Lines 11–13) are NS2 implementation for receiving a start message and a restart message, respectively. They take one input argument delay, and set the timer to expire after delay seconds. Regardless of the timer state, instproc sched{delay} cancels the timer by using instproc cancel{} in Line 8. In Line 9, it tells the timer to expire at delay seconds in future by invoking instproc after{ival args} of class Simulator. Shown in Lines 14–16, instproc after{...} employs command at of class Simulator to place an OTcl command in future.[1] From Line 9, instproc sched{delay} schedules an invocation of instproc

---

[1] As discussed in Section 4.2, the OTcl command "at{...}" places an AtEvent object on the simulation timeline, and returns the unique ID of the scheduled event to the caller.

Program 12.1 Timer related OTcl instprocs.

```
   //~/ns/tcl/mcast/timer.tcl
1  Class Timer
2  Timer instproc init { ns } {
3      $self set ns_ $ns
4  }
5  Timer instproc sched delay {
6      $self instvar ns_
7      $self instvar id_
8      $self cancel
9      set id_ [$ns_ after $delay "$self timeout"]
10 }
11 Timer instproc resched delay {
12     $self sched $delay
13 }


   //~/ns/tcl/lib/ns-lib.tcl
14 Simulator instproc after {ival args} {
15         eval $self at [expr [$self now] + $ival] $args
16 }


   //~/ns/tcl/mcast/timer.tcl
17 Timer instproc cancel {} {
18     $self instvar ns_
19     $self instvar id_
20     if [info exists id_] {
21         $ns_ cancel $id_
22         unset id_
23     }
24 }
25 Timer instproc destroy {} {
26     $self cancel
27 }
```

"timeout{}" at delay seconds in future and stores the unique ID corresponding to the timer expiration event in instvar id_.

Lines 17–27 in Program 12.1 show the details of instprocs cancel{} and destroy{} of class Timer. Both the instprocs act as an interface to receive a cancel message. Note that, id_ exists only when a timer expiration event is on the simulation timeline. Timer is cancelled only when id_ exists (i.e., the condition in Line 20 is true). In this case, Line 21 feeds id_ to instproc cancel{id_} (see Program 12.2) of the Simulator instance to remove the timer expiration event from the timeline. Finally, Line 22 unsets instvar id_ to indicate that the expiration event is no longer on the simulation timeline.

Program 12.2 shows the details of instproc cancel{...} of class Simulator and OTcl command cancel{uid} of class Scheduler. Instproc cancel{...}

Program 12.2 Instproc `cancel` of class `Simulator` and an OTcl command `cancel` of class `Scheduler`.

```
    //~/ns/tcl/lib/ns-lib.tcl
1   Simulator instproc cancel args {
2       $self instvar scheduler_
3       return [eval $scheduler_ cancel $args]
4   }


    //~/ns/common/scheduler.cc
5   int Scheduler::command(int argc, const char*const* argv)
6   {
7       ...
8       if (strcmp(argv[1], "cancel") == 0) {
9           Event* p = lookup(STRTOUID(argv[2]));
10          if (p != 0) {
11              cancel(p);
12              AtEvent* ae = (AtEvent*)p;
13              delete ae;
14          }
15      }
16      ...
17  }
```

takes one input argument `uid`, which is the unique ID of an event to be cancelled. Line 3 invokes command `cancel{uid}` of the Scheduler (stored in an instvar `scheduler_` of the Simulator), which removes the timer expiration event whose unique ID is "uid" (see Lines 9–13).

12.1.3 C++ Based Class Implementation

This section explains the C++ implementation of a timer. We first show the life cycle of a C++ timer based on C++ functions (in Table 12.1). Secondly, we briefly discuss the declaration of C++ abstract class `TimerHandler`, which represents timers in the C++ domain. Thirdly, we describe the details of three main components of a timer: (1) internal waiting mechanism, (2) interface functions, and (3) expiration actions. Fourthly, we demonstrate how a timer is cross-referenced with another object. Finally, we conclude this section by providing guidelines for implementing timers in NS2.

Timer Life Cycle

Based on Fig. 12.1 and Table 12.1, we redraw the life cycle of a `TimerHandler` object (i.e., a C++ timer object) in Fig. 12.2. The default state of a timer is `TIMER_IDLE`. Upon invoking functions `sched(delay)` or `resched (delay)`, the timer moves from state `TIMER_IDLE` to state `TIMER_PENDING`, where the

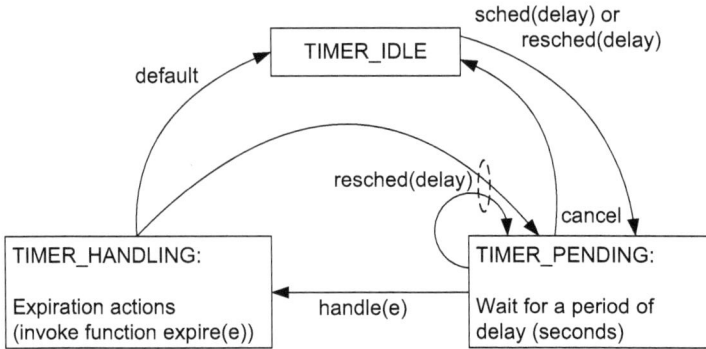

Fig. 12.2. Life cycle of a TimerHandler object.

timer starts a waiting period of delay seconds. When the timer expires, it moves to state TIMER_HANDLING and takes expiration actions by invoking function expire(e). After taking expiration actions, the timer moves to state TIMER_IDLE, and the cycle starts over again. Regardless of the state, function resched(delay) cancels the pending timer, and restarts the timer. In state TIMER_PENDING, we may cancel the timer by invoking function cancel(), which stops the active timer and changes the state of the timer to TIMER_IDLE.

Brief Overview of Class TimerHandler

Program 12.3 shows the declaration of abstract class TimerHandler, a C++ class which represents a timer. Line 7 defines three states of a TimerHandler object as members of TimerStatus enum data type: TIMER_IDLE, TIMER_PENDING, and TIMER_HANDLING. Class TimerHandler contains only two member variables: status_ in Line 12 and event_ in Line 13. Variable status_ stores the current timer state (or status). It takes a value in $\{0, 1, 2\}$, which corresponds to the values of members of the TimerStatus enum type shown in Line 7. The default state of a timer is TIMER_IDLE. Therefore, variable status_ is set to TIMER_IDLE at the timer construction (see Line 3). Another variable event_ (of class Event) represents a timer expiration event. It acts as a glue between a TimerHandler object and the Scheduler. The details of variable event_ will be discussed in the next section.

The key functions of class TimerHandler along with their descriptions are given below.

sched(delay) (Public) Start the timer and set the timer to expire at delay seconds in future.

_sched(delay) (Private) Place a timer expiration event on the simulation time line at delay seconds in future.

resched(delay) (Public) Restart the timer and set the timer to expire at delay seconds in future.

Program 12.3 Declaration of class `TimerHandler`.

```
//~/ns/common/timer-handler.h
1  class TimerHandler : public Handler {
2  public:
3     TimerHandler() : status_(TIMER_IDLE) { }
4     void sched(double delay);   // cannot be pending
5     void resched(double delay); // may or may not be pending
6     void cancel();              // must be pending
7     enum TimerStatus { TIMER_IDLE, TIMER_PENDING, TIMER_HANDLING};
8     int status() { return status_; };
9  protected:
10    virtual void expire(Event *) = 0;
11    virtual void handle(Event *);
12    int status_;
13    Event event_;
14 private:
15    inline void _sched(double delay) {
16       (void)Scheduler::instance().schedule(this, &event_, delay);
17    }
18    inline void _cancel() {
19       (void)Scheduler::instance().cancel(&event_);
20    }
21 };
```

cancel() (Public) Cancel the pending timer.

_cancel() (Private) Remove a timer expiration event from the simulation time line.

status() Return variable `status_`, the current state of the timer.

handle(e) Invokes function `expire(e)`. It is used by the Scheduler to dispatch a timer expiration event (see Chapter 4).

expire(e) Take expiration actions. It is a pure virtual function, and must be implemented by child instantiable classes of class `TimerHandler`.

Internal Waiting Mechanism

Class `TimerHandler` implements waiting mechanism through functions `_sched`(delay) and `_cancel(delay)`. Basically, these two functions place and remove `event_` on the simulation timeline. In Line 16 of Program 12.3, function `_sched(delay)` executes "schedule(this,code&event_,delay)", where "this" is the timer address, "event_" is an expiration dummy event (see Section 4.3.6), and "delay" is the duration until the timer expires. Function schedule(...) stores the address of timer "this" in variable `handler_` of the Event pointer `event_`, essentially setting `event_->handler_` to point to the `TimerHandler` object. Then, it places the object `event_` on the simulation

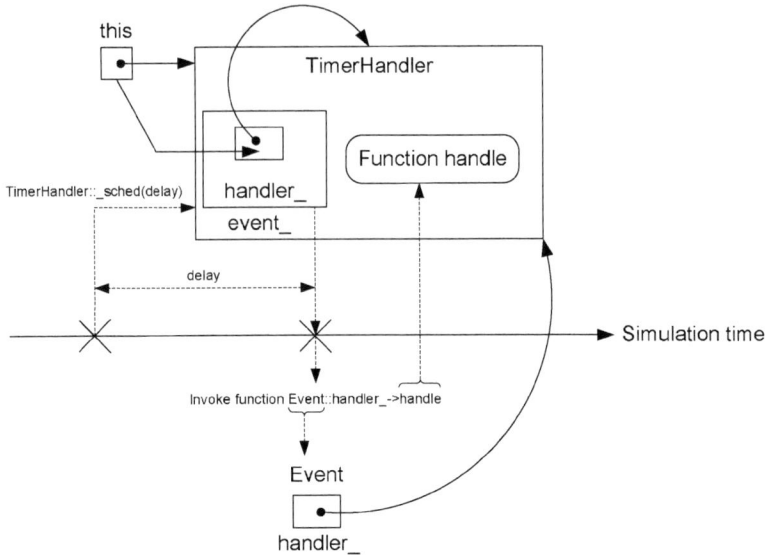

Fig. 12.3. A diagram which represents the timer waiting process (i.e., function _sched(delay)).

timeline at delay seconds in future. At the firing time, the Scheduler invokes function dispatch(e), which in turn executes event_->handler_->handle (...). Since variable handle_ of the dispatched event_ points to the Timer-Handler object (see Fig. 12.3), NS2 invokes function handle(e) associated with the TimerHandler object at the firing time. Function handle(e) of class TimerHandler in turn invokes function expire(e) (Line 6 of Program 12.4) which takes expiration actions specified by the derived classes of class TimerHandler.

Function _cancel() does the opposite of what function _sched(delay) does. It removes the timer expiration event from the simulation timeline. From Line 19 in Program 12.3, it invokes function cancel(&event_) of class Scheduler to remove the event "event" from the simulation timeline.

Expiration Actions

At the firing time, the Scheduler dispatches a timer expiration event by invoking function handle(e) of the associated timer (see also Fig. 12.3). The details of function handle(e) are shown in Program 12.4. Line 3 first checks whether the current status_ is TIMER_PENDING. If so, Line 5 will change the variable status_ to TIMER_HANDLING, and Line 6 will invoke function expire(e) to take expiration actions. After returning from the function expire(e), variable status_ is set by default to TIME_IDLE (Line 8). However, if status_ has already changed (e.g., when the timer is rescheduled;

status_$\neq$TIMER_HANDLING in Line 7), function handle(e) will not change variable status_.

Program 12.4 Function handle of class TimerHandler.

```
   //~/ns/common/timer-handler.cc
1  void TimerHandler::handle(Event *e)
2  {
3      if (status_ != TIMER_PENDING)
4          abort();
5      status_ = TIMER_HANDLING;
6      expire(e);
7      if (status_ == TIMER_HANDLING)
8          status_ = TIMER_IDLE;
9  }
```

In Line 10 of Program 12.3, function expire(e) is pure virtual. Therefore, derived instantiable classes of class TimerHandler are responsible for providing expiration actions by overriding this function. For example, class MyTimer below derives from class TimerHandler, and overrides function expire(e):

```
void MyTimer::expire(Event *e)
{
    printf("MyTimer has just expired!!\n");
}
```

which prints the statement "MyTimer has just expired!!" on the screen upon timer expiration.

Interface Functions to Start, Restart, and Cancel a Timer

Program 12.5 Function sched of class TimerHandler.

```
   //~/ns/common/timer-handler.cc
1  void TimerHandler::sched(double delay)
2  {
3      if (status_ != TIMER_IDLE) {
4          fprintf(stderr,"Couldn't schedule timer");
5          abort();
6      }
7      _sched(delay);
8      status_ = TIMER_PENDING;
9  }
```

The details of function `sched(delay)` of class `TimerHandler` is shown in Program 12.5. Function `sched(delay)` takes one input argument `delay`, and sets the timer to expire at `delay` seconds in the future by feeding `delay` into function `_sched(delay)` (Line 7). Note that function `sched(delay)` must be invoked when the `status_` of the timer is `TIMER_IDLE`. Otherwise, Lines 4-5 will show an error message and exit the program.

Program 12.6 shows the details of functions `resched(delay)` and `cancel()` of class `TimerHandler`. Function `resched(delay)` is very similar to function `sched(delay)`. In fact, when invoked with `status_` $\neq$ `TIMER_PENDING`, it does the same as function `sched(delay)` does (i.e., starts the timer). However, when `status_=TIMER_PENDING` (Line 3)–meaning `event_` was placed on the simulation timeline prior to the invocation–function `resched(delay)` removes the timer expiration event from the simulation time line, by invoking function `_cancel()`, and (re)starts the timer (Lines 4 and 5, respectively).

Program 12.6 Functions `resched` and `cancel` of class `TimerHandler`.

```
      //~/ns/common/timer-handler.cc
1   void TimerHandler::resched(double delay)
2   {
3       if (status_ == TIMER_PENDING)
4           _cancel();
5       _sched(delay);
6       status_ = TIMER_PENDING;
7   }

8   void TimerHandler::cancel()
9   {
10      if (status_ != TIMER_PENDING) {
11          ...
12          abort();
13      }
14      _cancel();
15      status_ = TIMER_IDLE;
16  }
```

Lines 8–16 of Program 12.6 show the details of function `cancel()` of class `TimerHandler`. Function `cancel()` invokes function `_cancel()` in Line 14 to remove the pending timer expiration event from the simulation timeline. Function `cancel()` must not be invoked, when `event_` is not on the simulation timeline (i.e., `status_` is either `TIMER_IDLE` or `TIMER_HANDLING`). Otherwise, NS2 will show an error message on the screen and exit the program (Lines 11-12).

Cross Referencing a Timer with Another Object

In most cases, the usefulness of a timer stands out when it is cross-referenced with another object. In this case, the object employs a timer as a waiting tool, which starts, restarts, and cancels the waiting process as necessary. The timer, on the other hand, informs the object of timer expiration, upon which the object may take expiration actions.

A typical cross-reference between a timer and an object can be created as follows:

(i) Declare a timer as a variable of an object class.
(ii) Declare a pointer to the object as a member of the timer class.
(iii) Define a non-default constructor for the timer class. Store the input argument of the constructor in its member pointer variable (which points to the associated object).
(iv) Instantiate a timer object from within the constructor of the associated object. Use the non-default constructor of the timer class defined above. Feed the pointer **this** (i.e., the pointer to the object) as an input argument to the constructor of the timer.

We now conclude this section with a simple timer example.

Example 12.1. Consider a process of counting the number of customers who enter a store during a day. Let class **Store** represent a convenience store (i.e., an object class), and let class **StoreHour** represent the number of opening hours of a day (i.e., a timer class). The opening hours is specified when the store is opened. The objective here is to count the number of visiting customers during a day, and print out the result when the store is closed.

Classes **Store** *and* **StoreHour**

From Program 12.7, class **Store** also has 3 variables. First, **hours_** (Line 17) contains opening hours of the store and is set to zero at the construction. Secondly, **count_** (Line 18) records the number of customers who have entered the store so far and is set to zero at the construction. Finally, variable **timer_** is a **StoreHour** object. Function **close()** (Lines 12–13) of class **Store** is invoked when the store is being closed. It prints out the opening hours and number of visiting customers for today on the screen. Declared in Line 1–8, class **StoreHour** has only one variable **store_** (Line 7) which is a pointer to a **Store** object.

Cross-Referencing **Store** *and* **StoreHour** *Objects*

The process of cross-referencing a **Store** object and a **StoreHour** object is shown in Fig. 12.4. The constructor of class **Store** constructs its variable **timer_** with the pointer **this** to the **Store** object (see Line 11). The constructor of class **StoreHour** stores the input pointer in variable **store_**. Since

Program 12.7 Declaration of classes `Store` and `StoreHour`.

```
   //store.h
1  class Store;
2  class StoreHour : public TimerHandler {
3  public:
4      StoreHour(Store *s) { store_ = s; };
5      virtual void expire( Event *e );
6  protected:
7      Store *store_;
8  };

9  class Store : public TclObject {
10 public:
11     Store() : timer_(this) { hours_ = -1; count_ = 0; };
12     void close(){
13         printf("The number of customers during
                   %2.2f hours is %d\n", hours_,count_);
14     };
15     int command(int argc, const char*const* argv);
16 protected:
17     double hours_;
18     int count_;
19     StoreHour timer_;
20 }
```

the input argument is the pointer to the `Store` object, the constructor of the `StoreHour` object essentially sets the variable `store_` to point back to the `Store` object.

Due to the cross-referencing, the compiler needs to recognize one of these two classes when declaring another. If Line 1 was removed, the compiler would not recognize class `Store` when compiling Line 7, and would show a compilation error message on the screen. After compiling Line 2, the compiler recognizes class `StoreHour` and can compile Line 19 without error.

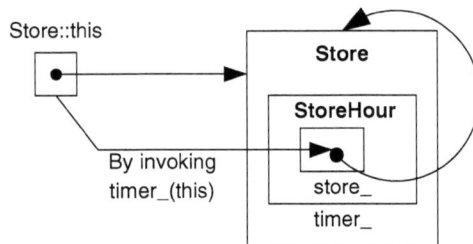

Fig. 12.4. A diagram which represents the process of cross-referencing a `Store` object and a `StoreHour` object.

Program 12.8 Function `expire` of class `StoreHour` as well as OTcl Commands `open` and `new-customer` of class `Store`.

```
    //store.cc
1   void StoreHour::expire(Event*) {
2       store_->close();
3   };

4   int Store::command(int argc, const char*const* argv)
5   {
6       if (argc == 3) {
7           if (strcmp(argv[1], "open") == 0) {
8               hours_ = atoi(argv[2]);
9               count_ = 0;
10              timer_.sched(hours_);
11              return (TCL_OK);
12          }
13      } else  if (argc == 2) {
14          if (strcmp(argv[1], "new-customer") == 0) {
15              count_++;
16              return (TCL_OK);
17          }
18      }
19      return TclObject::command(argc,argv);
20  }
```

It is also important to note that when compiling Lines 2–8, the compiler recognizes only `Store` class name. Any attempt to invoke functions (e.g., `close()`) of class `Store` will result in a compilation error. This is the reason why we need to separate C++ codes into header and C++ files. Again, since a header file is included at the top of a C++ file, the compiler first goes through the header file and recognizes all the variables and functions specified in the header file. With this knowledge, the compiler can compile the C++ file without error.

Defining Expiration Actions

Derived from class `TimerHandler`, class `StoreHour` overrides function `expire(e)` as shown in Lines 1–3 of Program 12.8. At the expiration, the timer (i.e., `StoreHour` object) simply invokes function `close()` of the associated `Store` object.

Creating OTcl Interface

We bind the C++ class `Store` to an OTcl class with the same name using a mapping class `StoreClass` shown in Program 12.9. Lines 4–20 in Program 12.8 also show OTcl interface commands `open{hours}` and `new-customer{}`. With opening hours **hours** as an input argument, the OTcl command **open{hours}** (Lines 8–11) is invoked when the store is opened. Line 8 stores the opening hours in variable **hours_**, Line 9 resets the number of visiting customers to zero, and Line 10 tells **timer_** to expire at "hours_" hours in future. The OTcl command **new-customer{}** is invoked as a customer enters the store. In Line 15, this command simply increases **count_** by one. Again, at the timer expiration, the timer invokes function **close()** through the pointer **store_** and prints out the opening hours (i.e., **hours_**) as well as the number of visiting customers (i.e., **count_**) for today (see function **expire(e)** in Line 2 of Program 12.8).

Program 12.9 A mapping class **StoreClass** which binds C++ and OTcl classes **Store**.

```
    //store.cc
1 static class StoreClass : public TclClass {
2 public:
3     StoreClass() : TclClass("Store") {}
4     TclObject* create(int, const char*const*) {
5         return (new Store);
6     }
7 } class_store;
```

Testing the Codes

After defining files **store.cc** and **store.h**, we include **store.o** to the **Make File** and run **make** at NS2 root directory to include classes **Store** and **StoreHour** into NS2 (see Section 2.7).

For, define a test Tcl simulation script in a file **store.tcl**,

```
    //store.tcl
1 set ns [new Simulator]
2 set my_store [new Store]
3 $my_store open 10.0
4 $ns at 1 "$my_store new-customer"
5 $ns at 5 "$my_store new-customer"
6 $ns at 6 "$my_store new-customer"
7 $ns at 8 "$my_store new-customer"
8 $ns at 11 "$my_store new-customer"
9 $ns run
```

We, run the script **store.tcl**, and obtain the following results:

```
>>ns store.tcl
The number of customers during 10.0 hours is 4
```

From the above script, when Line 2 creates a **Store** object, NS2 automatically creates a shadow C++ **Store** Object. Line 3 invokes command **open** with input argument 10.0, essentially opening the store for 10.0 hours. From Program 12.8, an OTcl command **open{10.0}** and tells the associated timer to expire at 10.0 hours in future, and clears the variable **count_**. Lines 4–8 invoke command **new-customer{}** at 1st, 5th, 6th, 8th, and 11th hours. Each of these lines increases the number of visiting customers (i.e., **count_**) by one. By the end of 11th hour in future, variable **count_** should be 5. However, the program shows that the number of visiting customers is 4. This is because the timer expires and invokes function **close()** at the 10th hour.

12.1.4 Guidelines for Implementing Timers in NS2

We now summarize the process of defining a new timer. Suppose that we would like to define a new timer class **StoreHour**. Suppose further that a **Store** object is responsible for starting, restarting, and canceling the **StoreHour** object, and for taking expiration actions. Then, the implementation of the above timer classes proceeds as follows:

From class StoreHour

- **Step 1:** Design class structure:
  - Derive class **StoreHour** from class **TimerHandler**.
  - Declare a pointer (e.g., **store_**) to class **Store**. The public function of class **Store** is accessible through the above pointer (e.g., **store_**) to class **Store**.
- **Step 2:** Bind the reference to class **Store** in the constructor.
- **Step 3:** Define expiration actions in function **expire(e)**.

From class Store

- **Step 1:** Design class structure:
  - Derive class **Store** from class **TclObject** *only if* an interface to OTcl is necessary.
  - Declare a **StoreHour** variable (e.g., **timer_**) as a member variable.
- **Step 2:** Instantiate the above **StoreHour** variable (e.g., **timer_**) with the pointer "**this**".

At runtime, we only need to instantiate a **Store** object. The internal mechanism of class **Store** will automatically create and configure a **StoreHour** object. Also, we do not need any global (or OTcl) reference to the **StoreHour** object, since the timer is usually manipulated by class **Store**.

12.2 Implementation of Random Numbers in NS2

This section demonstrates implementation of random number generators in NS2. In principle, NS2 employs so-called *Random Number Generator* (RNG) to generate random numbers. An RNG sequentially picks numbers from a stream of psudo-random numbers. A set of generated random numbers is characterized by the point where the RNG starts picking the numbers–called "seed". By default, NS2 sets the seed to 1. Therefore, the results obtained from every run are essentially the same.

Random numbers can also be transformed to conform to a given distribution. Such the transformation is carried out through instprocs in the OTcl domain, and through classes derived from class `RandomVariable` in the C++ domain. We will discuss the details of RNGs and the seeding mechanism in Sections 12.2.1 and 12.2.2, respectively. Section 12.2.3 shows the implementation of RNGs in NS2. Section 12.2.4 discusses different simulation scenarios, where RNGs are set differently. Section 12.2.5 explains the implementation of a C++ class `RandomVariable` which transforms random numbers according to a given distribution. Finally, Section 12.2.6 gives a guideline to define a new RNG and a new random variable in NS2.

12.2.1 Random Number Generation

NS2 generates random numbers by sequentially picking numbers from a stream of pseudo-random number (as discussed in Section 1.3.2). It uses the *combined multiple recursive generator* (MRG32k3a) proposed by L'Ecuyer [22] as a pseudo-random number generator. Generally speaking, an MRG32k3a generator contains streams of pseudo-random numbers from which the numbers picked sequentially seem to be random. In Fig. 12.5, an MRG32k3a generator provides 1.8×10^{19} independent streams, each of which consists of 2.3×10^{15} substreams. Each substream contains 7.6×10^{22} random numbers (i.e., the period of each substream is 7.6×10^{22}). In summary, an MRG32k3a generator can create 3.1×10^{57} numbers which appear to be random.

12.2.2 Seeding a Random Number Generator

As mentioned in Section 1.3.2, "seed" is one of the main ingredients of Random Number Generator (RNG). Loosely speaking, a seed specifies the location on a stream of pseudo-random numbers, where an RNG starts picking random numbers sequentially. When seeded differently, two RNGs start picking pseudo-random numbers from different locations, and therefore generate two distinct sets of random numbers. On the other hand, if seeded with the same number, two RNGs will start picking random numbers from the same location, and therefore, generate the same set of random numbers.

By default, NS2 always uses only one OTcl variable `defaultRNG` as a default RNG, and always seeds `defaultRNG` with 1. Therefore, the simulation

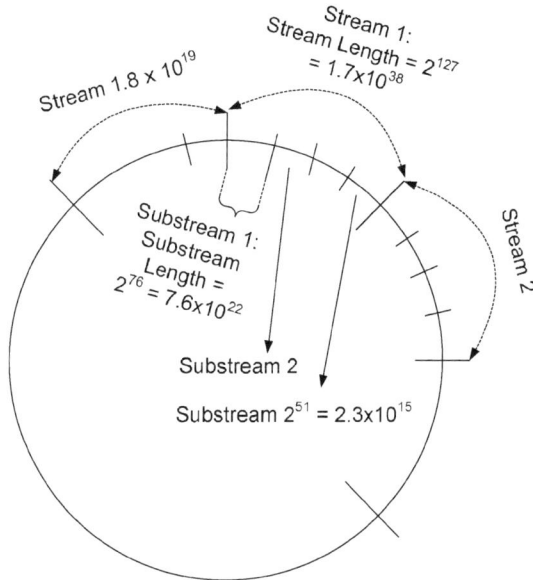

Fig. 12.5. Streams and substreams of an MRG32k3a generator.

results for every run are essentially the same. To collect independent simulation results, we must seed different runs differently.

Example 12.2. In the following, we run NS2 for three times to show NS2 seeding mechanism.

```
1 >>ns
2 >>$defaultRNG seed
3 1
4 >>$defaultRNG next-random
5 729236
6 >>$defaultRNG next-random
7 1193744747
8 >>exit

### RESTART NS2 ###
9 >> ns
10 >>$defaultRNG seed
11 1
12 >>$defaultRNG next-random
13 729236
14 >>$defaultRNG next-random
15 1193744747
16 >>exit
```

```
### RESTART NS2 ###
17 >>ns
18 >>$defaultRNG seed 101
19 >>$defaultRNG next-random
20 72520690
21 >>$defaultRNG next-random
22 308637100
23 >>exit
```

In the first run (Lines 1–8), variable `defaultRNG` (i.e., the default RNG) is used to generate two random numbers. In Line 2, instproc `seed` returns the current seed which is set (by default) to 1. Lines 4 and 6 use instproc `next-random{}` to generate two random numbers, 729236 and 1193744747, respectively. Finally, Line 8 exits the NS2 environment.

Lines 9–16 repeat the process in Lines 1–8. In Lines 10–11, we can observe that the seed is still 1. As expected, the first and the second random numbers generated are 729236 and 1193744747, respectively. These two numbers are the same as those in the first run. Essentially, the first run and the second run generate the same results. To generate different results, we need to seed the simulation differently.

Lines 17–22 show the last run, where the seed is set differently (to 101). The first and the second random number generated in this case are 72520690 and 308637100, respectively. These two numbers are different from those in the first two runs, since Line 15 sets the seed of `defaultRNG` to 101.

The key points about seeding the mechanism in NS2 are as follows:

- A seed specifies the starting location on a stream of psudo-random numbers, and hence characterizes an RNG.
- To generate two independent simulation results, each simulation must be seeded differently.
- At initialization, NS2 creates a variable `defaultRNG` as the default RNG, and seeds `defaultRNG` with 1. By default, NS2 generates the same simulation result for every run.
- When seeded with zero, an RNG replaces the seed with current time of the day and counter. Despite their tendency to be independent, two runs may pick the same seed and generate the same result. To ensure independent runs, we must seed the RNG manually.
- NS2 seeds a *new* RNG object to the beginning of the next random stream. Therefore, every RNG object is independent of each other.

12.2.3 OTcl and C++ Implementation

NS2 employs a C++ class `RNG` (which is bound to an OTcl class with the same name) to generate random numbers (see Program 12.10). In most cases, it is not necessary to understand the details of the MRG32k3a generator. This section shows only the key configuration and implementation in the OTcl and C++ domains. The readers may find the detailed implementation of an MRG32k3a generator in files ˜*ns*/tools/rng.cc,h.

Program 12.10 A mapping class `RNGClass` which binds OTcl and C++ classes `RNG`.

```
  //~/ns/tools/rng.cc
1 static class RNGClass : public TclClass {
2 public:
3     RNGClass() : TclClass("RNG") {}
4     TclObject* create(int, const char*const*) {
5         return(new RNG());
6     }
7 } class_rng;
```

OTcl Commands and Instprocs

In the OTcl domain, class `RNG` defines the following OTcl commands:

seed{}	Return the seed of RNG.
seed{n}	Set the the seed of RNG to be **n**.
next-random{}	Return a random number.
next-substream{}	Advance to the beginning of the next substream.
reset-start-substream{}	Return to the beginning of the current substream.
normal{avg std}	Return a random number normally distributed with average **avg** and standard deviation **std**.
lognormal{avg std}	Return a random number log-normally distributed with average **avg** and standard deviation **std**.

Defined in file ˜*ns*/tcl/lib/ns-random.tcl, the following instprocs generate random numbers with exponential distribution and uniform distribution:

exponential{mu} Return a random number exponentially distributed with mean mu.

uniform{min max} Return a random number uniformly distributed in [min,max].

integer{k} Return a random integer uniformly distributed in {0,1,...,k-1}.

C++ Functions

In the C++ domain, the key functions of class RNG include (see the details in files ˜ns/tools/rng.cc,h):

set_seed(n) If n = 0, set the the seed of the RNG to be current time and counter. Otherwise, set the seed to be n.

seed() Return the seed of the RNG.

next() Return a random int in {0,1,..., MAX_INT}.

next_double() Return a random double in [0,1].

reset_start_substream() Move to the beginning of the current substream.

reset_next_substream() Move to the beginning of the next substream.

uniform(k) Return a random int number uniformly distributed in {0,1,...,k-1}.

uniform(r) Return a random double number uniformly distributed in [0,r].

uniform(a,b) Return a random double number uniformly distributed in [a,b].

exponential(k) Return a random number exponentially distributed with mean k.

normal(avg,std) Return a random number normally distributed with average avg and standard deviation std.

lognormal(avg,std) Return a random number log-normally distributed with average avg and standard deviation std.

12.2.4 Randomness in Simulation Scenarios

In most cases, a simulation falls into one of the following three scenarios.

Deterministic Setting

This type of simulation is usually used for debugging. Its purpose is to locate programming errors in the simulation codes or to understand complex behavior of a certain network. In both cases, it is convenient to run the program under a deterministic setting and generate the same result repeatedly. By default, NS2 seeds the simulation with 1. The deterministic setting is therefore the default setting for NS2 simulation.

Single-Stream Random Setting

The simplest form of statistical analysis is to run a simulation for several times and compute statistical measures such as average and/or standard deviation. By default, NS2 always uses `defaultRNG` with seed "1" to generate random numbers. To statistically analyze a system, we need to generate several distinct sets of results. Therefore, we need to seed different runs differently. In a single-stream random setting, we need only one RNG. Hence, we may simply introduce the diversity to each run by seeding different runs with different values `<n>` (e.g., in Example 12.2, Line 18 seeds the default RNG with 101).

```
$defaultRNG <n>
```

which seeds the default RNG with a number `<n>`.

Multiple-Stream Random Setting

In some cases, we may need more than one independent random variable for a simulation. For example, we may need to generate random values of packet inter-arrival time as well as packet size. These two variables should be independent and should not share the same random stream. We can create two independent RNG using "`new RNG`". Since NS2 seeds each RNG with different random stream (see Section 12.2.2), the random processes with different RNGs are independent of each other.

Example 12.3. Suppose that the inter-arrival time and packet size are exponentially distributed with mean 5 and uniformly distributed within $[100, 5000]$, respectively. Print out the first 5 random values of inter-arrival time and packet size.

Tcl simulation script:

```
1 $defaultRNG seed 101
2 set arrivalRNG [new RNG]
3 set sizeRNG [new RNG]

4 set arrival_ [new RandomVariable/Exponential]
5 $arrival_ set avg_ 5
6 $arrival_ use-rng $arrivalRNG

7 set size_ [new RandomVariable/Uniform]
8 $size_ set min_ 100
```

```
9 $size_ set max_ 5000
10 $size_ use-rng $sizeRNG

11 puts "Inter-arrival time Packet size"
12 for {set j 0} {$j < 5} {incr j} {
13 puts [format "%-8.3f %-4d" [$arrival_ value] \
                        [expr round([$size_ value])]]
14 }
```

Results on the Screen:

```
Inter-arrival time    Packet size
1.048                 1880
7.919                 116
8.061                 3635
4.675                 2110
7.201                 1590
```

The details of the above Tcl simulation script are as follows. Lines 4 and 7 create an exponentially random variable[2] `arrival_` and a uniformly random variable `size_` whose parameters are defined in Lines 5–6 and Lines 8–10, respectively. Lines 11–14 print out five random numbers generated by `arrival_` and `size_`. In Section 12.2.5, we will see that the OTcl command "value" of class `RandomVariable` returns a random number and the OTcl command "use-rng" is used to specify an RNG for a random variable.

By default, `defaultRNG` is used to generate random numbers for both `arrival_` and `size_`. In this case, Lines 2 and 3 create two independent RNGs: `arrivalRNG` and `sizeRNG`. NS2 specifies these two variables as RNGs for `arrival_` and `size_` by using an OTcl command `use-rng` in Lines 6 and 10, respectively. Since the created RNG objects are independent, random variable `arrival_` and `size_` are independent of each other.

Exercise 12.4. From Example 12.3,

(i) Change the seed to "999". Re-run the script for a couple of times. Observe and explain the output.
(ii) Change the seed to "0". Re-run the script for a couple of times. Observe and explain the output.
(iii) Print out values of `arrival_` and `size_` for (i) and (ii), and show that they are exponentially and uniformly distributed (Hint: Set the seed properly).
(iv) Change the mean of `arrival_` to 10 and the interval of `size_` to $[400, 2000]$, and repeat (iii).
(v) Remove Line 6 and repeat (iii). Observe and explain the output.
(vi) Remove Lines 6 an 10 and repeat (iii). Observe and explain the output.

---

[2] We will discuss the details of random variables in the next section.

12.2.5 Random Variables

In NS2, a random variable is a module which generates random values whose statistics follow a certain distribution. It employs an RNG to generate random numbers and transforms the generated numbers to values which conform to a given distribution. This implementation is carried out in C++ abstract class RandomVariable whose diagram and declaration are shown in Fig. 12.6 and Program 12.11, respectively.

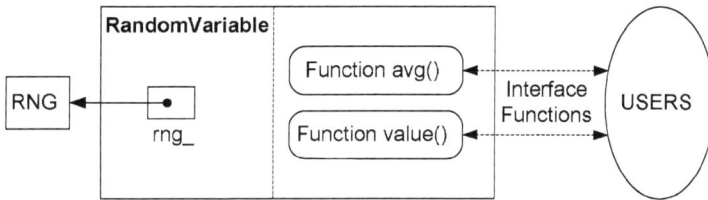

Fig. 12.6. A schematic diagram of class RandomVariable.

Consider the declaration of class RandomVariable in Program 12.11. Class RandomVariable contains a pointer rng_ (Line 9) to an RNG object (used to generate random numbers), and two pure virtual interface functions: value() in Line 3 and avg() in Line 4. Function value() generates random numbers, transforms the generated numbers to values conforming to a given distribution, and returns the transformed values to the caller. Function avg() returns the average value of the underlying distribution. Since these two functions are pure virtual, they must be overridden by all derived instantiable classes of class RandomVariable. The list of key built-in instantiable C++ classes as well as their bound OTcl classes is given in Table 12.2.

Program 12.11 Declaration of class RandomVariable.

```
   //~/ns/tools/ranvar.h
1  class RandomVariable : public TclObject {
2  public:
3     virtual double value() = 0;
4     virtual double avg() = 0;
5     int command(int argc, const char*const* argv);
6     RandomVariable();
7     int seed(char *);
8  protected:
9     RNG* rng_;
10 };
```

Table 12.2. Built-in C++ and OTcl random variable classes.

C++ class	OTcl class
UniformRandomVariable	RandomVariable/Uniform
ExponentialRandomVariable	RandomVariable/Exponential
ParetoRandomVariable	RandomVariable/Pareto
ParetoIIRandomVariable	RandomVariable/ParetoII
NormalRandomVariable	RandomVariable/Normal
LogNormalRandomVariable	RandomVariable/LogNormal
ConstantRandomVariable	RandomVariable/Constant
HyperExponentialRandomVariable	RandomVariable/HyperExponential
WeibullRandomVariable	RandomVariable/Weibull
EmpiricalRandomVariable	RandomVariable/Empirical

Random Number Generator

A RandomVariable object utilizes variable rng_ to generate random numbers. By default, every random variable uses the defaultRNG as its RNG. As shown in Program 12.12, the constructor (Lines 1–4) of class RandomVariable stores the default RNG returned from function RNG::defaultrng() in variable rng_.

To create multiple *independent* random variables, variable rng_ of each random variable must be independent of each other. From Example 12.3, this can be achieved by creating and binding a dedicated RNG to each random variable. As will be discussed in the next section, the process of binding an RNG to a random variable is carried out by using the OTcl command use-rng associated with a RandomVariable object.

OTcl Commands

Shown in Program 12.12, class RandomVariable defines the following two commands, which can be invoked from the OTcl domain:

- value{}: Returns a random number by invoking function value() (Lines 9–12).
- use-rng{rng}: Casts the input argument rng to type RNG*, and stores the cast object in variable rng_ (Lines 15–19).
  Note that an example use of the OTcl command use-rng{rng} is shown in Lines 6 and 10 in Example 12.3.

Since class RandomVariable is abstract, it does not provide a shadow class in the OTcl domain. However, all its derived classes do have shadow classes in the OTcl domain. Table 12.2 lists 10 built-in C++ and OTcl random variable classes.

Program 12.12 The constructor, OTcl command `value`, and OTcl command `use-rng` of class `RandomVariable`.

```
    //~/ns/tools/ranvar.cc
1   RandomVariable::RandomVariable()
2   {
3       rng_ = RNG::defaultrng();
4   }

    //~/ns/tools/ranvar.cc
5   int RandomVariable::command(int argc, const char*const* argv)
6   {
7       ...
8       if (argc == 2) {
9           if (strcmp(argv[1], "value") == 0) {
10              tcl.resultf("%6e", value());
11              return(TCL_OK);
12          }
13      }
14      if (argc == 3) {
15          if (strcmp(argv[1], "use-rng") == 0) {
16              rng_ = (RNG*)TclObject::lookup(argv[2]);
17              ...
18              return(TCL_OK);
19          }
20      }
21      ...
22  }
```

Exponential Random Variable

As an example, consider implementation of an exponentially random variable in Program 12.13. From Table 12.2, NS2 implements an exponentially random variable using the C++ class `ExponentialRandomVariable` and the OTcl class `RandomVariable/Exponential`.

Since an exponential random variable is completely characterized by an average value, class `ExponentialRandomVariable` has only one member variable `avg_` (Line 9), which stores the average value. At the construction (see Lines 18–20), class `ExponentialRandomVariable` binds variable `avg_` to instvar `avg_` in the OTcl domain. Functions `avg()` in Line 6 and `avgp()` in Line 5 return the value stored in `avg_` and the address of `avg_`, respectively. Function `setavg(d)` in Line 7 stores the value in "d" into variable "avg_". Function `value()` in Lines 21–23 returns a random number exponentially distributed with mean `avg_`. It invokes function `exponential(avg_)` of variable `rng_`, feeding variable `avg_` as an input argument to obtain an exponentially distributed random number.

Program 12.13 An implementation of class ExponentialRandomVariable.

```
   //~/ns/tools/ranvar.h
 1 class ExponentialRandomVariable : public RandomVariable {
 2   public:
 3     virtual double value();
 4     ExponentialRandomVariable();
 5     double* avgp() { return &avg_; };
 6     virtual inline double avg() { return avg_; };
 7     void setavg(double d) { avg_ = d; };
 8   private:
 9     double avg_;
10 };

   //~/ns/tools/ranvar.cc
11 static class ExponentialRandomVariableClass : public TclClass {
12 public:
13     ExponentialRandomVariableClass() : TclClass(
                            "RandomVariable/Exponential") {}
14     TclObject* create(int, const char*const*) {
15         return(new ExponentialRandomVariable());
16     }
17 } class_exponentialranvar;

18 ExponentialRandomVariable::ExponentialRandomVariable(){
19     bind("avg_", &avg_);
20 }

21 double ExponentialRandomVariable::value(){
22     return(rng_->exponential(avg_));
23 }
```

Exercise 12.5. Write a simulation script which generates random numbers exponentially distributed with mean 1.0. To verify the script, plot the probability density function.

Exercise 12.6. Write a simulation script which generates a random number normally distributed with mean 1.0 and standard deviation 0.05. To verify the script, plot the probability density function.

Exercise 12.7. Develop a new class for a discrete random variable whose probability mass function is $(0.1, 0.3, 0.3, 0.2, 0.1)$. Test the code by generating random numbers and verify the probability mass function.

12.2.6 Guidelines for Random Number Generation in NS2

We conclude this section, by providing the following guidelines for implementing randomness numbers in NS2:

- Determine the type of simulation: deterministic setting, single-stream random setting, or multi-stream random setting.
- Create RNG(s) according to the simulation type.
- If needed, create a random variable
  - Define the inheritance structure: C++, OTcl, and mapping classes.
  - Define function `avg()` which returns the average value of the distribution to the caller.
  - Define function `value()` which returns a random number conforming to the specified distribution.
- Specify an RNG for each random variable by using an OTcl command `use-rng` of class `RandomVariable`.

12.3 Built-in Error Models

An error model is an NS2 module which imposes error on packet transmission. Derived from class `Connector`, it can be inserted between two NsObjects. An error model simulates packet error upon receiving a packet. If the packet is simulated to be in error, the error model will either drop the packet or mark the packet with an error flag. If the packet is simulated not to be in error, on the other hand, the error model will forward the packet to its downstream object. An error model can be used for both wired and wireless networks. However, this section discusses the details of an error model through a wired class `SimpleLink` only.

Program 12.14 Class `ErrorModelClass` which binds C++ and OTcl classes `ErrorModel`.

```
      //~/ns/queue/errmodel.cc
1   static class ErrorModelClass : public TclClass {
2   public:
3       ErrorModelClass() : TclClass("ErrorModel") {}
4       TclObject* create(int, const char*const*) {
5           return (new ErrorModel);
6       }
7   } class_errormodel;
```

NS2 implements error models using a C++ class `ErrorModel` which is bound to an OTcl class with the same name (see Program 12.14). Class `ErrorModel` simulates Bernoulli error, where transmission is simulated to be either in error or not in error. NS2 also provides `ErrorModel` classes with more functionalities such as two-state error model. Tables 12.3 and 12.4 show NS2 built-in error models whose implementation is in the C++ and OTcl domain, respectively.

Table 12.3. Built-in error models which contain C++ and OTcl implementation.

C++ class	OTcl class	Description
TwoStateErrorModel	ErrorModel/TwoState	Error-free and error-prone states
ComplexTwoState MarkovModel	ErrorModel/Complex TwoStateMarkov	Contain two objects of class TwoStateErrorModel
MultiStateErrorModel	ErrorModel/MultiState	Error model with more than two states
TraceErrorModel	ErrorModel/Trace	Impose error based on a trace file
PeriodicErrorModel	ErrorModel/Periodic	Drop packets once every n packets
ListErrorModel	ErrorModel/List	Specify the a list of packets to be dropped
SelectErrorModel	SelectErrorModel	Selective packet drop
SRMErrorModel	SRMErrorModel	Error model for SRM
MrouteErrorModel	ErrorModel/Trace/Mroute	Error model for multicast routing
ErrorModule	ErrorModule	Send packets to classifier rather than target_
PGMErrorModel	PGMErrorModel	Error model for PGM
LMSErrorModel	LMSErrorModel	Error model for LMS

Table 12.4. Built-in OTcl error models defined in file ˜ns/tcl/lib/ns-errmodel.tcl.

OTcl class	Base class	Description
ErrorModel/Uniform	ErrorModel	Uniform error model
ErrorModel/Expo	ErrorModel/TwoState	Two state error model; Each state is represented by an exponential random variable.
ErrorModel/Empirical	ErrorModel/TwoState	Two state error model; Each state is represented by an empirical random variable.
ErrorModel/TwoStateMarkov	ErrorModel/Expo	ErrorModel/Expo model where the state residence time is exponential

12.3.1 OTcl Implementation: Error Model Configuration

In common with those of most objects, configuration interfaces of an error model are defined in the OTcl domain. Such a configuration includes parameter configuration and network configuration.

Parameter Configuration

There are two ways to configure an error model object: through bound variables, and through OTcl commands. Class `ErrorModel` binds the following C++ variables to OTcl instvars with the same name:

enabled_	Set to 1 if this error model is active, and set to 0 otherwise.
rate_	Error probability
delay_pkt_	If set to `true`, the error model will delay (rather than drop) the transmission of corrupted packets.
delay_	Delay time in case that `delay_pkt_` is set to `true`.
bandwidth_	Used to compute packet transmission time
markecn_	If set to `true`, the error model will mark error flag (rather than drop) in flag header of the corrupted packet.

The second configuration method is through the following OTcl commands whose input arguments are stored in `args`:

unit{arg}	Store `arg` in C++ variable `unit_`.
ranvar{arg}	Store `arg` in C++ variable `ranvar_`.
FECstrength{arg}	Store `arg` in C++ variable `FECstrength_`.
datapktsize{arg}	Store `arg` in C++ variable `datapktsize_`.
cntrlpktsize{arg}	Store `arg` in C++ variable `cntrlpktsize_`.
eventtrace{arg}	Store `arg` in C++ variable `et_`.

Among the above OTcl commands, commands `unit{}`, `ranvar{}`, and `FECstrength{}`, when taking no input argument, return values stored in `unit_`, `ranvar_`, and `FECstrength_`, respectively.

Network Configuration

As a `Connector` object, an error model can be inserted into a network to simulate packet errors. OTcl defines two pairs of instprocs to insert an error model into a `SimpleLink` object (see Section 7.1). Each pair consists of one instproc from class `SimpleLink` and one instproc from class `Simulator` as shown below (see Fig. 12.7):

- `SimpleLink::errormodule{em}`: Inserts an error model "em" right after instvar `head_` of a `SimpleLink` object.

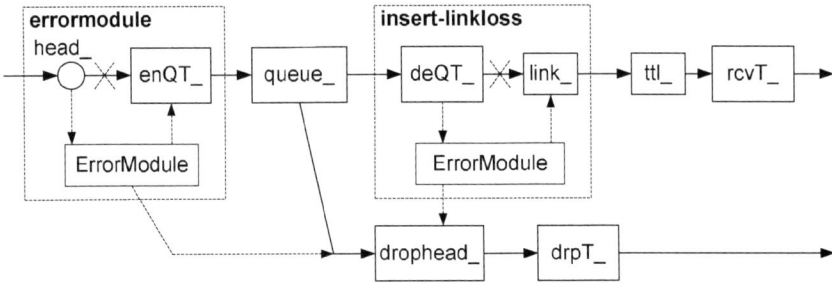

Fig. 12.7. Instprocs `errormodule` and `insert-linkloss` of class `SimpleLink`.

- `Simulator::lossmodel{lossobj from to}`: Executes "errormodule" from within the `SimpleLink` object which connects node "from" to node "to".
- `SimpleLink::insert-linkloss{em}`: Inserts an error model "em" right after instvar `queue_` (or instvar `deqT_` if it exists) of the `SimpleLink` object.
- `Simulator::link-lossmodel{lossobj from to}`: Executes the instproc "insert-linkloss{...}" from within the `SimpleLink` object which connects node "from" to node "to".

Program 12.15 shows the details of instproc `errormodule{em}` of class `SimpleLink`, which inserts the input error model (e.g., em) immediately after the link's head. Lines 6–7 store the input error model (i.e., em) in instvar

---

Program 12.15 Instproc `errormodule` of class `SimpleLink`, and instproc `add-to-head` of class `Link`.

```
    //~/ns/tcl/lib/ns-link.tcl
1   SimpleLink instproc errormodule args {
2       $self instvar errmodule_ queue_ drophead_
3       if { $args == "" } {
4           return $errmodule_
5       }
6       set em [lindex $args 0]
7       set errmodule_ $em
8       $self add-to-head $em
9       $em drop-target $drophead_
10  }

11  Link instproc add-to-head { connector } {
12      $self instvar head_
13      $connector target [$head_ target]
14      $head_ target $connector
15  }
```

errmodule_. Line 8 inserts the input error model next to the link's head by invoking instproc add-to-head{em}, and Line 9 sets the drop target of the input error model em to drophead_.

In Lines 11–15 of Program 12.15, instproc add-to-head{connector} inserts the input argument connector between link's head (i.e., the instvar head_) and target of the link's head (see Lines 13–14).

Program 12.16 shows the details of instproc insert-linkloss{em}, which inserts the input error model after instvar queue_ or instvar deqT_. Line 6 stores the input error model in variable em. Lines 7–9 delete instvar link_errmodule_ if it exists. Then Line 10 stores variable em in instvar link_errmodule_. If instvar deqT_ exists (i.e., trace is enabled), Lines 12–13 insert the input variable em immediately after instvar deqT_. Otherwise, Lines 15-16 insert the input variable em immediately after instvar queue_. Finally, Line 18 sets the drop target of the input variable em to instvar drophead_.

Program 12.16 An instproc insert-linkloss of class SimpleLink.

```
     //~/ns/tcl/lib/ns-link.tcl
1    SimpleLink instproc insert-linkloss args {
2        $self instvar link_errmodule_ queue_ drophead_ deqT_
3        if { $args == "" } {
4            return $link_errmodule_
5        }
6        set em [lindex $args 0]
7        if [info exists link_errmodule_] {
8            delete link_errmodule_
9        }
10       set link_errmodule_ $em
11       if [info exists deqT_] {
12               $em target [$deqT_ target]
13               $deqT_ target $em
14       } else {
15               $em target [$queue_ target]
16               $queue_ target $em
17       }
18       $em drop-target $drophead_
19 }
```

In most cases, a SimpleLink object is inaccessible from a Tcl simulation script. Therefore, class Simulator provides interface instprocs lossmodel{...} and link-lossmodel{...} to invoke instprocs errormodule{em} and insert-linkloss{em}, respectively, of class SimpleLink.

The details of both the instproc lossmodel{lossobj from to} and the instproc link-lossmodel{lossobj from to} of class Simulator are shown in Program 12.17, where they insert an error model "lossobj" into the link

Program 12.17 Instprocs `lossmodel`, `link-lossmodel`, and `link` of class `Simulator`.

```
    //~/ns/tcl/lib/ns-lib.tcl
1   Simulator instproc lossmodel {lossobj from to} {
2       set link [$self link $from $to]
3       $link errormodule $lossobj
4   }

5   Simulator instproc link-lossmodel {lossobj from to} {
6       set link [$self link $from $to]
7       $link insert-linkloss $lossobj
8   }

9   Simulator instproc link { n1 n2 } {
10      $self instvar Node_ link_
11      if { ![catch "$n1 info class Node"] } {
12      set n1 [$n1 id]
13      }
14      if { ![catch "$n2 info class Node"] } {
15      set n2 [$n2 id]
16      }
17      if [info exists link_($n1:$n2)] {
18          return $link_($n1:$n2)
19      }
20      return ""
21  }
```

which connect a node "`from`" to a node "`to`". Lines 2 and 6 invoke instproc `link{from to}` of class `Simulator`. In Line 18, this instproc returns the `Link` object which connects a node "`from`" to a node "`to`". Lines 3 and 7 then insert an error model into the returned `Link` object, by executing `errormodule{em}` and `insert-linkloss{em}`, respectively.

12.3.2 C++ Implementation: Error Model Simulation

The internal mechanism of an error model is specified in the C++ domain. As shown in Program 12.18, C++ class `ErrorModel` derives from class `Connector`. It employs packet forwarding/dropping capabilities (e.g., a variable `target_` and a function `recv(p,h)`) inherited from class `Connector`, and define error simulation mechanism.

Variables

Key variables of class `ErrorModel` are given below:

Program 12.18 Declaration of class `ErrorModel`.

```
   //~/ns/queue/errmodel.h
1  enum ErrorUnit { EU_TIME=0, EU_BYTE, EU_PKT, EU_BIT };

2  class ErrorModel : public Connector {
3  public:
4      ErrorModel();
5      virtual void recv(Packet*, Handler*);
6      virtual void reset();
7      virtual int corrupt(Packet*);
8      inline double rate() { return rate_; }
9      inline ErrorUnit unit() { return unit_; }
10 protected:
11     int enable_;
12     ErrorUnit unit_;
13     double rate_;
14     double delay_;
15     double bandwidth_;
16     RandomVariable *ranvar_;
17     int FECstrength_;
18     int datapktsize_;
19     int cntrlpktsize_;
20     double *cntrlprb_;
21     double *dataprb_;
22     Event intr_;
23     virtual int command(int argc, const char*const* argv);
24     int CorruptPkt(Packet*);
25     int CorruptByte(Packet*);
26     int CorruptBit(Packet*);
27     double PktLength(Packet*);
28     double* ComputeBitErrProb(int);
29 };

   //~/ns/queue/errmodel.cc
30 ErrorModel::ErrorModel() : firstTime_(1), unit_(EU_PKT),
                                ranvar_(0), FECstrength_(1)
31 {
32     bind("enable_", &enable_);
33     bind("rate_", &rate_);
34     bind("delay_", &delay_);
35 }
```

enable_   Set to 1 if this error model is active, and set to 0 otherwise.
rate_   Error probability
delay_   Time used to delay (rather than dropping) a corrupted packet
bandwidth_   Transmission bandwidth used to compute packet transmission
         time

unit_	Error unit (EU_TIME, EU_BYTE(default), EU_PKT, or EU_BIT)
ranvar_	Random variable which simulates error
FECstrength_	Number of bits in a packet which can be corrected
datapktsize_	Number of bytes in data payload
cntrlpktsize_	Number of bytes in packet header
dataprb_	An array whose i^{th} entry is the probability of having at most i corrupted data bits
cntrlprb_	An array whose i^{th} entry is the probability of having at most most i corrupted control bits
firstTime_	Indicate whether an error has occurred.
intr_	A queue callback object (see Section 7.3.3).

Variable rate_ specifies the error probability, while the variable unit_ indicates the unit of rate_. If unit_ is packets (i.e., EU_PKT), rate_ will represent packet error probability. If unit_ is bytes (i.e., EU_BYTE) or bits (i.e., EU_BIT), rate_ will represent byte error probability or bit error probability, respectively.

Functions

Key functions of class ErrorModel are given below:

rate()	Return the error probability stored in variable rate_.
unit()	Return the error unit stored in variable unit_.
PktLength(p)	Return the length (in error units) of the packet p.
reset()	Set the variable firstTime_ to 1.
recv(p,h)	Receive a packet p and a handler h.
corrupt(p)	Return 1/0 if the transmission is in error/not in error.
CorruptPkt(p)	Return 1/0 if the transmission is in error/not in error.
CorruptByte(p)	Return 1/0 if the transmission is in error/not in error.
CorruptBit(p)	Return the number of corrupted bits in error.
ComputeBitErrProb(size)	Computes the cumulative distribution of having i= $\{0, \cdots,$ FECstrength_$\}$ error bits.

Main Mechanism

The main mechanism of an ErrorModel object lies within the packet reception function recv(p,h) shown in Program 12.19. When receiving a packet, an ErrorModel object simulates packet error (by invoking function corrupt(p)

Program 12.19 Function `recv` of class `ErrorModel`.

```
    //~/ns/queue/errmodel.cc
1   void ErrorModel::recv(Packet* p, Handler* h)
2   {
3       hdr_cmn* ch = hdr_cmn::access(p);
4       int error = corrupt(p);
5       if (h && ((error && drop_) || !target_)) {
6           double delay = Random::uniform(8.0*ch->size()/bandwidth_);
7           if (intr_.uid_ < 0)
8               Scheduler::instance().schedule(h, &intr_, delay);
9       }
10      if (error) {
11          ch->error() |= error;
12          if (drop_) {
13                  drop_->recv(p);
14                  return;
15          }
16      }
17      if (target_) {
18              target_->recv(p, h);
19      }
20  }
```

in Line 4 of Program 12.19), and reacts to the error based on the underlying configuration. If an error occurs, Line 11 will mark an error flag in the common packet header. Then if `drop_` exists, Lines 13 and 14 will drop the packet and terminate the function. If the packet is not in error, on the other hand, function `recv(p,h)` will skip Lines 11–15, and will forward the packet to `target_` if it exists. A cautionary note: since a corrupted packet will also be forwarded to `target_` if `drop_` does not exist. *NS2 will not show any error but the simulation results might not be correct!*

Lines 6–8 in Program 12.19 are related to NS2 callback mechanism discussed in Section 7.3.3. Callback mechanism is an NS2 technique to have a downstream object invoke an upstream object along a downstream path. For example, after transmitting a packet, a queue needs to wait until the packet leaves the queue (i.e., wait for a *callback* signal to release the queue for the blocked state), before commencing another packet transmission. From Section 7.2, a `LinkDelay` object employs the Scheduler to inform the queue of packet departure (i.e., send a release signal) at the packet departure time.

A callback process is implemented by passing the handler (`h`) of an upstream object (e.g., the queue) along with packet (`p`) to a downstream object through function `recv(p,h)`. Upon receiving the handler, an NsObject reacts by either (1) passing the handler to its downstream object and hoping that the handler will be dealt with somewhere along the downstream path, or (2) immediately scheduling a callback event at a certain time.

According to Line 5 in Program 12.19, the `ErrorModel` object chooses to call back when both of the following conditions are satisfied:

(i) Handler "h" exists (i.e., non-zero), and
(ii) Either
    (a) Packet is in error and variable `drop_` exists, and/or
    (b) Variable `target_` does not exist.

Condition (i) occurs when an upstream object passes down the handler "h", and is waiting for a callback signal. Condition (ii) indicates the case where the `ErrorModel` object is responsible for sending a callback signal.[3] Condition (ii) consists of two following subconditions. One is the case where the packet will be dropped. Another is when `target_` does not exist. In these cases, the `ErrorModel` will be the last object in a downstream path which can deal with the packet, and is therefore, responsible for the callback mechanism.

When choosing to callback, Line 8 schedules a callback event after a delay time of "delay" seconds. NS2 assumes that an error can occur in any place in a packet with equal probability. Correspondingly, the time at which an error is materialized is uniformly distributed in $[0, txt]$, where txt is the packet transmission time (Line 6).

Simulating Transmission Errors

In the previous section, we have discussed how class `ErrorModel` forwards or drops (or marks with an error flag) packets based on the simulated error. In this section, we will discuss the details of function `corrupt(p)` which simulates transmission error. Taking a packet pointer `p` as an input argument, function `corrupt(p)` returns zero and one if the transmission is simulated not to be and to be in error, respectively.

Program 12.20 shows the details of function `corrupt(p)`. The function `corrupt(p)` always returns zero if the `ErrorModel` object is disabled (i.e., `enable_=0`; see Lines 4–5). Given that the `ErrorModel` object is enabled, function `corrupt(p)` will return a logic value (i.e., true or false) depending on whether the value returned from functions `CorruptPkt(p)` in Line 16, `CorruptByte(p)` in Line 10, `CorruptBit(p)` in Lines 13-14, and `CorruptTime(p)` in Line 8 is zero, when `unit_` is equal to EU_PKT, EU_BYTE, EU_BIT, and EU_TIME, respectively. Similar to function `corrupt(p)`, these functions return a zero and a non-zero value if the packet is not in error and is in error, respectively.

In some cases, the packet error process in a communication link can be modeled as having Bernoulli distribution. Suppose that `ranvar_` (Line 16 in Program 12.18) is a random variable which generates uniformly distributed

---

[3] If not, the `ErrorModel` object will assign the responsibility to its downstream object. In this case, handler "h" should be passed to the downstream object, by invoking `target_->recv(p,h)`.

Program 12.20 Functions corrupt CorruptPkt, CorruptByte, and PktLength of class ErrorModel.

```
    //~/ns/queue/errmodel.cc
 1  int ErrorModel::corrupt(Packet* p)
 2  {
 3      hdr_cmn* ch = HDR_CMN(p);
 4      if (enable_ == 0)
 5          return 0;
 6      switch (unit_) {
 7      case EU_TIME:
 8          return (CorruptTime(p) != 0);
 9      case EU_BYTE:
10          return (CorruptByte(p) != 0);
11      case EU_BIT:
12          ch = hdr_cmn::access(p);
13          ch->errbitcnt() = CorruptBit(p);
14          return (ch->errbitcnt() != 0);
15      default:
16          return (CorruptPkt(p) != 0);
17      }
18      return 0;
19  }

20  int ErrorModel::CorruptPkt(Packet*)
21  {
22      double u = ranvar_ ? ranvar_->value() : Random::uniform();
23      return (u < rate_);
24  }

25  int ErrorModel::CorruptByte(Packet* p)
26  {
27      double per = 1 - pow(1.0 - rate_, PktLength(p));
28      double u = ranvar_ ? ranvar_->value() : Random::uniform();
29      return (u < per);
30  }

31  double ErrorModel::PktLength(Packet* p)
32  {
33      if (unit_ == EU_PKT)
34          return 1;
35      int byte = hdr_cmn::access(p)->size();
36      if (unit_ == EU_BYTE)
37          return byte;
38      if (unit_ == EU_BIT)
39          return 8.0 * byte;
40      return 8.0 * byte / bandwidth_;
41  }
```

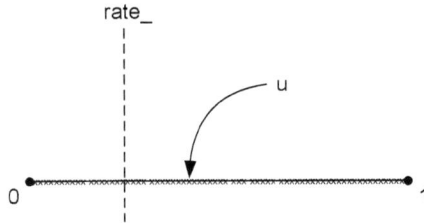

Fig. 12.8. Transforming uniform distribution to Bernoulli distribution.

random numbers "u" in the range [0,1]. From Fig. 12.8, "u" could be any point "×" in [0,1] with equal probability. Given a threshold rate_, "u" will be in [0,rate_) with probability rate_. In other words, to have probability of rate_ for an event (e.g., packet error), we need to generate a uniformly distributed random number "u", and assume the occurrence of the event if and only if u < rate_.

Lines 20–41 of Program 12.20 show the details of functions CorruptPkt(p), CorruptByte(p), and pktLength(p). Function CorruptPkt(p) in Lines 20-24 employs the above method (see Fig. 12.8) to simulate packet error. In other words, it generates uniformly distributed random numbers "u" and assumes that a packet is in error if and only if u < rate_.

For function CorruptByte(p), variable rate_ represents byte error probability. Line 27 translates byte error probability to packet error probability (per)[4] and simulates packet error in the same way as function CorruptPkt(p) does.

Function PktLength(p) in Lines 31–40 of Program 12.20 computes the length of a packet in the corresponding unit_. In particular, if unit_ is

- EU_PKT, function PktLength(p) will return 1 (see Line 34).
- EU_BYTE, function PktLength(p) will return the number of bytes in the packet stored in field size_ of common header (see Lines 35-37).
- EU_BITS, function PktLength(p) will return the number of bits in the packet (see Line 39).
- EU_TIME (if none of the above matches), function PktLength(p) will return the transmission time of the packet (see Line 40).

Program 12.21 shows the details of function CorruptBit(p) of class ErrorModel. When this function is called for the first time (i.e., firstTime_ is 1), Lines 5 and 6 precompute error probabilities for control header and data payload and store the probabilities in cntrlprb_ and dataprb_, respectively. The computation is achieved via function ComputeBitErrProb(size) which takes the size of control header (i.e., size=cntrlpktsize_) or data payload (i.e., size=datapktsize_) as its input argument. The values stored

---

[4] Packet error probability is $1 - (1-\texttt{rate\_})^n$, where rate_ is byte error probability and $n = \texttt{PktLength(p)}$ is number of bytes in a packet.

in `cntrlprb_[i]` and `dataprb_[i]` denote the probability that at most i bits are in error. Line 7 then sets `firstTime_` to zero so that function `CorruptBit` will skip Lines 5–7 when it is invoked again.

Function `CorruptBit(p)` computes packet error probability based on either `dataprb_` or `cntrlprb_`, not on the packet size specified in common header. In Line 10, its uses `cntrlprb_` and `dataprb_` as packet error probability, if the packet size specified in common header is not less than and less than `datapktsize_`, respectively. Since the value stored in `dptr[i]` is the probability that at most i bits are in error, Lines 11–12 increment i until the probability exceeds u and returns i to the caller. In this case, variable i is the number of corrupted bits.

The details of function `ComputeBitErrProb(size)` are shown in Program 12.21. This function takes the packet size as an input argument and returns an array `dptr` of `double` whose i^{th} entry contains the probability

---

Program 12.21 Functions `CorruptBit` and `ComputeBitErrProb` of class `ErrorModel`.

```
   //~/ns/queue/errmodel.cc
1  int ErrorModel::CorruptBit(Packet* p)
2  {
3      double u, *dptr; int i;
4      if (firstTime_ && FECstrength_) {
5          cntrlprb_ = ComputeBitErrProb(cntrlpktsize_);
6          dataprb_ = ComputeBitErrProb(datapktsize_);
7          firstTime_ = 0;
8      }
9      u = ranvar_ ? ranvar_->value() : Random::uniform();
10     dptr = (hdr_cmn::access(p)->size() >= datapktsize_)
                                     ? dataprb_ : cntrlprb_;
11     for (i = 0; i < (FECstrength_ + 2); i++)
12         if (dptr[i] > u) break;
13     return(i);
14 }

15 double* ErrorModel::ComputeBitErrProb(int size)
16 {
17     double *dptr; int i;
18     dptr = (double *)calloc((FECstrength_ + 2), sizeof(double));
19     for (i = 0; i < (FECstrength_ + 1) ; i++)
           dptr[i] = comb(size, i) * pow(rate_,
20                  (double)i) * pow(1.0 - rate_, (double)(size - i));
21     for (i = 0; i < FECstrength_ ; i++)
22         dptr[i + 1] += dptr[i];
23     dptr[FECstrength_ + 1] = 1.0;
24     return dptr;
25 }
```

of having at most i corrupted bits. Given packet size size, the probability of having exactly i corrupted bits is $\binom{\texttt{size\_}}{\texttt{i}} (\texttt{rate\_})^{\texttt{i}} (1 - \texttt{rate\_})^{\texttt{size-i}}$, as shown in Line 20, where rate_ is the bit error probability. Lines 21–23 compute the cumulative summation of dprt. Note that Line 23 sets dptr[FECstrength_ + 1] to 1.0 since a packet is considered to be in error if the number of corrupted bits is greater than FECstrength_.

12.3.3 Guidelines for Implementing a New Error Model in NS2

In order to implement a new error model in NS2, we need to follow the three steps below:

(i) Design and create an error model class in OTcl, C++, or both domains.
(ii) Configure the parameters of the error model object such as error probability (rate_), error unit (unit_), random variable (ranvar_).
(iii) Insert an error model into the network (e.g., by using instproc lossmodel{lossobj from to} or instproc link-lossmodel{lossobj from to} of class Simulator).

Example 12.8. Consider the simulation script in Program 9.1, which creates a network as shown Fig. 9.3. Include an error model with packet error probability 0.1 for the link connecting nodes n1 and n3.

Tcl Simulation Script:

```
1 set ns [new Simulator]
2 set n1 [$ns node]
3 set n2 [$ns node]
4 set n3 [$ns node]
5 $ns duplex-link $n1 $n2 5Mb 2ms DropTail
6 $ns duplex-link $n2 $n3 5Mb 2ms DropTail
7 $ns duplex-link $n1 $n2 5Mb 2ms DropTail

8 set em [new ErrorModel]
9 $em set rate_ 0.1
10 $em unit pkt
11 $em ranvar [new RandomVariable/Uniform]
12 $em drop-target [new Agent/Null]
13 $ns link-lossmodel $em $n1 $n3

14 set udp [new Agent/UDP]
15 set null [new Agent/Null]
16 set cbr [new Application/Traffic/CBR]
17 $ns attach-agent $n1 $udp
18 $ns attach-agent $n3 $null
```

```
19 $cbr attach-agent $udp
20 $ns connect $udp $null
21 $ns at 1.0 "$cbr start"
22 $ns at 100.0 "$cbr stop"
23 $ns run
```

where Lines 8–13 are included (into the simulation script in Program 9.1) in order to impose error on packet transmission. Note that the OTcl command unit{u} sets variable unit_ to the value corresponding to the input argument u. The possible values of u include "time", "byte", "pkt", and "bit".

Exercise 12.9. From Example 12.8, collect statistics for packets which are in error and not in error. Verify that the packet error probability is 0.1. Adjust simulation time if necessary. How long must your simulation be to ensure the convergence of 0.1 error probability ?

- Initially set link bandwidth to be 5 Mbps.
- Change the bandwidth to be 500 Kbps. What happen to the measured convergence time ? Explain why.

Exercise 12.10. Consider a two state error model, which consists of *good* and *bad* states. Packet transmission in a good state is always error free, while packet transmitted in a bad state is always corrupted. The time that an error model stays in good and bad states is exponentially distributed with means t_{good} and t_{bad}, respectively. Write a simulation script for the above two state error model with $t_{good} = 10$ sand $t_{bad} = 1$ s. Verify the results and show the convergence time.

12.4 Bit Operations in NS2

12.4.1 Bit Masking

Bit masking is a bit transformation technique, which can be used for various purposes. Given a *mask*, a *bit masking process* transforms an *original value* to a *masked value* (see Fig. 12.9). In this section, we will show two examples of bit masking: subnet masking and modulo masking.

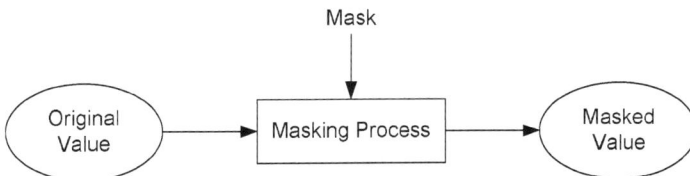

Fig. 12.9. Bit masking.

Subnet Masking

A 4-byte IP address can be divided into host address and network address. While a host address identifies a host (e.g., a computer), a network address characterizes a group of hosts. A host is given a host IP address as its identification and a 4-byte *subnet mask* which identifies its network. A subnet mask consists of all-one upper bits and all-zero lower bits (i.e., of format "$1\cdots10\cdots0$"). For a given host IP address and a subnet mask, the network IP address can be determined as follows:

$$\text{Network IP Address} = \text{Host IP Address \& Subnet Mask} \qquad (12.1)$$

where & is a bitwise "AND" operator.

Example 12.11. A class-C (i.e., subnet mask = 255.255.255.0) host IP address 10.1.2.3 has the network IP address of

$$(10.1.2.3)\&(255.255.255.0) = (10\&255).(1\&255).(2\&255).(3\&0) = 10.1.2.0 \tag{12.2}$$

In fact, all class-C IP addresses whose first three bytes are 10.1.2 have the same network address. Correspondingly, a class-C network address corresponds to 256 IP addresses.

From the above example, the original value (i.e., host IP address) 10.1.2.3 is *masked* (by using bitwise "and") with a *mask* 255.255.255.0 (i.e., class C subnet mask) such that the *masked value* (i.e., network IP address) is 10.1.2.0.

Modulo Masking

Modulo is a remainder computation process. Suppose $a = b \times c + d$. Then $a\%c = d$, where % is a modulo operator. Bit masking can also be used as a modulo operator with $c = 2^n$ where n is a positive integer.

To implement a modulo masking, the upper and lower bits of a modulo mask are set to contiguous zeros and contiguous ones, respectively (i.e., of format "$0\cdots01\cdots1$"), and the masking operation is a bitwise "AND" operation. Suppose, an original value is of format $xx..xx$, where x can be zero or one. The modulo masking applies bitwise "AND" to an original value and the modulo mask, and obtains the masked value as follows:

$$
\begin{aligned}
\text{original value} &= x\cdots xx\cdots x \\
\text{upper-bound mask} &= 0\cdots01\cdots1 \\
\text{masked value} &= 0\cdots0x\cdots x.
\end{aligned}
$$

Suppose the number of one-bits of a modulo mask is n. The bits whose positions are greater than n are removed during a masking process, and the masked value is bounded by 2^n-1. On the other hand, the bits whose positions are not greater than n are kept unchanged. These lower order bits in fact represent the remainder when the original value is divided by 2^n. Modulo masking is therefore equivalent to a modulo operation.

Exercise 12.12. Let a modulo mask be 64. Show that the modulo masking and modulo operation are equivalent for the following original values: 63, 64, 65, 127, 128, and 129.

Exercise 12.13. Consider a ball color-number matching experiment, where balls are fed one-by-one to an observer. Each ball is masked with a color and a number. The color can be either black or white, while the unique number is increased one-by-one as the balls are fed to the observer. From time to time, the observer is given a number and is asked to identify the color of one of the 64 most recently observed balls. Design a memory-friendly approach for the observation.

We summarize the masking components of subnet masking and modulo masking in Table 12.5. Note that, both subnet masking and modulo masking use a bitwise "AND" as their mask operation. Since their masks are different, the implications for their masked value are different.

12.4.2 Bit Shifting and Decimal Multiplication

Another important bit operation is bit shifting which is equivalent to decimal multiplication. If a binary value is shifted to the left by n bits, the corresponding decimal value will increase by 2^n times. Similarly, a binary number right shifted by n bits returns the quotient of the decimal value divided by 2^n.

To prove the above statement, consider an arbitrary value $y = \sum_{m=0}^{M} x_m 2^m$, where $x_m \in \{0,1\}, m = \{0,\cdots,M\}$. Let $y << n$ denote the value of y after being shifted to the left by n bits. Then

$$y << n = \left(\sum_{m=0}^{M} x_m 2^m\right) << n = (\underbrace{xx\cdots x}_{M \text{ bits}}) << n$$
$$= \left(\sum_{m=0}^{M} x_m 2^{m+n}\right) = (\underbrace{xx\cdots x}_{M \text{ bits}}\underbrace{00\cdots 0}_{n \text{ bits}}). \quad (12.3)$$

Table 12.5. Components of subnet masking and modulo masking.

Masking components	Subnet masking	Modulo masking
The mask	$1\cdots 10\cdots 0$	$0\cdots 01\cdots 1$
The mask operation	Bitwise "AND"	Bitwise "AND"
Masked value	Network IP address	Remainder

Suppose $y = \sum_{m=0}^{M} x_m 2^m$. We have

$$y \times 2^n = \left(\sum_{m=0}^{M} x_m 2^m \right) \times 2^n = \left(\sum_{m=0}^{M} x_m 2^{m+n} \right) \qquad (12.4)$$

which is the same as Eq. (12.3). This proves the first part (i.e., left-shifting) of the above statement. The second part of (i.e., right-shifting) the statement can be proven similarly and is omitted for brevity.

The relationship between bit shifting and decimal multiplication can be summarized as follows:

- An n-bit left shift results in multiplication of the decimal value by 2^n.
- An n-bit right shift returns the quotient when the decimal value is divided by 2^n.

Exercise 12.14. What are the values of 2, 3, 31, 45, and 56, when shifted to the left and right by 1, 2, and 3 bits ?

12.5 Chapter Summary

This chapter presents three major helper classes: timers, random number generators, and error models. The first helper class is `Timer`. Unless restarted or cancelled, class `Timer` waits for a certain time and takes expiration actions. Class `Timer` provides three main interface functions to start, restart, and cancel the waiting process. Class `Timer` is usually cross-referenced to another object, which contains an instruction on how to perform expiration actions. At the expiration (i.e., when `expire(e)` is invoked), the timer informs the object to execute the expiration actions. The object, on the other hand, may start, restart, or cancel the timer through its reference to the timer.

The second part of this chapter demonstrates how NS2 implements Random Number Generator (RNG) to generate random variables. By default, NS2 always seeds the simulation with 1–meaning NS2 is *deterministic* by default. To introduce randomness into simulation, we need to seed `defaultRNG` differently.

The last helper class is class `ErrorModel` which is a packet error simulation class. Derived from class `Connector`, it can be inserted into a network by using OTcl instprocs (e.g., `lossmodel{...}` and `insert-lossmodel{...}`). Class `ErrorModel` simulates packet error upon a packet reception. If the packet is simulated to be in error, it will either drop or mark the corrupted packet with an error flag. Otherwise, it will forward the packet to its downstream object.

This chapter also presents two main bit operations: bit masking and bit shifting. Bit masking is a bit transformation process which can be used for various purposes. This chapter gives two examples of bit basking. One is subnet masking, which is a process to determine a network for an IP address. Another is a modulo masking which can be used as a modulo operation. As another bit operation, bit shifting can be used for decimal multiplication or division. Shifting an original value to the left and right by n bits is equivalent to multiplying and dividing the original value by 2^n, respectively.

13

Processing an NS2 Simulation: Debugging, Tracing, and Result Compilation

Having discussed the main NS2 components as well as the methodology to configure a network, we now present the final three parts in network simulation. These three supplementary steps in network simulation are: *debugging, tracing,* and *compilation of simulation results*. Debugging is a process of removing programming errors. Variable tracing tracks changes in variables under consideration. Packet tracing records the details of packets passing through network checkpoints. Simulation result compilation collects information and computes relevant performance measures from the simulation. This chapter discusses the details of debugging, variable tracing, packet tracing, and result compilation in Sections 13.1, 13.2, 13.3, and 13.4, respectively. Finally, the chapter summary is given in Section 13.5.

13.1 Debugging: A Process to Remove Programming Errors

A programming error is usually referred to as a *bug*. The process of locating and fixing the error is usually called *debugging*. This section discusses two types of programming errors (i.e., bugs) and provide guidelines for debugging in NS2.

13.1.1 Types of Programming Errors

Based on the NS2 architecture, programming errors can be classified into compilation errors and runtime errors.

Compilation Errors (C++ Only)

This type of errors occurs during a compilation process, which consists of two phases. The first phase converts C++ files (with extension ".cc,h") into object files (with extension ".o"). In this phase, errors may occur if the compiler

T. Issariyakul, E. Hossain, *Introduction to Network Simulator NS2*,
DOI: 10.1007/978-0-387-71760-9_13, © Springer Science+Business Media, LLC 2009

is unable to understand the C++ codes. In this case, the compiler will show error messages on the screen, indicating where and why the errors occurred. Examples of C++ compilation errors include:

- Incorrect C++ syntax
- A use of undefined variables and/or functions

In the second phase, the compiler links the created object files and creates an executable **ns** file. An error in this phase is caused by improper linkage of C++ files. Again, the compiler will show error messages on the screen, indicating where and why the errors occur. Examples of C++ linking errors include:

- *Instantiate an object from an abstract class*: During a linking process, an error will occur if an object is instantiated from an abstract class, which leaves at least one pure virtual function unimplemented.
  A proper solution to this error is to provide implementation for the pure virtual function. However, for simplicity (but not for appropriateness), a user may provide empty implementation for the pure virtual function to remove the error.
- *Modifying a base class without creating the object files of the child classes*: This error usually occurs when the dependency in the **Makefile** is not properly defined. When a certain class is modified, the compiler does not recreate object files of the child classes. The solution is to define the dependency in the **Makefile** properly, or to remove all related object files before compiling the codes.

Note that OTcl is a scripting language. There is no need to compile OTcl code prior to the execution. Therefore, compilation errors do not occur in the OTcl domain.

Runtime Errors

This type of errors occurs during NS2 simulation. It is caused by improper OTcl and/or C++ programming. Since the OTcl domain implements error message trapping mechanism, an OTcl error message contains detailed and useful information. Each error message indicates where and why the error occurred. The C++ domain, on the other hand, does not implement error trapping. Generally the error messages in this case (e.g., segmentation fault) are fairly short and do not contain much information. Examples of OTcl runtime errors include

- Incorrect OTcl syntax
- Referring to instvars, instprocs, or commands which do not exist.

Examples of C++ runtime errors include

- *Segmentation fault*: This is usually caused due to invalid access to a memory content. For example, trying to access "a[6]" would cause a segmentation fault if "a" was declared as "int a[3];".

- *Not implementing a mandatory (non-pure virtual) function*: Apart from using a pure virtual function, NS2 provides another way to force a child class to implement a mandatory function. Here, NS2 may implement error-like actions (e.g., print out an error message) in the base class. If a child class does not implement this mandatory function, the function of the base class will be invoked and the error-like actions will be taken. Examples of this type of errors are the implementation of functions sendmsg(...) and sendto(...) of class Agent in file ~ns/common/agent.cc.

13.1.2 Debugging Guidelines

After identifying the type of programming errors (i.e., bugs), the next step is to locate the programming codes which cause the errors and to fix the errors. This section provides guidelines which facilitate the debugging process.

In general, two useful debugging tools are breakpoints and variable viewers. A breakpoint is the place where a program is intentionally stopped during an execution. By strategically placing breakpoints in the program, a programmer can easily find out the statement(s) responsible for an error. A variable viewer, on the other hand, allows the programmers to determine the values of variables, and analyze the cause of an error.

There are two debugging methods in NS2. The first method is to use debugging tools. For Tcl, NS2 supports Don Libs' debugger [23], while the standard GNU debugger [24] can be used to debug the C++ codes. The second method is to manually debug the program. Table 13.1 shows a list of OTcl and C++ commands which can be used for manual debugging.

Table 13.1. Debugging command in the OTcl C++ domains.

Tools	OTcl	C++
Breakpoints	"gets stdin"	"getchar","cin"
Variable viewer	"puts"	"printf","cout"
Simulation time	"Simulator::now"	"Scheduler::clock"
Cross-domain function invocation	Commands	"Tcl::evalf"
Cross-domain variable retrieval	Bound variables	Bound variables
Simulator object retrieval	N/A	"Simulator::instance()"
Scheduler object retrieval	N/A	"Scheduler::instance()"
Passing an OTcl value to the C++ domain	N/A	"TclObject::result"
OTcl to C++ variable conversion	N/A	"TclObject::name"
C++ to OTcl variable conversion	N/A	"TclObject::lookup"

Program   13.1   OTcl   commands   `show-target-class`   and `show-target-class` of C++ class `TcpAgent`.

```
    //~/ns/tcp/tcp.cc
1   int TcpAgent::command(int argc, const char*const* argv)
2   {
3       ...
4       Tcl& tcl = Tcl::instance();
5       if (strcmp(argv[1], "show-target-class") == 0) {
6           Simulator& sim = Simulator::instance();
7           tcl.evalf("puts [format \"%%1.1f OTcl class of
                TcpAgent::target is [%s info class]\" [%s now] ]",
                target_->name(),sim.name());
8           cout<<"Press RETURN to continue!!\n\n";
9           getchar();
10          return (TCL_OK);
11      }
12      if (strcmp(argv[1], "show-target-address") == 0) {
13          Scheduler& sch = Scheduler::instance();
14          tcl.evalf("%s target",this->name());
15          Connector *conn=(Connector*)TclObject::lookup(tcl.result());
16          cout<<sch.clock()<<" \[$tcp target\] returns OTcl reference
                string "<<tcl.result()<<" and C++ address "<<conn<<"\n"
17          cout<<sch.clock()<<" Variable TcpAgent::target_
                corresponds to OTcl reference string "<<target_->name()
                <<" and C++ address "<<target_<<"\n";
18          cout<<"Press RETURN to continue!!\n\n";
19          getchar();
20          return (TCL_OK);
21      }
22      ...
23  }
```

To debug NS2 codes, it is usually useful to identify objects and/or the types of objects. In this example, we develop 2 non-built-in OTcl commands–namely `show-target-class` and `show-target-address` below. Shown in Program 13.1, these two OTcl commands show the class and address, respectively, of the target of a `TcpAgent` object.

Example 13.1. Consider Example 10.1 which implements the network in Fig. 9.3. Let us insert the following lines immediately before Line 15 in Example 10.1:

```
1 puts "The reference string for \$tcp is $tcp"
2 puts "Press RETURN to start the simulation!!"
3 gets stdin
4 $ns at 3.1 "$tcp show-target-class"
5 $ns at 5.1 "$tcp show-target-address"
```

```
6 $ns at 10.1 "$ns halt"
7 $ns run

>>ns tcp-dbg.tcl
The reference string for $tcp is _o55
Press RETURN to start the simulation!!
<RETURN>
3.1 OTcl class of TcpAgent::target is Classifier/Hash/Dest
Press RETURN to continue!!
<RETURN>
5.1 [$tcp target] returns OTcl reference string _o12 and C++
  address 0xd655c0
5.1 Variable TcpAgent::target_ corresponds to OTcl reference
  string _o12 and ++ address 0xd655c0
Press RETURN to continue!!
<RETURN>
```

Here, the lines with <RETURN> are actually blank lines, where the program is paused and waits for a <RETURN> keystroke.

Lines 5–11 in Program 13.1 shows the details of OTcl commands show-target-class{}. Line 6 retrieves the Simulator object and stores it in variable sim. In Line 7, function evalf(...) of class Tcl evaluates the Tcl statement in the same manner as printf(...) (see also Fig. 13.1). It puts the values stored in target_->name() and sim->name() as the first and second arguments, respectively, and passes the entire statement to the Tcl interpreter. Here, function name() defined in class TclObject is used to translate the C++ variables target_ and sim to OTcl reference strings.

Lines 12–21 in Program 3.1 shows the details of the OTcl command show-target-address{}. Line 13 first retrieves the Scheduler object and stores it in variable sch. Line 14 asks the Tcl to interpret "_o55 target", where _o55 is the OTcl reference string corresponding to the current TCP object. Line 15 uses function result() to obtain OTcl reference string of the target of the TcpAgent object. It obtains the C++ object corresponding to the string using function lookup(...). The obtained C++ object is then cast to a Connector pointer and stored in variable conn.

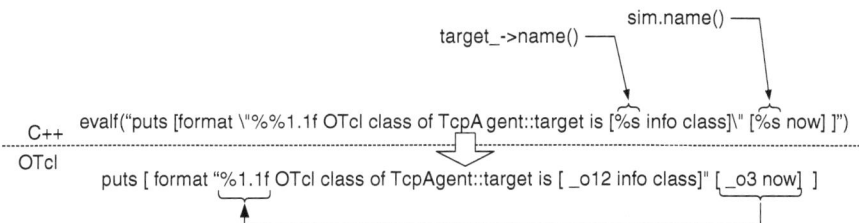

Fig. 13.1. Details of Line 7 in Program 13.1.

13.2 Variable Tracing

Variable tracing is responsible for recording changes in variables of a TclObject under consideration. Variable tracing consists of 6 main components: a TclObject, an `InstVar` object, a `TracedVar` object, a tracer, a Tcl channel, and a trace file. The relationship among the above six components for variable tracing are shown in Fig. 13.2. Here, the solid lines represent the configuration after NS2 initialization, while the dotted lines are the relationship which are created later by NS2 users. The details of the relationship are as follows:

- Class `TclObject` declares a pointer `instvar_` to an `InstVar` object (Line 6 in Program 13.2). This object is the head of the link list which contains all bound OTcl instvars of the TclObject.
- Class `Instvar` holds information (e.g., the name of the bound OTcl variable) about a bound instvar. It also contain a pointer `traced_var` to a `TracedVar` object.
- A `TracedVar` object is responsible for keeping track of the change in its value and reporting the change to a tracer.
- Receiving a report from a `TracedVar` object, a tracer records the changes in variable. In most cases, the record is written to a Tcl channel which is attached to a traced file. Unless a traced is explicitly given, NS2 uses the TclObject as a tracer for its `TracedVar` object.

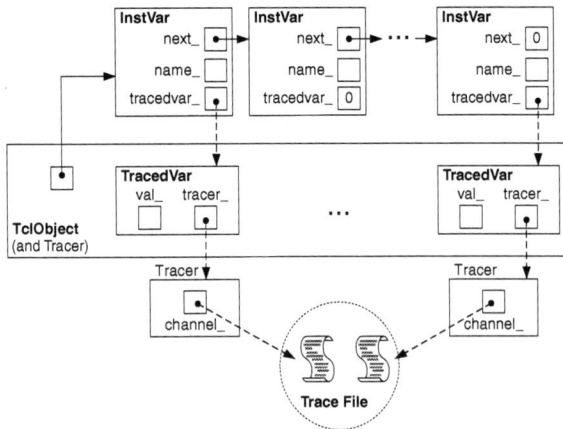

Fig. 13.2. An architecture of variable tracing.

13.2.1 Activation Process for Variable Tracing

Variable tracing can be activated in the OTcl domain using the following three steps:

Program 13.2 Declaration of classes `TclObject` and `Instvar`

```
   //~/tclcl/tclcl.h
1  class TclObject {
2  public:
3      ....
4      virtual void trace (TracedVar*);
5  protected:
6      int traceVar(const char* varName, TclObject* tracer);
7      InstVar* instvar_;
8      ....
9  }

   //~/tclcl/Tcl.cc
10 class InstVar {
11 protected:
12     InstVar(const char* name);
13     const char* name_;
14     TracedVar* tracedvar_;
15 public:
16     virtual ~InstVar();
17     InstVar* next_;
18     inline const char* name() { return name_; }
19     inline TracedVar* tracedvar() { return tracedvar_; }
20     inline void tracedvar(TracedVar* v) { tracedvar_ = v; }
21 };
```

(i) Specify a `TracedVar` object.
(ii) Create a trace file.
(iii) Attach the created trace file to a tracer.

Let `obj`, `traced_var`, and `tracer` be a `TclObject`, a `TracedVar` object, and a tracer, respectively. Then, the above three steps are carried out by the following three OTcl statements (respectively):

```
obj trace $traced_var $tracer
set fh [open "filename" w]
$tracer attach $fh
```

Optionally, we can use the TclObject `obj` as a tracer, by replacing the first and last statements above with the following statements (respectively):

```
obj trace $traced_var
obj attach $fh
```

Note that variable tracing can be applied only to a traceable `TracedVar` object, which complies with the following two criteria:

• The instvar `$traced_var` must be bound to the C++ domain.

- The C++ class of the bounded variable must derive from class `TracedVar`.
- The bound C++ variable must be a member of the class of the TclObject `obj`.

Example 13.2. Suppose we would like to trace variable `t_seqno_` of a `TcpAgent` object `$tcp` in Example 10.1, and store the trace in file `trace.txt`. We may include the following codes into the Tcl simulation script:

```
1 $tcp trace t_seqno_
2 set trace_ch [open "trace.txt" w]
3 $tcp attach $trace_ch
```

Here, Line 1 indicates the need to trace the `TracedVar` object `t_seqno_` of tcp using tcp as a tracer. Line 2 creates a trace file "`trace.txt`", and Line 3 attaches the file to the tracer `$tcp`. These three lines inform NS2 to record all the changes in variable `t_seqno_` associated with `$tcp` in file "`trace.txt`". After simulation, the following trace file whose format complies with Fig. 13.3 is created:

```
...
4.06820 0 0 2 0 t_seqno_ 14
4.06986 0 0 2 0 t_seqno_ 15
4.07153 0 0 2 0 t_seqno_ 9
4.07153 0 0 2 0 t_seqno_ 10
4.08468 0 0 2 0 t_seqno_ 14
...
```

Time	Source Address	Source Port	Destination Address	Destination Port	Name of the TracedVar	Value

Fig. 13.3. Trace format defined in class `TcpAgent`.

13.2.2 Instvar Objects

Again, an `InstVar` object acts as a reference to an instvar in the OTcl domain. Since a TclObject can have several bound instvars, a TclObject resorts to a link-list structure of `InstVar` objects (see Figure 13.2). Here, a TclObject only needs to maintain a pointer to the head of the link-list.

The details of class `InstVar` are shown in Lines 10–21 of Program 13.2. Class `InstVar` has 3 main variables: `next_`, `name_`, and `tracedvar_`. The pointer `next_` provides a support to create a link list. The variable `name_` contains the OTcl instvar name. The pointer `tracedvar_` points to a `TracedVar` object. We note here that not all OTcl instvars need to be traceable. As a result, not all `InstVar` object in the link list has it pointer `tracedvar_` configured. From Fig. 13.2, the second `InstVar` object in the link list does not have it pointer point a `TracedVar` object.

Class InstVar has three main functions: name(), tracedvar(), and tracedvar(v). These functions are used to configures internal variables of class InstVar.

13.2.3 TracedVar Objects

A TracedVar object is a member variable of a TclObject, which is equipped with tracing capability. Overloading the basic operators (e.g., "+", "-", "*", "/"), TracedVar objects are used in place of ordinary C++ variables. Each overloaded operator executes the basic operation and reports the change in value to a tracer.

NS2 implements TracedVar objects through an abstract class TracedVar. From Program 13.3, class TracedVar has four main variables:

name_ The name of this TracedVar object in the OTcl domain
owner_ A pointer to the TclObject which declares this TracedVar object as its variable
tracer_ A pointer to the TclObject responsible for keeping track of this TracedVar object
next_ A pointer to the next TracedVar object

Class TracedVar has one pure virtual function and six regular functions. The pure virtual function value(...) returns the value of the TracedVar object[1] (e.g., see the implementation of function value(...) of class TracedInt below). The other six functions act as interface functions to set and retrieve variables of class TracedVar.

NS2 has two built-in classes derived from class TracedVar: TracedInt and TracedDouble.[2] These two classes comply with the following mechanism. From Program 13.3, class TracedInt declares an int variable val_ (Line 28) to store its current value. It informs a tracer of changes in its value by feeding itself as an input argument of function trace(...) of the tracer. It also uses a single point of value assignment–function assign(newval) in Lines 24–25. When an overloaded operator is invoked, the basic operation is executed and the execution result is stored in variable val_ using function assign(newval). The statement tracer_->trace(this) invoked from within function assign(newval) ensures that all the changes in val_ are recorded by the tracer.

13.2.4 Tracers

A tracer is an object of class TclObject which is responsible for recording changes in the value of a TracedVar object. Class TclObject defines a

---

[1] Since the base class TracedVar does not define a variable to store the value, the function value(...) must be declared as pure virtual.

[2] For brevity, the following discussion is based on class TracedInt only.

Program 13.3 Declaration of classes `TracedVar` and `TracedInt`, and function `assign` of class `TracedInt`.

```
   //~/tclcl/tracedvar.h
1  class TracedVar {
2  public:
3      TracedVar();
4      virtual ~TracedVar() {}
5      virtual char* value(char* buf, int buflen) = 0;
6      inline const char* name() { return (name_); }
7      inline void name(const char* name) { name_ = name; }
8      inline TclObject* owner() { return owner_; }
9      inline void owner(TclObject* o) { owner_ = o; }
10     inline TclObject* tracer() { return tracer_; }
11     inline void tracer(TclObject* o) { tracer_ = o; }
12     TracedVar* next_;
13 protected:
14     TracedVar(const char* name);
15     const char* name_;
16     TclObject* owner_;
17     TclObject* tracer_;
18 };

19 class TracedInt : public TracedVar {
20 public:
21     TracedInt() : TracedVar() {}
22     TracedInt(int v) : TracedVar(), val_(v) {}
23     virtual ~TracedInt() {}
24     inline int operator++() { assign(val_ + 1); return val_; }
25     inline int operator=(int v) { assign(v); return val_; }
26 protected:
27     virtual void assign(const int newval);
28     int val_;
29 };

   //~/tclcl/tracedvar.cc
30 void TracedInt::assign(int newval)
31 {
32     if (val_ == newval)
33         return;
34     val_ = newval;
35     if (tracer_)
36         tracer_->trace(this);
37 }
```

function `trace(v)` to record the value of a `TracedVar` object `*v`. The function `trace(v)` is invoked from within every overloading operator of `TracedVar` objects to reports changes in a variable `*v`. Since function `trace(v)` is defined in class `TclObject`, every TclObject can be used as a tracer.

As shown in Lines 1–4 of Program 13.4, function `trace(v)` of class `TclObject` simply prints an error message on the screen (Line 3). This implementation (weakly) forces the derived classes of class `TclObject` to implement function `trace(v)` without declaring the function as pure virtual. If a derived class does not implement function `trace(v)`, it can still instantiate an object and operate normally. However, an error message (Line 3) will be shown on the screen, if function `trace(v)` is invoked.

As an example, consider class `TcpAgent` in Program 13.4. Here, function `trace(v)` of class `TcpAgent` in Lines 5–8 of Program 13.4 simply invokes

---

Program 13.4 Function `trace` of classes `TclObject` and `TcpAgent`, and function `traceVar` of class `TcpAgent`.

```
     //~/tclcl/Tcl.cc
1    void TclObject::trace(TracedVar*)
2    {
3        fprintf(stderr, "SplitObject::trace called in
                            the base class of %s\n",name_);
4    }

     //~/ns/tcp/tcp.cc
5    void TcpAgent::trace(TracedVar* v)
6    {
7            traceVar(v);
8    }

9    void TcpAgent::traceVar(TracedVar* v)
10   {
11       Scheduler& s = Scheduler::instance();
12       char wrk[TCP_WRK_SIZE];
13       double curtime = &s ? s.clock() : 0;
14       if (v == &cwnd_)
15           ...
16       else if (v == &t_rtt_)
17           ...
18       else
19           snprintf(wrk, TCP_WRK_SIZE,
20               "%-8.5f %-2d %-2d %-2d %-2d %s %d\n",
21               curtime, addr(), port(), daddr(), dport(),
22               v->name(), int(*((TracedInt*) v)));
23       (void)Tcl_Write(channel_, wrk, -1);
24   }
```

function traceVar(v). Based on the input TracedVar pointer v, function tracedVar(v) of class TcpAgent stores a string in a local variable wrk, and invokes function Tcl_Write(...) to write the string wrk to a Tcl_Channel object channel_ (see Line 23). In most cases, channel_ (defined in class Agent; see Line 4) is attached to a trace file, and Tcl_Write(...) simply prints the string to the attached trace file.

13.2.5 Connections Among a TclObject, a TracedVar Object, a Tracer, and a Trace File

A Connection from a TracedVar Object to Tracer

A connection from a TracedVar object to a tracer can be created using the OTcl command trace{...} (see Section 13.2.1) whose syntax is shown below

    $obj trace $traced_var[$tracer]

Again, if the optional argument $tracer$ is not present, the TclObject $obj will be used as a tracer.

The OTcl command informs the TclObject (i.e., $obj) to trace its variable whose OTcl bound instvar name is $traced_var trace{...} The details of the OTcl command trace{...} is shown in Lines 1–12 of Program 13.5. Lines 5 and 7 sets a variable tracer to be this and the second input argument of the OTcl command (i.e., argv[3]), if it exists, respectively. Then Line 8 invokes function traceVar(argv[2],tracer) to create a connection from a TraceVar object to a tracer input TracedVar object.

Function traceVar(varName,tracer)[3] is shown in Lines 13–24 of Program 3.5. This function creates a connection from a TracedVar object whose OTcl instvar name is varName to a tracer object tracer. Lines 15-16 locates an entry of the InstVar linked list whose name_ matches with the string varName. When the component is found, Line 17 ensures that the matching instvar contains a reference to a TracedVar object (see Line 8 in Program 13.2). Then, Line 18 sets the variable tracer_ of the matched TracedVar object to the value as specified in the input argument tracer (see the detail of function tracer({...}) of class TracedVar in Line 11 in Program 13.3). Line 18 tells the member pointer tracer_ of the located InstVar *p to point to the same as the input argument tracer. Line 19 informs the tracer (i.e., the input argument) to trace the TracedVar object associated with the InstVar object *p (see function trace(v) of class TcpAgent in Program 13.4). Note that, prior to the Lines 13–24 in Program 13.5, the TracedVar object must be created, and the address of the created TracedVar object must be assigned to variable tracedvar_ (declared in Line 12 in Program 13.2) of all the linked list components.

---

[3] This function is different from that in Program 13.4, since the input arguments are different.

Program 13.5 Command `trace` and function `TraceVar` of class `TclObject`.

```
//~/tclcl/Tcl.cc
1    int TclObject::command(int argc, const char*const* argv)
2    {
3        if (argc > 2) {
4            if (strcmp(argv[1], "trace") == 0) {
5                TclObject* tracer = this;
6                if (argc > 3)
7                    tracer = TclObject::lookup(argv[3]);
8                return traceVar(argv[2], tracer);
9            }
10       }
11       return (TCL_ERROR);
12   }

13   int TclObject::traceVar(const char* varName, TclObject* tracer)
14   {
15       for (InstVar* p = instvar_; p != 0; p = p->next_) {
16           if (strcmp(p->name(), varName) == 0) {
17               if (p->tracedvar()) {
18                   p->tracedvar()->tracer(tracer);
19                   tracer->trace(p->tracedvar());
20                   return TCL_OK;
21               }
22           }
23       }
24   }
```

Connection Between a Tracer and a Trace File

A tracer usually employs a Tcl channel to record changes in a `TracedVar` object in a trace file. For example, `TcpAgent` defines a Tcl channel `channel_` in its base class `Agent` (See Line 4 in Program 13.6). This Tcl channel is usually attached to a trace file "`file`" via an OTcl command `attach{file}` of class `TcpAgent`.

The details of the OTcl command `attach{file}` are shown in Lines 7–21 in Program 13.6. Here, Line 12 converts the input file name to a string id. Line 13 retrieves an OTcl file reference corresponding to `id`, and stores it in variable `channel_`. After this point, a connection to a trace file is created within a tracer, and the tracer is able to pass variable changing messages to the attached trace file.

13.2.6 Trace File Format

The trace file format defines how the variable details are recorded in a trace file. The format is defined in function `trace(v)` of class `TclObject` (i.e., a

---

Program 13.6 Declaration and OTcl command `attach` of class `Agent`.

```
    //~/ns/common/agent.h
1   class Agent : public Connector {
2   ...
3   protected:
4       Tcl_Channel channel_;
5   ...
6   }

    //~/ns/tcp/tcp.cc
7   int TcpAgent::command(int argc, const char*const* argv)
8   {
9       ...
10      if (strcmp(argv[1], "attach") == 0) {
11          int mode;
12          const char* id = argv[2];
13          channel_ = Tcl_GetChannel(tcl.interp(), (char*)id, &mode);
14          if (channel_ == 0) {
15              tcl.resultf("trace: can't attach %s for writing", id);
16              return (TCL_ERROR);
17          }
18          return (TCL_OK);
19      }
20      ...
21  }
```

---

tracer) under a `printf`-like environment. For example, the default trace format of a `TcpAgent` object is defined in its function `trace(v)` in (Fig. 13.3). Defined in Lines 19-22 of Program 13.4, the trace file format for class `TcpAgent` is shown in Fig. 13.3, where each field of the trace format is separated by a space.

Example 13.3. Consider Example 13.2. An operator "++" of a `TracedVar` object is overloaded by a function `operator++()` defined in Line 24 of Program 13.3 (see also Fig. 13.4). Function `operator++()` invokes function `assign(val_+1)` to increment the value stored in its variable `val_` by 1, store the incremented value in variable `val_`, and the change in the variable `val_`. From within the function `assign(newval)`, a `TracedVar` object (e.g., `t_seqno_`) executes "`tracer_->trace(this)`" where `tracer_` and `this` are the associated tracer and the address of the `TracedVar` object, respectively. Since, in this case, the variable `tracer` is a `TcpAgent` object, function `trace(v)` in Line 5–8 of Program 13.4 is invoked, and a trace string is printed to the trace file according to the format specified in Lines 19–22 of Program 13.4.

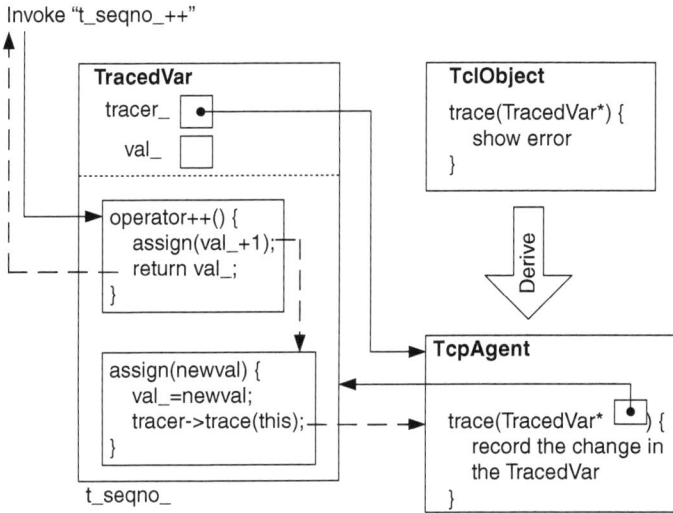

Fig. 13.4. The mechanism of function `operator++()` of class `TcpAgent`.

13.3 Packet Tracing

Packet tracing records packet details when they pass through network checkpoints, where a `Trace` object is install. This section discuss packet tracing mechanism through Example 13.4 below:

Example 13.4. Consider Example 9.1. A typical way to activate packet tracing which records changes in a file "`out.tr`" is to insert the following OTcl statements after Line 4 in Example 9.1.

```
set f [open out.tr w]
$ns trace-all $f
```

where the upper line creates a variable "f" which is a reference to file "`out.tr`". The lower line informs NS2 to activate the packet tracing mechanism and to record the details of packets flowing through the packet tracing objects in file "`out.tr`".

After adding the following statements at the end of the Tcl scripting file,

```
$ns at 0.0 "$cbr start"
$ns at 10.0 "$ns halt"
$ns run
```

we run the Tcl simulation script, and the trace result will be stored in file "`out.tr`". The followings are a part of the file "`out.tr`".

```
//out.tr
```

```
+ 0 0 2 cbr 210 ------- 0 0.0 2.0 0 0
- 0 0 2 cbr 210 ------- 0 0.0 2.0 0 0
r 0.002336 0 2 cbr 210 ------- 0 0.0 2.0 0 0
+ 0.00375 0 2 cbr 210 ------- 0 0.0 2.0 1 1
- 0.00375 0 2 cbr 210 ------- 0 0.0 2.0 1 1
. . .
```

We shall discuss the packet tracing OTcl configuration method and C++ internal mechanism implementation in Sections 13.3.1 and 13.3.2. Sections 13.3.3 discusses the details of the packet tracing helper class **Base Trace**. Various types of packet tracing objects are presented in Section 13.3.4. Finally, the packet trace format is shown in Section 13.3.5.

13.3.1 OTcl Configuration Interfaces

This section demonstrates how a packet tracing object is inserted into a network. We show packet trace configuration through a simplex-link with a drop-tail queue only. The readers are encouraged to look through the NS2 codes and find out more about packet tracing.

The key packet tracing configuration OTcl instprocs are given below:

- `Simulator::trace-all{file}`: Store the file handle `file` in instvar `traceAllFile_` of the Simulator (see Program 13.7).
- `Simulator::simplex-link{n1 n2 bw delay qtype}`: Creates a **Simple Link** object from node `n1` to node `n2` (see Section 7.1). If the instvar `traceAllFile_` of the Simulator exists, the instproc `trace-queue{n1 n2 file}` is invoked to configure the link as shown in Fig. 7.1.
- `Simulator::trace-queue{n1 n2 file}`: Creates and configures packet tracing objects in the link which connects Node `n1` to Node `n2` as shown in Fig. 7.1. Associates the tracing object to the file handle "`file`".
- `Simulator::create-trace{type file src dst}`: Creates a tracing object of type "`type`", attaches the file handle "`file`" to the created packet tracing object, and return the created tracing object to the caller.
- `Simulator::flush-trace{}`: Flushes the buffer of all packet tracing objects in the simulation.

Program 13.7 Instproc `trace-all` of class Simulator.

```
    //~/ns/tcl/lib/ns-lib.tcl
  1 Simulator instproc trace-all file {
  2     $self instvar traceAllFile_
  3     set traceAllFile_ $file
  4 }
```

- SimpleLink::trace{ns file}": Create packet tracing objects enqT_, deqT_, drpT_, and rcvT_, and configure them as shown in Fig. 7.1.

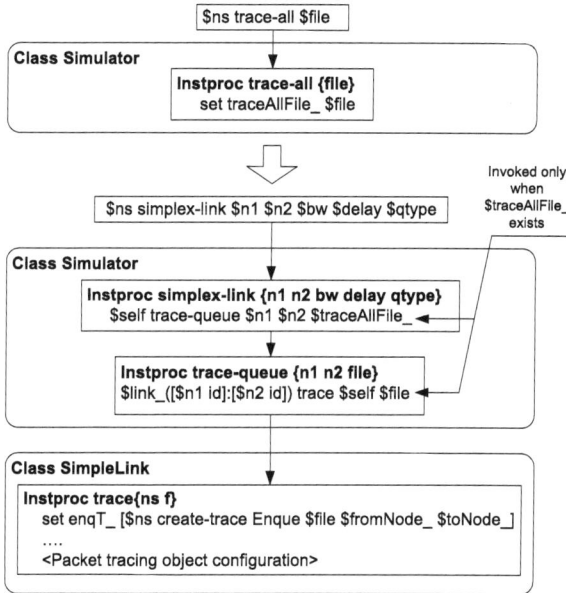

Fig. 13.5. The packet tracing configuration process of a SimpleLink object.

As an example, consider a packet tracing configuration for a SimpleLink object as shown in Fig. 13.5. The process starts with an activation of packet tracing through a statement $ns trace-all $file. This statement stores the input file handle file in a flag instvar traceAllFile_, which indicates whether the packet tracing is enabled. When other objects are created, packet tracing objects are inserted if instvar traceAllFile_ is not Null.

The next step is to create a SimpleLink object through an instproc "simplex-link{...}" of class Simulator. If instvar traceAllfile_ is not Null, instproc trace-queue{...} of class Simulator will be invoked. Instproc trace-queue{...} executes the statement "trace{ns file}" of the SimpleLink object, where ns and file are the Simulator and a trace file handle, respectively. From within the instproc trace{ns file}, packet tracing objects (e.g., enqT_), which are responsible for recording packet details, are created and configured.

$ns trace-all $file

where $ns and $file are the Simulator and a file handle, respectively (see Example 13.4).

Program 13.7 shows the details of instproc "`trace-all{file}`" of class
`Simulator`. This instproc stores the input file handle `file` in a flag instvar `traceAllFile_`, which indicates whether the packet tracing is enabled.
When other objects are created, packet tracing objects are inserted if instvar
`traceAllFile_` is not Null.

Similar to that of `Queue` and `QueueHandler` objects, a connection of `ARQTx`
and `ARQHandler` is created by the constructor of the `ARQTx` object. From Line
27 in Program 14.2, the `ARQTx` constructor instantiates an `ARQHandler` object,
`arqh_`, feeding itself as an input argument. The constructor of can `ARQHandler`
object (Line 5 in Program 14.1) stores the input argument in its variable
`arq_tx_`, crating a two-way connection between `ARQTx` and `ARQHandler` objects.

Instproc `trace-all{file}` of Class Simulator

Defined in Program 13.7, instproc `trace-all{file}` of class Simulator takes
a file handle `file` as an input argument (Line 1), and stores `file` in instvar
`traceAllFile_` (Line 3). The syntax of the instproc `trace-all{...}` is as
follows:

Instproc `simplex-link{...}` of Class Simulator

Instproc `simplex-link{...}` of class `Simulator` is used to create a link between two nodes (e.g., `n1` and `n2`; see Section 7.1 for the details of class
`SimpleLink`). Program 13.8 shows the part of instproc `simplex-link{...}`
which is related to packet tracing. Lines 6–7 create a `SimpleLink` object connecting node `n1` to node `n2`. The bandwidth and delay of the connecting
link are "bw" bps and "delay" seconds, respectively. The queue associated
with the link is of type "qtype". Line 9 stores instproc `traceAllFile_` in
a local variable `trace`, and Lines 10-12 execute "`$self trace-queue{n1 n2
trace}`" associated with the Simulator, if instvar `traceAllFile_` is not Null.

Instproc `trace-queue{n1 n2 file}` of Class Simulator

Lines 14–17 in Program 13.8 show the details of function `trace-queue{n1
n2 file}`. Again, class `Simulator` has an instance associative array `link_`.
The index of `link_` is of format "sid:did", where `sid` and `did` are node
IDs attaching to its beginning point and its ending point, respectively. The
instproc `trace-queue{n1 n2 file}` invokes instproc `trace{ns file}` associated with the `SimpleLink` object `link_([$n1 id]:[$n2 id])` (see Line 16),
to create and configure packet tracing components of the `SimpleLink` object.

Note that the input argument `$file` is a file handle. When `$ns trace-all
$file` is invoke the handle `$file` is stored in the instvar `traceAllFile_` of
the Simulator (see Program 13.7). Instproc `simplex-link{...}` (Line 9) retrieves the instvar `traceAllFile_`, and stores it in a local variable `trace`.

Program 13.8 Instprocs `simplex-link` and `trace-queue` of class Simulator.

```
//~/ns/tcl/lib/ns-lib.tcl
1 Simulator instproc simplex-link { n1 n2 bw delay qtype args } {
2       $self instvar link_
3       set sid [$n1 id]
4       set did [$n2 id]
5       ...
6       set q [new Queue/$qtype]
7       set link_($sid:$did) [new SimpleLink $n1 $n2 $bw $delay $q]
8       ...
9       set trace [$self get-ns-traceall]
10      if {$trace != ""} {
11              $self trace-queue $n1 $n2 $trace
12      }
13 }

14 Simulator instproc trace-queue { n1 n2 {file ""} } {
15      $self instvar link_ traceAllFile_
16      $link_([$n1 id]:[$n2 id]) trace $self $file
17 }
```

Essentially, this variable `trace` contains the file handle `$file`. This local variable (or equivalently the file handle) is then fed as the third input argument of instproc `trace-queue{...}` and as the second input argument of the instproc `trace{...}`, respectively.

Instproc trace{ns f} of Class SimpleLink

As shown in Program 13.9, instproc `trace{ns f}` of class `SimpleLink` takes two input arguments: the Simulator "ns" and a file handle "f". Line 5 stores the input file handle "f" in instvar `trace_`. Lines 6–9 create packet tracing objects `enqT_`, `deqT_`, `drpT_`, and `rcvT_`, by using the instproc `create-trace{...}` associated with the input Simulator ns. Lines 11–19 configure the created packet tracing objects as indicated in Fig. 7.1.

Instproc create-trace{type file src dst} of Class Simulator

In Program 13.10, instproc `create-trace{type file src dst}` creates and configures a packet tracing object whose type is "type". Line 3 first creates a packet tracing object with type specified in "type". Lines 4 and 5 configure member variables `src_` and `dst_`, respectively, of the created packet tracing object "p". Line 6 stores the created packet tracing object in instvar "alltrace_" of the Simulator. Lines 7–9 attach "file" to the created packet tracing object. Finally, Line 10 returns the created packet tracing object to the caller.

Program 13.9 Instproc `trace` of class `SimpleLink`.

```
    //~/ns/tcl/lib/ns-link.tcl
1   SimpleLink instproc trace { ns f {op ""} } {
2       $self instvar enqT_ deqT_ drpT_ queue_ link_ fromNode_ toNode_
3       $self instvar rcvT_ ttl_ trace_
4       $self instvar drophead_       ;# idea stolen from CBQ and Kevin
5       set trace_ $f
6       set enqT_ [$ns create-trace Enque $f $fromNode_ $toNode_ $op]
7       set deqT_ [$ns create-trace Deque $f $fromNode_ $toNode_ $op]
8       set drpT_ [$ns create-trace Drop $f $fromNode_ $toNode_ $op]
9       set rcvT_ [$ns create-trace Recv $f $fromNode_ $toNode_ $op]
10      $self instvar drpT_ drophead_
11      set nxt [$drophead_ target]
12      $drophead_ target $drpT_
13      $drpT_ target $nxt
14      $queue_ drop-target $drophead_
15      $deqT_ target [$queue_ target]
16      $queue_ target $deqT_
17      $self add-to-head $enqT_
18      $rcvT_ target [$ttl_ target]
19      $ttl_ target $rcvT_
20  }
```

Program 13.10 Instproc `create-trace` of class `Simulator`.

```
    //~/ns/tcl/lib/ns-lib.tcl
1   Simulator instproc create-trace { type file src dst {op ""} } {
2       $self instvar alltrace_
3       set p [new Trace/$type]
4       $p set src_ [$src id]
5       $p set dst_ [$dst id]
6       lappend alltrace_ $p
7       if {$file != ""} {
8           $p attach $file
9       }
10      return $p
11  }
```

Instproc `flush-trace{}` of Class Simulator

Program 13.11 shows the details of instproc `flush-trace{}` of class `Simulator`. The Simulator stores all packet tracing objects in instvar `alltrace_`. Therefore, Lines 3–7 invoke the OTcl command `flush{}` of all the packet tracing objects stored in instvar `alltrace_`.

Program 13.11 Instproc `flush-trace` of class `Simulator`.

```
    //~/ns/tcl/lib/ns-lib.tcl
1  Simulator instproc flush-trace {} {
2      $self instvar alltrace_
3      if [info exists alltrace_] {
4          foreach trace $alltrace_ {
5              $trace flush
6          }
7      }
8  }
```

13.3.2 C++ Main Packet Tracing Class Trace

In NS2, packet tracing objects are implemented using class `Trace` declared in Program 13.12, which is bound to an OTcl class with the same name (see Line 15–23). From Line 1, class `Trace` derives from class `Connector`, and can be inserted between two NsObjects to record the details of a packet passing through it. As a connector, a packet tracing object receives a packet `*p` by having its upstream object invoke its function `recv(p,h)`. Upon receiving a packet, it records the details of the packet in a trace file, and forwards the packet to its downstream object.

Main C++ Variable of Class Trace

Class `Trace` consists of four main variables: `src_`, `dst_`, `type_`, and `pt_`. Variables `src_` (Line 3) and `dst_` (Line 4) specify the beginning and the ending addresses of a `Trace` object. Variable `type_` in Line 10 indicates the type of the tracing object. Despite its `int` type, the true meaning of this variable is the `char` equivalent. For example, the types of objects which trace packet enquing and dequing are "+" and "-", which correspond to decimal values of 43 and 45, respectively. Finally, pointer `pt_` in Line 9 is a reference to a `BaseTrace` object, which provides the basic functionalities for packet tracing. We shall discuss the details of class `BaseTrace` later in Section 13.3.5.

Main C++ Functions of Class Trace

Class `Trace` has three following main functions: the constructor, function `recv(p,h)`, and function `format(tt,s,d,p)`.

The Constructors

Lines 24–29 and 30–34 show the constructors of C++ class `Trace` and OTcl class `Trace`, respectively. The OTcl constructor simply stores the input argument in its instvar `type_` (Line 33). Similarly, the C++ constructor stores

Program 13.12 Declaration of class `Trace` which is bound to the OTcl class with the same name, and their constructors.

```
    //~/ns/trace/trace.h
1   class Trace : public Connector {
2   protected:
3       nsaddr_t src_;
4       nsaddr_t dst_;
5       virtual void format(int tt, int s, int d, Packet* p);
6   public:
7       Trace(int type);
8       ~Trace();
9       BaseTrace *pt_;
10      int type_;
11      int command(int argc, const char*const* argv);
12      static int get_seqno(Packet* p);
13      void recv(Packet* p, Handler*);
14  };

    //~/ns/trace/trace.cc
15  class TraceClass : public TclClass {
16  public:
17      TraceClass() : TclClass("Trace") { }
18      TclObject* create(int argc, const char*const* argv) {
19          if (argc >= 5)
20              return (new Trace(*argv[4]));
21          return 0;
22      }
23  } trace_class;

24  Trace::Trace(int type) : Connector(), pt_(0), type_(type)
25  {
26      bind("src_", (int*)&src_);
27      bind("dst_", (int*)&dst_);
28      pt_ = new BaseTrace;
29  }

    //~/ns/tcl/lib/ns-trace.tcl
30  Trace instproc init type {
31      $self next $type
32      $self instvar type_
33      set type_ $type
34  }
```

the input argument in variable `type_` (Line 24). It also binds variables `src_` and `dst_` to instvars with the same name (Lines 26–27), and creates a new `BaseTrace` object `*pt_` (Line 28).

Function `recv(p,h)`

Function `recv(p,h)` is the main packet reception function. The details of function `recv(p,h)` is shown in Program 13.13. Line 3 invokes function `format(type_,src_,dst_,p)` to store the details of packet `*p` in the internal variable `wrk_` of the associated `BaseTrace` object `*pt_`. Line 4 executes "`pt_->dump()`" to print the packet details to an attached trace file. If the `Trace` object contains a non-Null downstream object, Line 8 will forward packet `p` to the downstream object. Otherwise, Line 6 will deallocate packet `p`.

Function `Format(tt,s,d,p)`

Shown in Programs 13.14–13.15, function `format(tt,s,d,p)` stores the packet details in the internal variable `wrk_` of the associated `BaseTrace` object `*pt_`. Taking the packet tracing type "tt", a source node ID "s", a destination node ID "d", and a pointer to an incoming packet "*p" as input arguments, this function proceeds as follows. Line 7 stores the packet type in a local variable `name`. Lines 9–21 create a flag string and store it in a local variable `flag`. Address and port of source node and destination node are retrieved in Lines 22–25. Finally, Lines 26–45 print out a packet tracing string to variable `pt_->wrk_`.[4] The packet trace format will be discussed in greater detail in Section 13.3.5.

Main OTcl Commands of a Packet Tracing Object

There are three OTcl main commands associated with class `Trace`: `flush{}`, `detach{}`, and `attach{file}`. In Program 13.16, command `flush{}` (Lines

Program 13.13 Function `recv` of class `Trace`.

```
    //~/ns/trace/trace.cc
1 void Trace::recv(Packet* p, Handler* h)
2 {
3     format(type_, src_, dst_, p);
4     pt_->dump();
5     if (target_ == 0)
6         Packet::free(p);
7     else
8         send(p, h);
9 }
```

---

[4] As we shall see in Section 13.3.3, function `buffer()` of class `BaseTrace` simply returns variable `wrk_`.

Program 13.14 Function format of class Trace.

```
//~/ns/trace/trace.cc
1   void Trace::format(int tt, int s, int d, Packet* p)
2   {
3       hdr_cmn *th = hdr_cmn::access(p);
4       hdr_ip *iph = hdr_ip::access(p);
5       hdr_tcp *tcph = hdr_tcp::access(p);
6       packet_t t = th->ptype();
7       const * name = packet_info.name(t);
8       int seqno = get_seqno(p);
9       char flags[NUMFLAGS+1];
10      for (int i = 0; i < NUMFLAGS; i++)
11          flags[i] = '-';
12      flags[NUMFLAGS] = 0;
13      hdr_flags* hf = hdr_flags::access(p);
14      flags[0] = hf->ecn_ ? 'C' : '-';
15      flags[1] = hf->pri_ ? 'P' : '-';
16      flags[2] = '-';
17      flags[3] = hf->cong_action_ ? 'A' : '-';
18      flags[4] = hf->ecn_to_echo_ ? 'E' : '-';
19      flags[5] = hf->fs_ ? 'F' : '-';
20      flags[6] = hf->ecn_capable_ ? 'N' : '-';
21      flags[7] = 0;
22      char *src_nodeaddr = Address::instance().
                                print_nodeaddr(iph->saddr());
23      char *src_portaddr = Address::instance().
                                print_portaddr(iph->sport());
24      char *dst_nodeaddr = Address::instance().
                                print_nodeaddr(iph->daddr());
25      char *dst_portaddr = Address::instance().
                                print_portaddr(iph->dport());
        ...
```

5–10) clears the buffer of the attached Tcl channel by invoking `pt_->flush(ch)`, where `ch` is the attached Tcl channel. The OTcl command `detach{}` does not clear the channel buffer, but simply sets the pointer to the attached Tcl channel to Null (see Line 12). Finally, the OTcl command `attach{file}` sets the input file handle `file` as a trace file (Lines 19–20).

13.3.3 C++ Helper Class BaseTrace

One of the main variables of class `Trace`, `pt_`, is a pointer to an object of class `BaseTrace`, a packet tracing helper class. Class `BaseTrace` acts as an interface from a packet tracing object to a Tcl channel. Shown in Program 13.17, class `BaseTrace` is bound to an OTcl class with the same name. Class `BaseTrace` has two main variables: `channel_` (Line 14) and `wrk_` (Line

Program 13.15 Function `format` of class `Trace` (Cont.).

```
//~/ns/trace/trace.cc
    ...
26      sprintf(pt_->buffer(),
27          "%c "TIME_FORMAT" %d %d %s %d %s %d %s.%s %s.%s
28          tt,
29          pt_->round(Scheduler::instance().clock()),
30          s,
31          d,
32          name,
33          th->size(),
34          flags,
35          iph->flowid(),
36          src_nodeaddr,
37          src_portaddr,
38          dst_nodeaddr,
39          dst_portaddr,
40          seqno,
41          th->uid(),
42          tcph->ackno(),
43          tcph->flags(),
44          tcph->hlen(),
45          tcph->sa_length() );
46      delete [] src_nodeaddr;
47      delete [] src_portaddr;
48      delete [] dst_nodeaddr;
49      delete [] dst_portaddr;
50 }
```

15). While `channel_` is an interface to a Tcl channel, `wrk_` is a buffer which stores a trace string to be written to the Tcl channel. At the construction, the Tcl channel `channel_` is set to Null, and the trace string `wrk_` is allocated with memory space which can hold upto 1026 characters.

Key functions of class `BaseTrace` include `channel(...)`, `buffer()`, `flush` `(channel)`, and `dump()`. The operations of the first three functions are fairly straightforward, and are omitted for brevity. Function `dump()` shown in Lines 28–37 of Program 13.17 is responsible for dumping a trace string stored in `wrk_` to the Tcl channel. Here, Line 30 retrieves the length of the string `wrk` and stores the length in a local variable "n". Line 32 attaches an end-of-line character to `wrk_`. Line 33 attaches zero to `wrk_` indicating the end of the string. Line 34 writes `wrk_` to the Tcl channel `channel_` using function `Tcl_Write(...)`. Finally, Line 35 clears the value stored in `wrk_`.

In common with class `Trace`, class `BaseTrace` has three main OTcl commands: `flush{}`, `detach{}`, and `attach{file}`. These three commands per-

Program 13.16 Function command of class Trace.

```
    //~/ns/trace/trace.cc
1   int Trace::command(int argc, const char*const* argv)
2   {
3       Tcl& tcl = Tcl::instance();
4       if (argc == 2) {
5           if (strcmp(argv[1], "flush") == 0) {
6               Tcl_Channel ch = pt_->channel();
7               if (ch != 0)
8                   pt_->flush(ch);
9               return (TCL_OK);
10          }
11          if (strcmp(argv[1], "detach") == 0) {
12              pt_->channel(0) ;
13              return (TCL_OK);
14          }
15      } else if (argc == 3) {
16          if (strcmp(argv[1], "attach") == 0) {
17              int mode;
18              const char* id = argv[2];
19              Tcl_Channel ch = Tcl_GetChannel(tcl.interp(),
                    (char*)id,&mode);
20              pt_->channel(ch);
21              if (pt_->channel() == 0) {
22                  tcl.resultf("trace: can't attach %s
                        for writing", id);
23                  return (TCL_ERROR);
24              }
25              return (TCL_OK);
26          }
27      }
28      return (Connector::command(argc, argv));
29  }
```

form the same action as those in class Trace. We will omit the details of these three OTcl commands for brevity.

13.3.4 Various Types of Packet Tracing Objects

NS2 employs different types of packet tracing objects to trace packets at different places. For example, a Trace/Enque object is placed immediately before a queue to trace packets which enter the queue. The type (i.e., variable type_) of a Trace/Enque object is "+", which is equivalent to 43 in decimal. When a packet passes through a Trace/Enque object, a line beginning with "+" is appended to the Tcl Channel.

Program 13.17 Declaration, an OTcl binding class, the constructor of class BaseTrace, and function dump of class BaseTrace.

```
   //~/ns/trace/basetrace.h
1  class BaseTrace : public TclObject {
2  public:
3      BaseTrace();
4      ~BaseTrace();
5      virtual int command(int argc, const char*const* argv);
6      virtual void dump();
7      inline Tcl_Channel channel() { return channel_; }
8      inline void channel(Tcl_Channel ch) {channel_ = ch; }
9      inline char* buffer() { return wrk_ ; }
10     void flush(Tcl_Channel channel) { Tcl_Flush(channel); }
11     #define PRECISION 1.0E+6
12     #define TIME_FORMAT "%.15g"
13 protected:
14     Tcl_Channel channel_;
15     char *wrk_;
16 };


   //~/ns/trace/basetrace.cc
17 class BaseTraceClass : public TclClass {
18 public:
19         BaseTraceClass() : TclClass("BaseTrace") { }
20         TclObject* create(int argc, const char*const* argv) {
21             return (new BaseTrace());
22         }
23 } basetrace_class;

24 BaseTrace::BaseTrace() : channel_(0),
25 {
26   wrk_ = new char[1026];
27 }

28 void BaseTrace::dump()
29 {
30     int n = strlen(wrk_);
31     if ((n > 0) && (channel_ != 0)) {
32         wrk_[n] = '\n';
33         wrk_[n + 1] = 0;
34         (void)Tcl_Write(channel_, wrk_, n + 1);
35         wrk_[n] = 0;
36     }
37 }
```

Among all built-in OTcl packet tracing classes, the most common ones include

- `Trace/Enque` ("+"): Trace packet arrival (usually at a queue)
- `Trace/Deque` ("−"): Trace packet departure (usually at a queue)
- `Trace/Drop` ("d"): Trace packet drop (delivered to a drop-target)
- `Trace/Recv` ("r"): Trace packet reception at a certain node

where the characters in the parentheses are attributed to each packet tracing object class.

Program 13.18 Constructors of classes `Trace`, `Trace/Enque`, and `Trace/Deque`.

```
      //~/ns/tcl/lib/ns-trace.tcl
 1   Class Trace/Enque -superclass Trace
 2   Trace/Enque instproc init {} {
 3       $self next "+"
 4   }
 5   Trace/Deque instproc init {} {
 6       $self next "-"
 7   }

     //~/ns/trace/trace.h
 8   static class DequeTraceClass : public TclClass {
 9   public:
10       DequeTraceClass() : TclClass("Trace/Deque") { }
11       TclObject* create(int args, const char*const* argv) {
12       if (args >= 5)
13               return (new DequeTrace(*argv[4]));
14           return NULL;
15       }
16   } dequetrace_class;
```

Among these four classes, only class `Trace/Deque` has an implementation in the C++ domain. Other three classes are not bound to the OTcl domain. The main difference among the above four packet tracing objects lie in their constructors. As shown in Program 13.18, OTcl class `Trace/Enque` derives from the OTcl class `Trace` (Line 1), while class OTcl `Trace/Deque` is mapped to the C++ OTcl class `DequeTrace` (Lines 8–16). Lines 3 and 6 show that classes `Trace/Enque` and `Trace/Deque` are constructed with characters "+" and "−", respectively. In Line 24 of Program 13.12, this character is stored in the variable `type_` of the packet tracing object.

As an example, consider the process of creating a `Trace/Enque` object in Fig. 13.6. The process starts when a statement "`new Trace/Enque`" is executed. From within an OTcl constructor, the type "+" is repeatedly fed to the

OTcl | C++

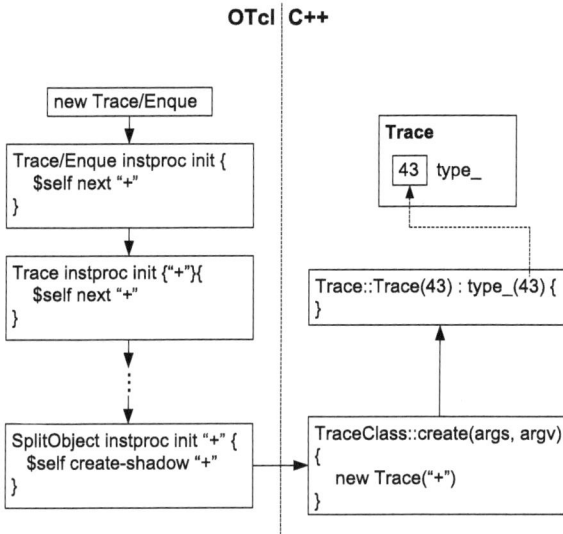

Fig. 13.6. Construction of a `Trace/Enque` object.

constructor up the hierarchy by the statement "$self next "+"". When class `SplitObject` is reached, instproc `create-shadow{...}` is invoked with an input argument "+". Instproc `create-shadow{...}` invokes function `create()` of class `TraceClass` in the C++ domain. From Line 24 in Program 13.12, the constructor of class `Trace` is invoked, and type "+" is fed as an input argument. Since the constructor takes an integer as an input argument, the ascii code "+" is converted into a decimal value "43". Finally, the constructor stores the input argument (i.e., "43" in this case) in the variable `type_`.

13.3.5 Packet Trace Format

Packet trace format is defined in function `format(...)` (Programs 13.14-13.15). In a normal case, each line of a trace file follows the format in Fig. 13.7. There are 12 fields in each line of a trace file (i.e., a Tcl channel):

Type Identifier	Time	Source Node	Destination Node	Packet Name	Packet Size	Flags	Flow ID	Source Address	Destination Address	Sequence Number	Packet Unique ID

Fig. 13.7. Packet tracing file format

- Type Identifier: depends on the type (i.e., variable `type_`) of packet tracing object which generates the string. Most widely used type identifiers are shown below. The complete list of type identifiers is given in file ~*ns*/tcl/lib/ns-trace.tcl.

- "+" which represents a packet enque event,
- "-" which represents a packet deque event,
- "r" which represents a packet reception event,
- "d" which represents a packet drop (e.g., sent to `dropHead_`) event, and
- "c" which represents a packet collision at the MAC level.
- Time: at which the packet tracing string is created.
- Source Node and Destination Node: denote the IDs of the source and the destination nodes of the tracing object.
- Packet Name: Name of the packet type (as specified in Program 8.9).
- Packet Size: Size of the packet in bytes.
- Flags: A 7-digit flag string is defined in Lines 9–21 of Program 13.14. Each flag digit is set to "-" if the corresponding flag is disabled. Otherwise, it will be set as follows. The first is set to "E" if an ECN (Explicit Congestion Notification) echo is enabled. The second is set to "P" if the priority in the IP header is enabled. The fourth is set to "A" if the corresponding TCP takes an action on a congestion (e.g., closes the congestion window). The fifth is set to "E" if the congestion has occurred. The sixth is set to "F" if the TCP fast start is used. Finally, the seventh is set to "N", when the transport layer protocol is capable of using Explicit Congestion Notification (ECN).
- Flow ID: Flow ID specified in the field `fid_` of an IP packet header.
- Source Address and Destination Address: The source and destination addresses of a packet specified in an IP packet header. For a flat addressing scheme, the format of these two fields is "a.b", where "a" is the address and "b" is the port.
- Sequence Number: The sequence number specified in packet header. Specified by a transport layes protocol.
- Packet Unique ID: A unique ID stored in a common packet header.

13.4 Compilation of Simulation Results

One of the main objectives of network simulation is to study network performance. Compilation of simulation results refers to a process of collecting information from simulation and compute performance measures under consideration. There are three main approaches to collect simulation results in NS2: through C++ codes, through Tcl codes, and through a trace file.

- **Through C++ codes**: This refers to an approach which inserts C++ codes into the original NS2 codes. As mentioned earlier in this book, the modification of C++ code results in a quick simulation. However, programmers require a fair amount of knowledge in the C++ architecture to collect results from the simulation.

- **Through Tcl codes**: This method is perhaps the most convenient way to collect the results. The programmers do not need to know the details of the C++ architecture. They only need to know the variable binding structure of classes under consideration.
- **Through trace file**: This method consists of two main steps. In the first step, a trace file is created during simulation. The second step is to retrieve the relevant information from the trace file. In most cases, a scripting language (e.g., AWK) can be used to extract the necessary information from a trace file (see Appendix A). Although this approach is widely demonstrated in the NS2 tutorial in the internet, advanced users are not encouraged to use this approach due to the following reasons. First, the OTcl command "`trace-all`" consumes a significant amount of resources (e.g., memory, simulation time), and dramatically slows down the simulation. Secondly, a generated trace file usually contains too much information. In most cases, an NS2 user need to learn another scripting language (e.g., AWK) to extract relevant information from a trace file. Finally, the trace file may not contain the required information. For example, information on instantaneous buffer occupancy is not available in a trace file.

Example 13.5. Consider Example 10.1 which creates the network in Fig. 9.3. Insert an error model with error probability 0.05 into the link connecting Node 1 and Node 3. Suppose the maximum TCP transmission window size is set to 20.

- **Through C++ result codes**: Find out the number of times TCP transmission window is reduced.
- **Through Tcl codes**: Plot the dynamic variation of TCP transmission window.
- **Through trace file**: Compute the average interval between two TCP packets entering the link layer buffer.

Constructing a Network

An error model can be inserted into the network by inserting the following OTcl codes immediately after Line 7 of Example 10.1:

```
set em [new ErrorModel]
$em set rate_ 0.005
$em unit pkt
$em ranvar [new RandomVariable/Uniform]
$em drop-target [new Agent/Null]
$ns lossmodel $em $n1 $n3
```

The maximum TCP transmission window is set to 20 by the following statement after Line 10 in Example 10.1: " `$tcp set window_ 20`".

Collection of Results Through C++ Codes

TCP shrinks its transmission window when function `slowdown(how)` of class `TcpAgent` is invoked. Therefore, we may declare a variable `num_slowdowns_` of class `TcpAgent` in file `~ns/tcp/tcp.h`, initialize it to zero in the constructor, and add the two following lines in function `slowdown(how)`:

```
num_slowdowns_++;
printf("Total number of TCP window reduction is %d \n", now,
  num_slowdowns_);
```

After recompiling NS2, we run the script "tcp.tcl" and obtain the following results:

```
>> ns tcp.tcl
Total number of TCP window reduction is 1
Total number of TCP window reduction is 2
Total number of TCP window reduction is 3
...
Total number of TCP window reduction is 36
```

In this simulation, TCP shrinks its transmission window 36 times.

Collection of Results Through Tcl Codes

Transmission window size of a TCP connection is the minimum of instvars `cwnd_` and `window_` of a `Agent/TCP` object. Since these two variables are available in the OTcl domain, we may collect samples of TCP window size by inserting the following Tcl script after Line 14 in Example 10.1.

```
1   set f_cwnd [open cwnd.tr w]
2   proc plot_tcp { } {
3       global f_cwnd tcp ns
4       if { [$tcp set cwnd_] < [$tcp set window_] } {
5               puts $f_cwnd "[$ns now] [$tcp set cwnd_]"
6       } else {
7               puts $f_cwnd "[$ns now] [$tcp set window_]"
8       }
9       $ns at [expr [$ns now] + 0.2] plot_tcp
10 }
11 $ns at 0.01 "plot_tcp"
```

The above statements put time and TCP transmission window size in file "cwnd.tr" every 0.2 seconds. Line 1 above creates a Tcl channel `f_cwnd` which is bound to the file `cwnd.tr`. Lines 2–10 define a procedure `plot_tcp{}`. Lines 11 invokes procedure `plot_tcp` at 0.01 second. Within the procedure `plot_tcp{}`, Lines 5 and 7 print instvars `cwnd_` and `window_`, whichever is less, on the Tcl channel `f_cwnd`. Line 9 schedules an invocation of procedure

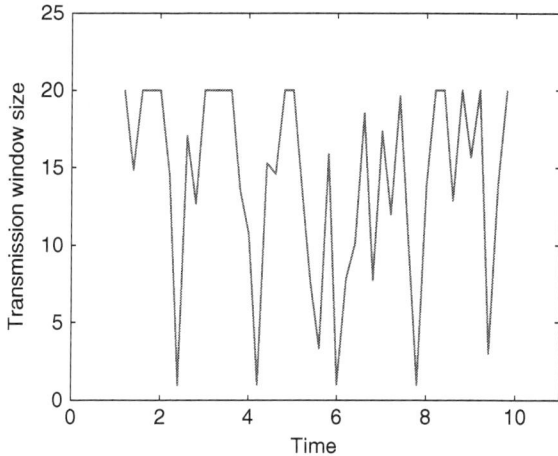

Fig. 13.8. Dynamics of TCP transmission window for Example 13.5.

`plot_tcp` at 0.2 seconds in future. This invocation continuously prints out simulation time and TCP transmission window size to the Tcl channel until the simulation terminates.

After running the above Tcl simulation script, the file `cwnd.tr` is created. The first and the second columns of file `cwnd.tr` are the time and the corresponding TCP transmission window, respectively. We now plot Fig. 13.8, using the first and second columns as X axis and Y axis, respectively. Since we set instvar `window_` to be 20, TCP transmission window can never exceed 20. We can also observe frequent decreases in TCP transmission window size due to packet losses.

Collection of Results Through Trace File

The first step in this process is to enable tracing in the Tcl simulation script. Again, this step can be carried out by inserting the following lines after Line 4 in Example 10.1.

```
set f_trace [opon trace.tr w]
$ns trace-all $f_trace
```

The second step is to process the trace file. In this case, there is only one TCP flow in the simulation and we can measure the interval between two TCP packets entering a queue, which connect Node 1 (with ID 0) to Node 3 (with ID 1), using the AWK script file `avg.awk` in Program 13.19. By executing the AWK script, we will see the following result on the screen:

```
>> awk -f avg.awk trace.tr
Average TCP packet inter-arrival time is 0.001703
```

Program 13.19 An AWK script which computes average interval between two TCP packets entering a link layer buffer of Node 1.

```
//avg.awk
1 BEGIN{ started = 0 }
2 /^+/ { time = $2;
3 if (started == 1) {
4 if ($3==0 && $4==2 && $5 == "tcp") {
5 interval = time-old_time;
6 old_time = time;
7 cum_interval += interval;
8 total_samples ++;
9 }
10 } else {
11 started = 1; old_time = time;
12 }
13 }
14 END { avg_interval = cum_interval/total_samples;
15 printf("Average TCP packet inter-arrival time is %f\n",
 avg_interval);
16 }
```

Line 1 in Program 13.19 initializes variable started to zero. Lines 2–13 collect samples of the inter-arrival time of TCP packets. Line 2 indicates the actions to be executed for all the lines beginning with "+" in the subsequent curly braces. From Line 4, the samples are collected only for the source node 0, the destination node 2, and protocol tcp. Finally, Lines 14–16 compute and print the average TCP packet inter-arrival time on the screen.

13.5 Chapter Summary

Two of the most important aspects in a network simulation are debugging and compilation of simulation results. Debugging refers to a procession of removing compilation and run-time errors in both C++ and OTcl domains. This chapter provides guidelines and necessary commands for debugging. Although originally designed to facilitate the understanding of network dynamics, NS2 tracing could also be useful in the debugging process. NS2 supports two types of tracing. Variable tracing records the changes in value of a variable (in most cases in a file), while packet tracing stores the details of packets passing through network checkpoints (again in most cases in a file).

There are three major ways to collect simulation results. First, collecting simulation results through C++ codes is a quick and easy way. However, the users may require a fair amount of knowledge on the C++ architecture. Also, since this method involves the modification of C++ code, it could mess up the original NS2 source codes. The upside of this approach is that it gives users a access to most NS2 components. At runtime, the simulation proceeds very fast since the modification is carried out using the C++ compiler. Secondly, collecting simulation results through Tcl codes allows the programmer to collect the results from the OTcl domain in a simple way. In this case, the users do not need to understand the entire architecture of NS2, but they need to know how the variables in C++ and OTcl domains are bound together. Since NS2 is written mostly in C++, some variables are inaccessible from the OTcl domain. This approach may not be able to collect all required performance measures. Despite its necessity, this approach does not provide an access to NS2 internal variables (which might be needed in some case). Proceeding by the interpreter, this approach can take long runtime compared to the first approach. Finally, collecting simulation results through a trace file consists of two steps: 1) running simulation to create a trace file and 2) processing the created trace file. Despite its prevalance in the on-line tutuorial, this method is not recommended since it takes too much simulation resource and might not give the users required information. In fact, the recommended method is the first one (using C++).

14

Developing New Modules for NS2

So far, we have explained the details of the basic components of NS2 including their functionalities, internal mechanisms, and configuration methods. In this final chapter, we demonstrate how new NS2 modules are created, configured, and incorporated through two following examples. One is an Automatic Repeat reQuest (ARQ) protocol, which is a mechanism to improve transmission reliability of a communication link by means of packet retransmission. Another is a packet scheduler which arranges the transmission sequence of packets from multiple incoming data flows.

14.1 Automatic Repeat reQuest (ARQ)

Automatic Repeat reQuest (ARQ) is a method of handling communication errors by packet retransmission. An ARQ transmitter (i.e., a transmitting node which implements an ARQ protocol) is responsible for transmitting data packets and retransmitting the lost packets. An ARQ receiver (i.e., a receiving node which implements an ARQ protocol), on the other hand, is responsible for receiving packets and (implicitly or explicitly) informing the transmitter of the transmission result. It returns an ACK (acknowledgement) message or a NACK (negative acknowledgement) message to the transmitter if a packet is successfully or unsuccessfully (respectively) received. Based on the received ACK/NACK pattern, the ARQ transmitter decides whether to retransmit the lost packet or to transmit a new packet.

This section focuses on a limited-persistence stop-and-wait ARQ protocol. This type of ARQ protocols is characterized by two following properties. With limited-persistence, an ARQ transmitter gives up the retransmission if the transmission fails consecutively for a certain number of times. Another property is "stop-and-wait". Here, an ARQ transmitter transmits a packets and waits for an acknowledgement from the corresponding ARQ receiver before commencing another (lost or new) packet transmission.

T. Issariyakul, E. Hossain, *Introduction to Network Simulator NS2*,
DOI: 10.1007/978-0-387-71760-9_14, © Springer Science+Business Media, LLC 2009

In the following, we first design NS2 modules for a limited-persistence stop-and-wait ARQ protocol with an error-free and delay-free (i.e., immediate) feedback channel in Section 14.1.1. Sections 14.1.2 and 14.1.3 demonstrate C++ and OTcl implementations, respectively. Finally, Section 14.1.4, we extend the ARQ model for an error-free feedback channel with non-zero processing and propagation delay. Implementation of an ARQ protocol with an error prone feedback channel is left as an exercise for the readers.

14.1.1 The Design

Figure 14.1 shows an architecture of a link with an ARQ protocol. Here, the feedback channel is assumed to be error-free and the feedbacks are assumed to be immediate. The link is constructed by inserting an error module and an ARQ module into a `SimpleLink` object. From Fig. 7.1, a `SimpleLink` object consists of four main instvars: `queue_` which models the packet buffering, `link_` which models the service time of the queue and the link propagation delay, `ttl_` which models *time-to-live* of a packet, and `drophead_` which acts as a common dropping point for a `SimpleLink` object. An error model `link_errmodule_` is inserted into a `SimpleLink` object by an OTcl command `link-lossmodel{...}` of class `Simulator`.

Based on this basic `SimpleLink` object with an error model, we incorporate the three following ARQ components (i.e., instvars) to implement an ARQ module. The first component is an ARQ transmitter (instvar `tARQ_`), which transmits, retransmits, and drops the packet based on the underlying ARQ protocol. The second and third components (instvars `acker_` and `nacker_`) are responsible for transmitting ACK and NACK messages, respectively, to the ARQ transmitter `tARQ_`.

NS2 employs a queue blocking and callback mechanism to model packet forwarding in a `SimpleLink` object. The process starts when a `Queue` object `queue_` receives and forwards a packet as well as its queue handler (whose class is `QueueHandler`) to its downstream object, and *blocks* itself. Here, the `Queue` object stops transmitting packets until the head-of-the-line packet is completely transmitted (when it is unblocked by the downstream object).

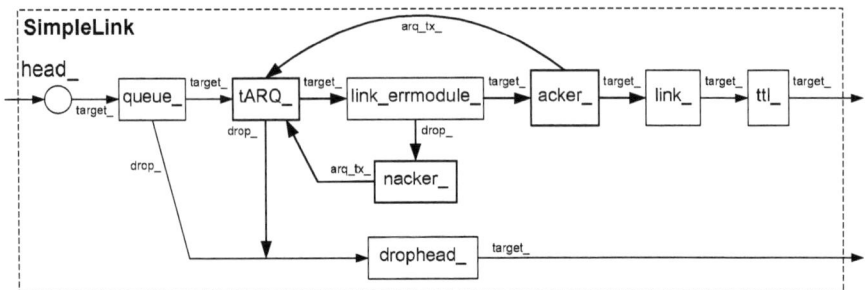

Fig. 14.1. Architecture of a `SimpleLink` object with an ARQ module.

In the absence of instvar `link_errmodule_` and the ARQ-related instvars, instvar `link_` (of class `LinkDelay`) is responsible for unblocking the `Queue` object. It does so by placing the input queue handler on the simulation time line at the time where the packet is completely transmitted (i.e., has left the `Queue` object). At the firing time, the queue handler is dispatched, and the `Queue` object is unblocked. At this moment, the `Queue` object is allowed to transmit another packet.

Instvar `link_errmodule_` is of C++ class `ErrorModel` (see Section 12.3) and is responsible for simulating packet errors. A packet will be forwarded to the variables `drop_` and `target_`, respectively, depending on whether it is in error or not.

We now incorporate the instvars `tARQ_`, `acker_`, and `nacker_` into the packet forwarding mechanism as follows:

- *ACK/NACK message passing*: The key components of the ACK/NACK message passing mechanism are instvars `acker_` and `nacker_`, which are responsible for creating and forwarding ACK and NACK messages (respectively) to an ARQ transmitter. From Fig. 14.1, these two components are attached to variables `target_` and `drop_`, respectively, of instvar `link_errmodule_`. A packet will be forwarded to the instvars `acker_` and `nacker_`, respectively, depending on whether the packet is in error or not in error, respectively. Instvar `acker_` informs the ARQ transmitter of a transmission success, and forwards the received packet to the instvar `link_`, while `nacker_` drops the corrupted packet, and informs the ARQ transmitter of transmission failure.
- *A callback mechanism*: In case of a `SimpleLink` object, instprocs `link_` and `link_errmodule_` are responsible for the callback mechanism. When inserting ARQ components, the callback mechanism is modified as follows. Instvars `link_` and `link_errmodule_` call back to an ARQ transmitter (i.e., `tARQ_`) which in turns calls back to a `Queue` object (i.e., `queue_`). Upon receiving a packet and a queue handler from the `Queue` object, the ARQ transmitter stores the queue handler in its member variable, and transmits the received packet as well as *its* handler to the downstream object. Depending on whether the packet is in error or not in error, the `link_errmodule_` and `link_` (respectively) will place a callback event on the simulation timeline. At the same time, `nacker_` and `acker_` will inform the ARQ transmitter of the transmission result. At the firing time (when the packet is completely transmitted), the ARQ transmitter determines whether the packet was successfully transmitted or not. Then, it decides whether to retransmit the lost packet or to fetch another packet from the upstream `Queue` object based on the received ACK/NACK messages.

14.1.2 C++ Implementation

In Fig. 14.1, the ARQ-related instvars include an ARQ transmitter `tARQ_`, an ACK message transmitter `acker_`, and a NACK message transmitter nacker_. These instvars are implemented in the C++ classes ARQTx, ARQAcker, and ARQNacker, respectively, which are bound to OTcl classes with the same name. Implementations of these three classes are shown in Programs 14.1–14.4.

Program 14.1 Declaration of classes ARQTx and ARQHandler

```
   //arq.h
1  class ARQTx;
2  enum ARQStatus {IDLE,SENT,ACKED,RTX,DROPPED};
3  class ARQHandler : public Handler {
4  public:
5       ARQHandler(ARQTx& arq) : arq_tx_(arq) {};
6       void handle(Event*);
7  private:
8       ARQTx& arq_tx_;
9  };
10 class ARQTx : public Connector {
11 public:
12      ARQTx();
13      void recv(Packet*, Handler*);
14      void nack(Packet*);
15      void ack();
16      void resume();
17 protected:
18      ARQHandler arqh_;
19      Handler* qh_;
20      Packet* pkt_;
21      ARQStatus status_;
22      int blocked_;
23      int retry_limit_;
24      int num_rtxs_;
25 };
```

Class ARQTx

Class ARQTx derives from class Connector, and can be used to connect two NsObjects.[1] The main C++ variables of class ARQTx are shown below:

num_rtxs_ Current number of packet retransmission; It is increased by one for every transmission failure, and is reset to zero when a new packet arrives (e.g., due to a packet drop or a transmission success).

---

[1] In Fig. 14.1, we use an ARQTx object tARQ_ to connect a Queue object queue_ with an ErrorModule object link_errmodule_.

Program 14.2 Functions of classes `ARQTx` and `ARQHandler`

```
     //arq.cc
26   void ARQHandler::handle(Event*){arq_tx_.resume();}
27   ARQTx::ARQTx() : arqh_(*this)
28   {
29       num_rtxs_ = 0; retry_limit_ = 0;
30       qh_ = 0;pkt_ = 0;
31       status_ = IDLE; blocked_ = 0;
32       bind("retry_limit_", &retry_limit_);
33   }
34   void ARQTx::recv(Packet* p, Handler* h)
35   {
36       qh_ = h; status_ = SENT;
37       blocked_ = 1;
38       send(p,&arqh_);
39   }
40   void ARQTx::ack()
41   {
42       num_rtxs_ = 0; status_ = ACKED;
43   }
44   void ARQTx::nack(Packet* p)
45   {
46       num_rtxs_++;
47       if( num_rtxs_ <= retry_limit_) {
48               pkt_ = p; status_ = RTX;
49       } else {
50               pkt_ = p; status_ = DROPPED;
51       }
52   }
53   void ARQTx::resume()
54   {
55       blocked_ = 0;
56       if ( status_ == ACKED ) {
57           status_ = IDLE; qh_->handle(0);
58       } else if ( status_ == RTX ) {
59           status_ = SENT; blocked_ = 1;
60           send(pkt_,&arqh_);
61       } else if ( status_ == DROPPED ) {
62           status_ = IDLE;
63           drop(pkt_); qh_->handle(0);
64       }
65   }
```

retry_limit_ The retry limit; The ARQ protocol will retransmit the lost packet as long as num_rtxs_<=retry_limit_.

blocked_ Indicates whether the ARQTx object is blocked (see the callback mechanism in Section 7.3.3).

arqh_ A handler which is passed to the downstream object

qh_ A handler of an upstream Queue object

status_ Current status of the ARQTx object defined in Line 2 of Program 14.1

pkt_ A pointer to the packet which is being transmitted.

Class ARQTx defines four following functions: recv(p,h), ack(), nack(p), and resume(). Function recv(p,h) is a main packet reception function. Since the upstream object of an ARQTx object is of class Queue, the input handler *h is of class QueueHandler. Function recv(p,h) stores the input handler in its variable qh_, forwards the input packet as well as its handler arqh_ (of class ARQHandler) to the downstream object, and sets the variable status to SENT.

Functions ack() and nack(p) are ACK and NACK message reception functions, respectively, whose flowcharts are shown in Fig. 14.2. Function ack() resets num_rtxs_ to zero and sets status_ to ACKED. Function nack(p) stores the input packet in variable pkt_, and increases the value of variable num_rtxs_ by 1. If num_rtxs_ is greater than retry_limit_, it will set status_ to DROPPED and reset num_rtxs_ to zero. Otherwise, it will set status_ to RTX.

Function resume() is invoked when the ARQHandler object arqh_ is dispatched (see Line 26 in Program 14.2). It takes actions based on the value stored in the variable status_ (see the flow chart in Fig. 14.3). In particular,

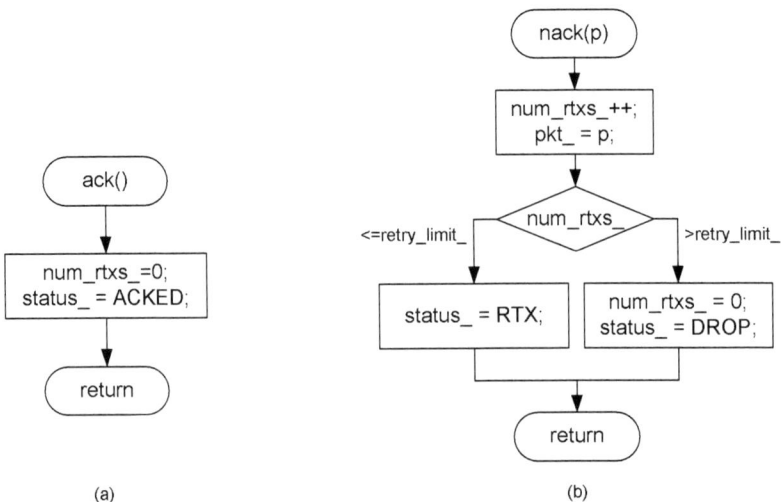

Fig. 14.2. Flowchart of functions (a) ack() and (b) nack(p) of class ARQTx.

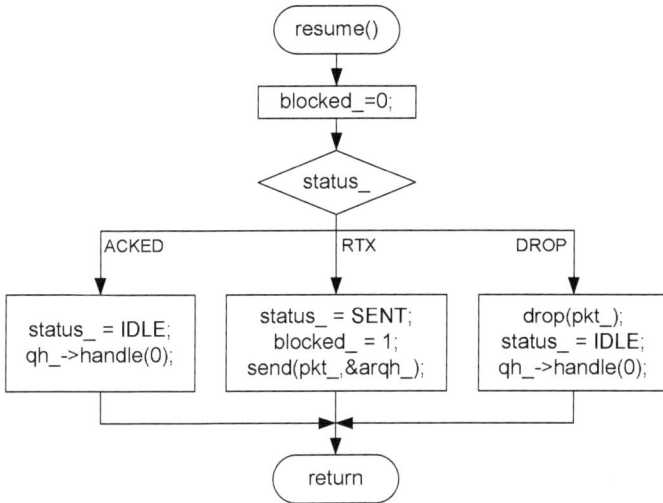

Fig. 14.3. Flowchart of function `resume()` of class `ARQTx`.

if `status_` is `ACKED`, function `resume()` will set `status_` to be `IDLE` and fetch another packet from the upstream `Queue` object by invoking `qh_->handle(0)` (Line 57 in Program 14.2). If `status_` is `DROPPED`, it will drop the packet stored in `pkt_`, set `status_` to be `IDLE` and fetch another packet from the upstream `Queue` object by invoking `qh_->handle(0)` (Lines 62–63 in Program 14.2). Finally, if `status_` is `RTX`, function `resume()` will block the ARQ transmitter and forward packet `*pkt_` as well as its handler `*arqh_` to the downstream object (Lines 59–60 in Program 14.2).

Classes `ARQRx`, `ARQAcker`, and `ARQNacker`

Another part of ARQ implementation is an ARQ receiver, which is responsible for reacting to packet transmission from the ARQ transmitter. Here, ARQ receivers are represented by a C++ class `ARQRx`, which contains a pointer `arq_tx_` (see Line 72 in Fig. 14.3) to an ARQ transmitter (i.e., an `ARQTx` object). This pointer is initialized to zero at the object construction, and is associated with an ARQ transmitter by OTcl command `attach-ARQTx` (Lines 104–107 in Program 14.4). Class `ARQRx` declares function `recv(p,h)` as pure virtual (see Line 70 in Program 14.3) to force its derived classes to implement this function.

There are two classes derived from class `ARQRx`: classes `ARQAcker` and `ARQNacker`. These two classes are responsible for sending ACK and NACK messages, respectively, to the associated ARQ transmitter. Upon receiving a packet, class `ARQAcker` (see Lines 110–114 in Program 14.4) informs the associated ARQ transmitter of successful packet delivery by invoking function `ack()` associated with the pointer `arq_tx_` (see the detail of function `ack()`

Program 14.3 Declaration of classes ARQRx, ARQAcker, and ARQNacker, and their OTcl classes

```
   //arq.h
66 class ARQRx : public Connector {
67 public:
68     ARQRx();
69     int command(int argc, const char*const* argv);
70     virtual void recv(Packet*, Handler*)=0;
71 protected:
72     ARQTx* arq_tx_;
73 };
74 class ARQAcker : public ARQRx {
75 public:
76     ARQAcker() {};
77     virtual void recv(Packet*, Handler*);
78 };
79 class ARQNacker : public ARQRx {
80 public:
81     ARQNacker() {};
82     virtual void recv(Packet*, Handler*);
83 };

   //arq.cc
84 static class ARQAckerClass: public TclClass {
85 public:
86     ARQAckerClass() : TclClass("ARQAcker") {}
87     TclObject* create(int, const char*const*) {
88         return (new ARQAcker);
89     }
90 } class_arq_acker;
91
92 static class ARQNackerClass: public TclClass {
93 public:
94     ARQNackerClass() : TclClass("ARQNacker") {}
95     TclObject* create(int, const char*const*) {
96         return (new ARQNacker);
97     }
98 } class_arq_nacker;
```

Program 14.4 Functions of classes `ARQRx`, `ARQAcker`, and `ARQNacker`

```
      //arq.cc
99    ARQRx::ARQRx() { arq_tx_ = 0; }
100   int ARQRx::command(int argc, const char*const* argv)
101   {
102       Tcl& tcl = Tcl::instance();
103       if (argc == 3) {
104           if (strcmp(argv[1], "attach-ARQTx") == 0) {
105               arq_tx_ = (ARQTx*)TclObject::lookup(argv[2]);
106               return(TCL_OK);
107           }
108       } return Connector::command(argc, argv);
109   }
110   void ARQAcker::recv(Packet* p, Handler* h)
111   {
112       arq_tx_->ack();
113       send(p,h);
114   }
115   void ARQNacker::recv(Packet* p, Handler* h)
116   {
117       arq_tx_->nack(p);
118   }
```

in Lines 40–43 of Program 14.2). Then, it sends out the received packet to its downstream NsObject. Similarly, class `ARQNacker` (see Lines 115–118 in Program 14.4) informs the associated ARQ transmitter of transmission failure by invoking function `nack(p)` associated with the pointer `arq_tx_` (see function `nack(p)` in Lines 44–52 of Program 14.12).

14.1.3 OTcl Implementation

In the OTcl domain, we need to create `ARQTx`, `ARQAcker`, and `ARQNack` objects– `tARQ_`, `acker_`, and `nacker_`, respectively, and insert them into a `SimpleLink` object as shown in Fig. 14.1. Program 14.5 shows two OTcl instprocs developed for this purpose.

- `SimpleLink::link-arq{limit}`: This instproc creates the ARQ-related instances and configures the `SimpleLink` object as shown in Fig. 14.1. Lines 4–6 create instvars `tARQ_`, `acker_`, and `nacker_`. Line 7 stores the input argument "`limit`" in variable `retry_limit_` of `tARQ_`. Lines 8 and 9 associate `acker_` and `nacker_`, respectively, with `tARQ_`. Finally, Lines 10–15 configure the rest of components as shown in Fig. 14.1.
- `Simulator::link-arq{limit from to}`: This instproc is an interface instproc which creates and configures ARQ modules of the link connecting Node "`from`" to Node "`to`". The input argument `limit` here is used as the retry limit of the ARQ module.

Program 14.5 OTcl Instprocs for an ARQ Module

```
    //~ns/tcl/lib/ns-link.tcl
 1  SimpleLink instproc link-arq { limit } {
 2      $self instvar link_ link_errmodule_ queue_ drophead_
 3      $self instvar tARQ_ acker_ nacker_
 4      set tARQ_ [new ARQTx]
 5      set acker_ [new ARQAcker]
 6      set nacker_ [new ARQNacker]
 7      $tARQ_ set retry_limit_ $limit
 8      $acker_ attach-ARQTx $tARQ_
 9      $nacker_ attach-ARQTx $tARQ_
10      $queue_ target $tARQ_
11      $tARQ_ target $link_errmodule_
12      $link_errmodule_ target $acker_
13      $acker_ target $link_
14      $tARQ_ drop-target $drophead_
15      $link_errmodule_ drop-target $nacker_
16  }


    //~ns/tcl/lib/ns-lib.tcl
17  Simulator instproc link-arq {limit from to} {
18      set link [$self link $from $to]
19      $link link-arq $limit
20  }
```

Example 14.1. We now setup an experiment to show the impact of retry limit
of a limited-persistence stop-and-wait ARQ protocol on TCP throughput. Our
experiment is based on Example 10.1. We insert an error module with 0.3
error probability in the link connecting Node **n1** to Node **n3**, and implement a
limited-persistence ARQ over this lossy link, and plot TCP throughput versus
the retry limit.

Tcl Simulation Script

We insert the following codes in the Tcl simulation script file "tcp.tcl" in
Example 10.1:

```
    //tcp.tcl
 1  set em [new ErrorModel]
 2  $em set rate_ 0.3
 3  $em unit pkt
 4  $em ranvar [new RandomVariable/Uniform]
 5  $em drop-target [new Agent/Null]
 6  $ns link-lossmodel $em $n1 $n3

 7  $ns link-arq 3 $n1 $n3
```

```
8   proc show_tcp_seqno {} {
9       global tcp
10      puts "The final tcp sequence number is
                          [$tcp set t_seqno_]"
11  }

12  $ns at 0.0 "$ftp start"
13  $ns at 100.0 "show_tcp_seqno"
14  $ns at 100.1 "$ns halt"
15  $ns run
```

Here, Lines 1–6 create an error module with packet error probability 0.3, and insert the created error module immediately after instvar `queue_` of the link connecting Node **n1** and Node **n3**. Line 7 creates and configures ARQ-related components with retry limit of 3. We run the simulation for 50.1 seconds and collect the results when the simulation time is 50.0. After running the script file "`tcp.tcl`" above, the following result appear on the screen:

```
>> ns tcp.tcl
>> The final tcp sequence number is 37587
```

TCP throughput in packets per second is computed as the final TCP sequence number divided by the simulation time. We vary the retry limit (in Line 7 above) to $\{0, 1, 2, 3\}$, and plot TCP throughput in Fig. 14.4. Clearly, increasing retry limit increases link reliability, and therefore, increases TCP throughput.

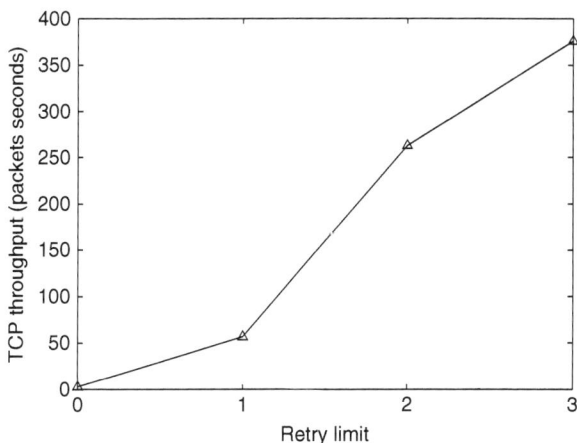

Fig. 14.4. Impact of retry limit of a limited persistent ARQ protocol on TCP throughput.

14.1.4 ARQ Under a Delayed (Error-Free) Feedback Channel

We have developed an NS2 module for an ARQ protocol with an immediate and error-free feedback. In practice, a feedback channel would be non-immediate and/or error prone. This section extends the modules developed earlier for a non-immediate error-free feedback channel. The extension for a non-immediate and *error-prone* feedback channel is left for the reader as an exercise (Exercise 14.4).

Program 14.6 shows the details of class ARQRx modified to support a delayed feedback channel. The idea is to defer the generation of ACK/NACK message for delay_ seconds, where the variable delay_ is bound to an inst-

---

Program 14.6 Modification of class ARQRx for a limited-persistence ARQ protocol with a delayed feedback channel.

```
   //arq.h
1  class ARQRx : public Connector {
2  public:
3      virtual void recv(Packet*, Handler*);
4      virtual void handle(Event*);
5      virtual void resume()=0;
6  protected:
7      ARQTx* arq_tx_;
8      Packet *pkt_;
9      Handler *handler_;
10     double delay_;
11     Event event_;
12 };

   //arq.cc
13 ARQRx::ARQRx()
14 {
15     arq_tx_ = 0; pkt_ = 0; handler_ = 0;
16     bind("delay_", &delay_);
17 }
18 void ARQRx::handle(Event *e) {resume();}
19 void ARQRx::recv(Packet* p, Handler* h)
20 {
21     pkt_ = p; handler_ = h;
22     Scheduler::instance().schedule(this, &event_, delay_);
23 }
24 void ARQAcker::resume()
25 {
26     arq_tx_->ack();
27     send(pkt_,handler_);
28 }
29 void ARQNacker::resume() {arq_tx_->nack(pkt_);}
```

var with the same name in the OTcl domain (see Line 16). Defined in Lines 19–23, function recv(p,h) invokes function schedule(this,&event_,delay_) of class Scheduler to defer the generation of ACK/NACK message. At the firing time, the Scheduler dispatches the scheduled event by invoking function handle(e) of the ARQRx object. In Line 18, function handle(...) invokes the pure virtual function resume() to resume the pending actions. Defined in classes ARQAcker (Lines 24–28) and ARQNacker (Line 29), function resume() simply sends either an ACK message or a NACK message, respectively, to the attached ARQTx object. Note that class ARQRx also defines a variable event_ of class Event which is used as an ACK/NACK reception dummy event (see also Section 4.3.6).

In the OTcl domain, we only need to include the two following lines into instproc link-arq{limit} of class SimpleLink (e.g., after Line 6 in Program 14.5):

```
$acker_ set delay_ [$self delay]
$nacker_ set delay_ [$self delay]
```

Here, the link delay in the forward direction (returned from $self delay) is used as the ARQ feedback delay for both ACK and NACK generators (i.e., acker_ and nacker_, respectively).

Example 14.2. Compare the TCP throughputs for the cases with an immediate feedback channel and a delayed feedback channel in the link layer ARQ protocols. Here, we use the results in Example 14.1 as a benchmark. When rerunning the Tcl simulation script in Example 14.1 under the ARQ protocol with a delayed feedback channel, the following result should appear on the screen:

```
>> ns tcp.tcl
>> The final tcp sequence number is 20596
```

which is less than 37587 in Example 14.1. The readers are encourage to experiment with different input parameters (e.g., feedback delay or retry limit) to gain more insights into the impact of link layer ARQ protocols on TCP performance.

Exercise 14.3. Why class TimerHandler was not used to implement the delayed feedback channel?

Exercise 14.4. Based on Examples 14.1 and 14.2, modify the ARQ protocol as follows.

(i) Remove the variable nacker_, and use a timer-based retransmission mechanism: A packet is assumed to be lost unless an ACK message is received within a timeout period.

(ii) Develop the codes for an ARQ protocol with an *error prone* delayed feedback channel.

14.2 Packet Scheduling for Multi-Flow Data Transmission

Packet scheduling is a mechanism to arrange transmission sequence of incoming packets in a node. For example, a *round-robin* (RR) packet scheduler transmits packets from different flows in sequence. This section shows the implementation of a round-robin packet scheduler in NS2.

14.2.1 The Design

Figure 14.5 shows the architecture of a packet scheduler in NS2. Here, the packet scheduler is implemented in an OTcl class LinkSch, which is modified from class SimpleLink.

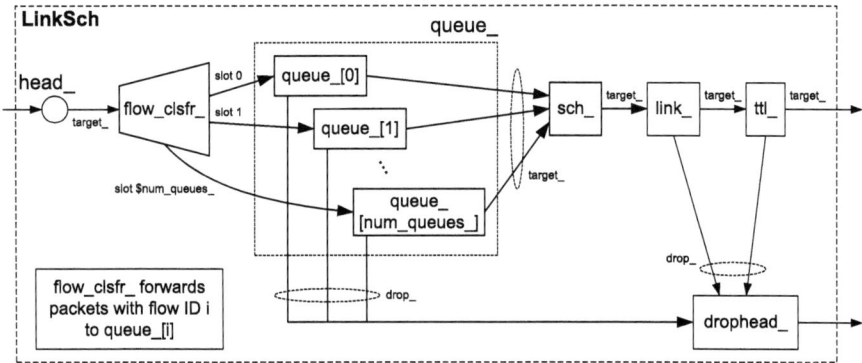

Fig. 14.5. Architecture of a LinkSch object.

Key Differences between Class LinkSch and Class SimpleLink

Class LinkSch defines two addition components–a flow classifier flow_clsfr_ and a packet scheduler sch_, and modifies one component of class SimpleLink– it is instvar queue_.

- Flow classifier flow_clsfr_ examines packet header and forwards packets with the same flow ID to the same forwarding NsObject.[2]
- Packet scheduler sch_ takes a packet from one of the attached upstream data flows and forwards the packet to its downstream object. It complies with an underlying packet scheduling protocol to take a packet from a certain flow.

---

[2] Flow classifiers are implemented in a C++ class FidHashClassifier. However, we do not use the built-in C++ class, since we would like to show how to implement a new C++ class.

- Modified instvar `queue_`: Instvar `queue_` is scalar in class `SimpleLink`. However, it is used as an associative array in class `LinkSch`. Its index and value are, respectively, the flow ID and the `Queue` object which stores packets of the corresponding flow ID. Since each element of `queue_` contains packets with the same flow ID, we use the terms "flow ID" and "queue ID" interchangeably.

Packet Flow Mechanism

When a packet enters a `LinkSch` object, it is sorted by a flow classifier, `flow_clsfr_`, which forwards packets from the same flow to the same queue. In particular, the flow classifier forwards packets with flow ID "i" to instvar "`queue_[i]`", as shown in Fig. 14.5. Each element of `queue_` forwards packets to the packet scheduler `sch_` according to the underlying mechanism defined in the C++ domain (e.g., class `Queue`). Based on an underlying scheduling mechanism, the packet scheduler `sch_` takes a packet from one of these queues and forwards it to a `LinkDelay` object, `link_`.

Callback Mechanism

The packet scheduler breaks a callback connection between a `Queue` object `queue_` and a `LinkDelay` object `link_` into two connections. One is between instvar `queue_` and instvar `sch_` and another is between instvar `sch_` and instvar `link_`. Instead of calling back to instvar `queue_`, instvar `link_` reports (i.e., calls back) to the packet scheduler `sch_` to indicate that it is ready to receive another packet. Upon receiving a call back message, the packet scheduler selects the next transmission flow based on its underlying scheduling discipline, and fetches another packet from the selected flow (i.e., an element of `queue_`).

Every element of `queue_` deactivates the queue blocking mechanism, since they do not need to wait before sending a packet to a packet scheduler. Rather, such a blocking (i.e., waiting) mechanism is implemented in the packet scheduler. Under this call back mechanism, instvar `link_` calls back to the packet scheduler (rather than a `Queue` object) to indicate the completion of packet transmission.

14.2.2 C++ Implementation

In the C++ domain, we define two new NS2 components – flow classifiers and packet schedulers in C++ classes `FlowClassifier` and `PktScheduler`, respectively. Class `FlowClassifier` defines how a flow classifier forwards packets with the same flow ID to the same NsObject. Class `PktScheduler` is a base class from which more specific packet scheduler classes derive. As an example, we develop a C++ class `RRScheduler` which is a derived class of class `PktScheduler` to represent round-robin packet schedulers.

Program 14.7 C++ Implementation of class `FlowClassifier`

```
    //classifier-flow.h
1   class FlowClassifier : public Classifier {
2   protected:
3       int classify(Packet *p);
4   };

    //classifier-flow.cc
5   static class FlowClassifierClass : public TclClass {
6   public:
7       FlowClassifierClass() : TclClass("Classifier/Flow") {}
8       TclObject* create(int, const char*const*) {
9           return (new FlowClassifier());
10      }
11  } class_flow_classifier;
12  int FlowClassifier::classify(Packet *p)
13  {
14      return hdr_ip::access(p)->flowid();
15  }
```

Here, we bind the C++ class `FlowClassifier` to an OTcl class `Classifier/Flow`, but do not bind the C++ class `PktScheduler`. However, we bind a C++ class `RRScheduler`, a child class of class `PktScheduler`, to an OTcl class `PktScheduler/RR`.

Flow Classifiers

A flow classifier is represented by a C++ class `FlowClassifier` implementation of which is shown in Program 14.7. Class `FlowClassifier` is bound to an OTcl class `Classifier/Flow` (see Lines 5–10). Derived from class `Classifier`, class `FlowClassifier` overrides function `classify(p)` by returning the flow ID specified in the header of packet `p*` (Line 14).

Packet Schedulers

The main responsibility of a packet scheduler is to determine transmission sequence of the attached upstream `Queue` objects. In this section, we assume that each `Queue` object holds packets of the same flow ID and the packet scheduler determines the transmission sequence based on the flow ID only.

Packet schedulers are implemented using a C++ class `PktScheduler`, declaration and implementation of which are shown in Programs 14.8 and 14.9, respectively. From Program 14.8, class `PktScheduler` has one constant and four key variables:

Program 14.8 Declaration of a C++ class `PktScheduler`

```
   //pkt-sched.h
16 #define MAX_FLOWS 10
17 class PktScheduler : public Connector {
18 public:
19     PktScheduler();
20     virtual void handle(Event*);
21     virtual void recv(Packet*, Handler*);
22 protected:
23     void send(int fid, Handler* h);
24     virtual void resume();
25     int getFlowID(Packet* p) {return hdr_ip::access(p)->flowid();};
26     virtual int nextID() = 0;
27     Handler* qh_[MAX_FLOWS];
28     Packet* pkt_[MAX_FLOWS];
29     int blocked_;
30     int active_flow_id_;
31 };
```

MAX_FLOW	The maximum number of queues which can be attached to the packet scheduler.
blocked_	Set to "1" if the packet scheduler is in the "blocked" state, and set to "0" otherwise.
active_flow_id_	The flow ID of the packet being transmitted
pkt_[i]	The HOL packet of the queue corresponding to flow "i"
qh_[i]	The QueueHandler object of the queue corresponding to flow "i"

Class `PktScheduler` has two main tasks. One is to determine the transmission sequence for all attached data flow. Another is to insert itself in the middle of a callback connection between a `Queue` object and a `LinkDelay` object. While the first task is implemented in function `nextID()`, the second task is attributed to functions `recv(p,h)` and `resume()`. Taking no input argument, function `nextID()` returns the next transmitting flow ID based on the underlying scheduling discipline. Class `PktScheduler` declares this function as pure virtual, and leaves the detailed implementation to its derived classes (Line 26 in Program 14.8). As an example, we will show how a round-robin packet scheduler implements this function later in this section.

The details of functions `recv(p,h)` and `resume()` are shown in Lines 41–50 and 57–67 of Program 14.9. Function `recv(p,h)` is the main packet reception function. Function `recv(p,h)` first determines the flow ID of packet *p by invoking function `getFlow(p)` in Line 43. Line 44 stores the input packet *p and the input handler *h in variable `pkt_[fid]` and `qh_[fid]`, respectively, where `fid` is the ID of the packet *p. If the `PktScheduler` object is not blocked, Line 46 will send the head of the line packet stored in `pkt_[fid]` to

Program 14.9 Functions of a C++ class `PktScheduler`

```
    //pkt-sched.cc
32  PktScheduler::PktScheduler()
33  {
34      int i;
35      for (i=0;i<handle(0);
32  PktScheduler::PktScheduler()
33  {
34      int i;
35      for (i=0;i<MAX_FLOWS;i++){
36          pkt_[i] = 0;
37          qh_[i]=0;
38      }
39      blocked_ = 0;active_flow_id_ = -1;
40  }
41  void PktScheduler::recv(Packet* p, Handler* h)
42  {
43      int fid = getFlowID(p);
44      pkt_[fid] = p;qh_[fid] = h;
45      if (!blocked_) {
46          send(fid,this);
47          blocked_ = 1;
48          active_flow_id_ = fid;
49      }
50  }
51  void PktScheduler::send(int fid_idx, Handler* h)
52  {
53      Connector::send(pkt_[fid_idx],h);
54      pkt_[fid_idx] = 0;
55  }
56  void PktScheduler::handle(Event*) { resume();}
57  void PktScheduler::resume()
58  {
59      qh_[active_flow_id_]->handle(0);
60      int index = nextID();
61      blocked_ = 0;
62      if (index >= 0) {
63          send(index,this);
64          blocked_ = 1;
65          active_flow_id_ = index;
66      }
67  }
```

the downstream object, and reset `pkt_[fid]` to zero (see Lines 51–55). Note that the `PktScheduler` object passes its address (i.e., `this`) rather than the input handler to the downstream object. Line 47 blocks the `PktScheduler` object and Line 48 stores the flow ID of the packet under transmission in variable `active_flow_id_`.

Upon receiving a packet `*p` and a handler `*h`, a `LinkDelay` object schedules a packet departure event. Since the received handler belongs to the above `PktScheduler` object, at the firing time, function `handle(e)` of the `PktScheduler` object is invoked. In Line 56, function `handle(e)` simply invokes function `resume()`, the details of which are shown in Lines 57–67. Line 59 first fetches a packet (for which transmission has just been finished) from the flow by invoking function `handle(e)` of `queue_[active_flow_id_]`. Line 60 determines the next transmitting flow based on the underlying scheduling discipline. Finally, Lines 61–66 forward the selected packet to the downstream object (similar to Lines 45–49).

As an example, consider an implementation of round-robin schedulers, which transmit packets from each flow sequentially. We implement this type of schedulers using a C++ class `RRScheduler` which is bound to an OTcl class `Scheduler/RR`. The declaration and implementation of the C++ class `RRScheduler` are shown in Program 14.10. Class `RRScheduler` has one variable `current_id_` and one function `NextID()`. The variable `current_id_` records the most recently selected flow ID. Based on the round-robin scheduling principle, class `RRScheduler` overrides function `nextID()` by returning the next ID, whose corresponding `Queue` object contains at least one packet. If the `Queue` objects do not contain any packet, this function will return −1 (see Lines 95–108).

14.2.3 OTcl Implementation

In the OTcl domain, we put together the components of a link with a scheduler, as shown in Fig. 14.5. Again, the major differences of class `LinkSch` and class `SimpleLink` lie in the instvars `flow_clsfr_`, `queue_`, and `sch_` of class `LinkSch`. Instvar `flow_clsfr_` is a flow classifier (whose OTcl class is `Classifier/Flow`). It forwards incoming packets with flow ID "i" to the NsObject stored in the slot number "i". Instvar `queue_` is an array of `Queue` objects. Here, `queue_[fid]` is installed in the slot corresponding to flow ID "fid". Finally, instvar `sch_` is a round-robin packet scheduler instantiated from an OTcl class `PktScheduler/RR`.

Programs 14.11–14.12 show the OTcl implementation of a link with a scheduler. The implementation involves two OTcl classes: `LinkSch` and `Simulator`. Similar to class `SimpleLink`, class `LinkSch` derives from class `Link`. In addition to those defined in class `SimpleLink`, the following instvars are defined in class `LinkSch`:

Program 14.10 C++ implementation of c Class `RRScheduler`

```
    //pkt-sched.h
68  class RRScheduler : public PktScheduler {
69  public:
70      RRScheduler() ;
71  private:
72      virtual int nextID();
73      int current_id_;
74  };

    //pkt-sched.cc
75  RRScheduler::RRScheduler()
76  {
77      current_id_ = -1;
78  }
79  static class RRSchedulerClass: public TclClass {
80  public:
90      RRSchedulerClass() : TclClass("PktScheduler/RR") {}
91      TclObject* create(int, const char*const*) {
92          return (new RRScheduler());
93      }
94  } class_rr_scheduler;
95  int RRScheduler::nextID()
96  {
97      int count = 0;
98      current_id_++;current_id_ %= MAX_FLOWS;
99      while((pkt_[current_id_] == 0)&&(count<MAX_FLOWS)){
100         current_id_++;current_id_ %= MAX_FLOWS;
101         count++;
102     }
103     if (count == MAX_FLOWS)
104         return -1;
105     else{
106         return current_id_;
107     }
108 }
```

       `sch_` A round-robin scheduler whose class is `Classifier/Flow`

`flow_clsfr_` A flow classifier

`num_queues_` The number of queues which are attached to the packet
             scheduler `sch_`

Class `LinkSch` also has two main instprocs `init{...}` (i.e., the constructor) and `add-flow{...}`. As shown in Lines 1–30 of Program 14.11, instproc,
`init{src dst bw delay num_queues}` creates a `LinkSch` object connecting
a Node `src` and a Node `dst`. The bandwidth and delay of the `LinkSch` object

Program 14.11 OTcl implementation of a link with a round-robin scheduling.

```
     //~ns/tcl/lib/ns-link.tcl
1    Class LinkSch -superclass Link
2    LinkSch instproc init {src dst bw delay num_queues} {
3        $self next $src $dst
4        $self instvar link_ queue_ head_ toNode_ ttl_
5        $self instvar drophead_
6        $self instvar num_queues_ sch_ flow_clsfr_
7        set ns [Simulator instance]
8        set head_ [new Connector]
9        set drophead_ [new Connector]
10       set link_ [new DelayLink]
11       set ttl_ [new TTLChecker]
12       set flow_clsfr_ [new Classifier/Flow]
13       set sch_ [new PktScheduler/RR]
14       set num_queues_ $num_queues
15       $head_ set link_ $self
16       $drophead_ target [$ns set nullAgent_]
17       $head_ target $flow_clsfr_
18       for {set i 0} {$i < $num_queues_} {incr i} {
19           set queue_($i) [new Queue/DropTail]
20           $queue_($i) target $sch_
21           $queue_($i) drop-target $drophead_
22       }
23       $sch_ target $link_
24       $link_ target $ttl_
25       $link_ drop-target $drophead_
26       $link_ set bandwidth_ $bw
27       $link_ set delay_ $delay
28       $ttl_ target [$dst entry]
29       $ttl_ drop-target $drophead_
30   }
```

are bw bps and delay seconds, respectively. In regards to packet scheduling, Lines 12 and 13 create a flow classifier and a round-robin packet scheduler, respectively. Lines 18–22 create "num_queues" Queue objects, and configure each of the created Queue objects to point to the packet scheduler sch_ and the common dropping point drophead_. The connection from a flow classifier flow_clsfr_ to the created Queue object is not created here. Rather, it is created by using instproc add-flow{fid}, which simply installs queue_[fid] in the slot number fid of the flow classifier flow_clsfr_ (see Lines 43–48 in Program 14.12).

To facilitate the construction and configuration of LinkSch objects, we also develop two interface instprocs in class Simulator. The first instproc, sch-link{...} is an interface to create a LinkSch object (see Lines 35–42 in

Program 14.12 OTcl implementation of a link with a round-robin scheduling (Cont.).

```
31 LinkSch instproc add-flow { fid } {
32     $self instvar queue_ flow_clsfr_
33     $flow_clsfr_ install $fid $queue_($fid)
34 }
35 Simulator instproc sch-link { n1 n2 bw delay num_queues} {
36     $self instvar link_ queueMap_ nullAgent_ useasim_
37     set sid [$n1 id]
38     set did [$n2 id]
39     set link_($sid:$did) [new LinkSch
                              $n1 $n2 $bw $delay $num_queues]
40     set pushback 0
41     $n1 add-neighbor $n2 $pushback
42 }
43 Simulator instproc add-flow { n1 n2 prio } {
44     $self instvar link_
45     set sid [$n1 id]
46     set did [$n2 id]
47     $link_($sid:$did) add-flow $fid
48 }
```

Program 14.12). The second instproc, add-flow{fid} is an interface to create a connection from a flow classifier to the queues in the LinkSch object (see Lines 43–48 in Program 14.12).

Example 14.5. Consider Example 10.1 and Fig. 9.3. Replace the TCP flow with "num_queues" TCP flows whose flow ID are 0, 1, 2, and so on. Apply a round robin packet scheduling discipline to the link connecting the Node n1 and the Node n3.

Tcl Simulation Script

```
   //rr.tcl
1  set num_queues [lindex $argv 0]
2  set ns [new Simulator]
3  set n1 [$ns node]
4  set n2 [$ns node]
5  set n3 [$ns node]
6  $ns duplex-link $n1 $n2 5Mb 2ms DropTail
7  $ns duplex-link $n2 $n3 5Mb 2ms DropTail
8  $ns sch-link $n1 $n3 5Mb 2ms 10
9  $ns simplex-link $n3 $n1 5Mb 2ms DropTail
10 for {set i 0} {$i < $num_queues} {incr i} {
11     $ns add-flow $n1 $n3 $i
12     set tcp($i) [new Agent/TCP]
```

```
13      set sink($i) [new Agent/TCPSink]
14      set ftp($i) [new Application/FTP]
15      $tcp($i) set fid_ $i
16      $ns attach-agent $n1 $tcp($i)
17      $ns attach-agent $n3 $sink($i)
18      $ftp($i) attach-agent $tcp($i)
19      $ns connect $tcp($i) $sink($i)
20      $ns at 0.0 "$ftp($i) start"
21 }
22 proc show_tcp_seqno {} {
23      global tcp num_queues
24      for {set i 0} {$i < $num_queues} {incr i} {
25          puts "The final tcp($i) sequence number is
26                  [$tcp($i) set t_seqno_]"
27      }
28 }
29 $ns at 100.0 "show_tcp_seqno"
30 $ns at 100.1 "$ns halt"
31 $ns run
```

The above Tcl simulation script "rr.tcl" for this example takes the number of TCP flows as an input argument, and simulates the transmission of these TCP flows under a round-robin packet scheduler, and shows the final sequence number of every TCP flow.

Line 1 takes an input argument from the shell and stores it in variable num_queues. Lines 8–9 replace a bi-directional link between Node 1 and Node 3 with a uni-directional LinkSch object from Node 1 to Node 3 and a uni-directional SimpleLink object from Node 3 to Node 1. The "for" loop in Lines 10–21 creates and configures TCP flows. Each packet created by the "ith" element of variable tcp is assigned with flow ID "i" (by Line 15). In p Line 10, packets with flow ID "i" will be forwarded to queue_[i] of the LinkSch object created in Line 8.

By running the simulation for 1 TCP flow and 3 TCP flows, the following results are shown on the screen.

```
>> ns rr.tcl 1
The final tcp(0) sequence number is 60110

>> ns rr.tcl 3
The final tcp(0) sequence number is 20052
The final tcp(1) sequence number is 20051
The final tcp(2) sequence number is 20051
```

The TCP throughput is computed by the final sequence number divided by the simulation time. Since the simulation time here is 100 seconds (see

Line 28), the throughput of TCP flow "0" is 610.1 packets/sec and 200.52 packets/sec when the number of TCP flows is 1 and 3, respectively.

With a round-robin scheduler, each element of the array `queue_` has equal chance to transmit packets. In principle, each TCP flow should have the same throughput performance (as shown above). Also, the throughput in case of n TCP flows should be approximately n times less than that in case of single TCP flow. From the above result, the per-flow TCP throughput in case of 3 TCP flows is almost the same as each other, and is approximately one third of TCP throughput in case of single TCP flow (i.e., $(60110/100)/3 = 601.10/3 = 200.37$).

Next, we run the above Tcl simulation script for 1 to 10 TCP flows. We compare the average TCP throughput and the *fair share* TCP throughput in Fig. 14.6. Here, we define the *fair share* TCP throughput for n TCP flows as γ/n, where γ is the TCP throughput in case of single TCP flow. We observe that both average and fair share throughput are almost inline with each other. We also observe that TCP throughput for each flow is very similar to each other. These two observations validate the round-robin operation, which treats every TCP flow equally.

Exercise 14.6. A Weighted Fair Queue (WFQ) packet scheduler gives fair access to every data flow. Under a WFQ packet scheduler, each data flow gains channel access in proportion to its weight. The algorithm for WFQ-based packet scheduling can be found in [25]. Develop a module for a WFQ packet scheduler. Validate the module by plotting the results in a graph.

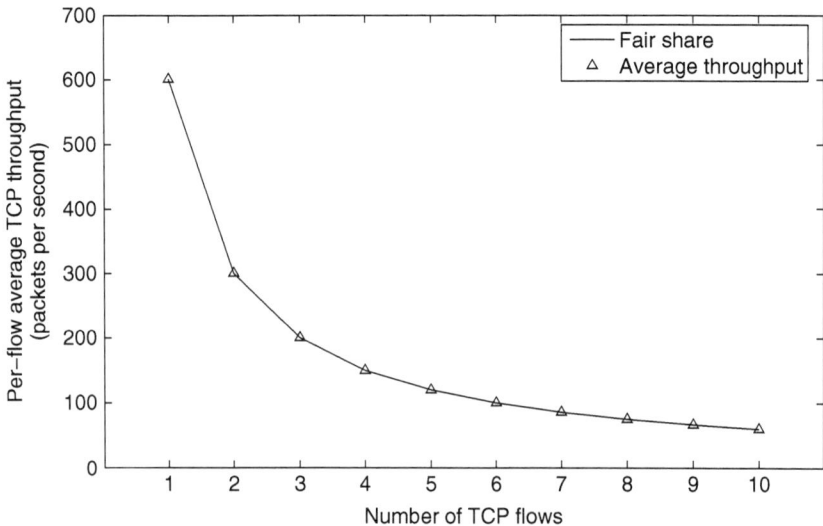

Fig. 14.6. Impact of number of TCP flows on per-flow throughput under round-robin packet scheduling.

14.3 Chapter Summary

This final chapter has demonstrated how new modules are created, configured, and incorporated into NS2. Two examples provided here include: an Automatic Repeat reQuest (ARQ)-based error recovery module and a packet scheduling module. In most of the cases, we need to develop NS2 codes in both C++ and OTcl domains. In the C++ domain, the main task is to define the internal mechanisms of the new NS2 components. The main task in the OTcl domain, on the other hand, are to integrate the developed NS2 components into the existing NS2 modules, and to instantiate and configure the newly developed modules from a Tcl simulation script.

A

Programming Essentials

This appendix covers the basic elements of the programming languages, which are essential for developing NS2 simulation programs. These include Tcl/OTcl which is the basic building block of NS2 and AWK which can be used for post simulation analysis.

A.1 Tcl Programming

Tcl is a general purpose scripting language. While it can do anything other languages could possibly do, its integration with other languages has proven even more powerful. Tcl runs on most of the platforms such as Unix, Windows, and Mac. The strength of Tcl is its simplicity. It is not necessary to declare a data type for variable prior to the usage. At runtime, Tcl interprets the codes line by line and converts the string into appropriate data type (e.g., integer) on the fly.

A.1.1 Program Invocation

Tcl can be invoked from a shell command prompt with the following syntax:

```
tclsh [<filename> <arg0> <arg1> ...]
```

where `tclsh` is mandatory. Other input arguments are optional. When the above command is invoked without input argument, the shell enters Tcl environment where it waits for the Tcl statements line by line. If `<filename>` is specified, Tcl will interpret the text specified in the file whose name is `<filename>` line by line. In addition, if `<arg0> <arg1>` ... are specified, they will be placed in a `list` variable (see Section A.1.3) argv. In the main program, `<argi>` can be retrieved by executing "`lindex $argv $i`".

A.1.2 A Simple Example

To get a feeling about the language, we look at Example A.1 below:

Example A.1. The following Tcl script, "`convert.tcl`", converts tempera-
tures from Fahrenheit to Celsius. The conversion starts at 0 degree (Fahrenheit), proceeds with a step of 25 degrees (Fahrenheit), and stops when the
temperature exceeds 140 degrees (Fahrenheit). The program prints out the
converted temperature in Celsius as long as the temperature in Fahrenheit
does not exceed 140 degrees.

```
# convert.tcl
# Fahrenheit to Celsius Conversion
1  proc tempconv {} {
2     set lower 0
3     set upper 140
4     set step 25
5     set fahr $lower
6        while {fahr < $upper} {
7            set celsius [expr 5*($fahr - 32)/9]
8            puts "Fahrenheit / Celsius : $fahr / $celsius"
9            set fahr [expr $fahr + $step]
10       }
11 }
```

The details of the above example are as follows. The symbol **#** here denotes
the beginning of a line comment. The reserved word **proc** in Line 1 declares
a procedure **tempconv{}** which takes no input argument. The procedure also
defines four local variables (i.e., **lower**, **upper**, **step**, and **fahr**) and assigns
values to them using the reserved word **set** followed by the name and its assigned value (Lines 2–5). Note here that, to refer to the value of a variable, the
reserved character **$** is used in front of the variable (e.g., **set fahr $lower**).
The keyword **expr** in Line 9 informs the Tcl interpreter to interpret the following string as a mathematical expression. The **while** loop in Lines 6–10
controls the iteration of the procedure through the test expression enclosed in
a double quotation mark. The command **puts** in Line 8 prints out the string
contained within the quotation mark. If the name of the script is **convert.tcl**,
the script can be executed by typing the following on a shell prompt:

```
>>tclsh convert.tcl
Fahrenheit / Celsius : 0 / -17.778
Fahrenheit / Celsius : 25 / -3.889
Fahrenheit / Celsius : 50 / 10
Fahrenheit / Celsius : 75 / 23.889
Fahrenheit / Celsius : 100 / 37.778
Fahrenheit / Celsius : 125 / 51.667
```

Alternatively, since NS2 is written in Tcl, the following invocation would lead to the same result.

```
>>ns convert.tcl
```

A.1.3 Variables and Data Types

Data Types

As an interpreter, Tcl does not need to define data type of variables. Instead, it stores everything in string and interprets them based on the context.

Example A.2. Consider the following Tcl codes:

```
# vars.tcl
1 set a "10+1"
2 set b "5"
3 set c $a$b
4 set d [expr $a$b]
5 puts $c
6 puts $d
7 unset c
8 puts $c
```

After executing the Tcl script "vars.tcl", the following result should appear on the screen:

```
>>tclsh vars.tcl
10+15
25
```

Here, variable c is simply a string "10+15", whereas variable d is 25 obtained by numerically evaluating the string "10+15" stored in variable c. Therefore, we may conclude that everything is treated as a string unless specified otherwise [26].

Variable Assignment and Retrieval

Tcl stores a value in a variable using the reserved word "set". The value stored in a variable can be retrieved by placing a character "$" in front of a variable name. In addition, a reserved word "unset" is used to clear the value stored in a variable.

Example A.3. Insert the following two lines into the end of the codes in Example A.2.

```
7 unset c
8 puts $c
```

After executing the Tcl script "vars.tcl", the following result should appear on the screen:

```
>>tclsh vars.tcl
10+15
25
can't read "c": no such variable
 while executing
"puts c"
 (file "var.tcl" line 8)
```

Clearly after being unset, variable c stores nothing. Printing the variable would result in a runtime error.

Bracketing

There are four type of bracketing in Tcl. These are used to group a series of strings. Tcl interprets strings inside different types of bracket differently. Suppose a variable $var stores a value 10. Tcl interprets a statement "expr $var + 1" with four different bracketing differently.

- Curly braces ({expr $var + 1}): Tcl interprets this statement as it is.
- Quotation marks ("expr $var + 1"): Tcl interpolates the variable var in the string. This statement would be interpreted as "expr 10 + 1".
- Square brackets ([expr $var + 1]): Tcl regards a square bracket in the same way that C++ regards a parenthesis. It interprets the string in a square bracket before interpreting the entire line. This statement would be interpreted as "11".
- Parentheses ((expr $var + 1)): Tcl uses a parentheses for indexing an array and for invoking built-in mathematical function.

Example A.4. Insert the following two lines into the end of the codes in Example A.2.

```
7 puts -nonewline {{}: }
8 puts {expr $c}
9 puts -nonewline {"": }
10 puts "expr $c"
11 puts -nonewline {[]: }
12 puts [expr $c]
```

After executing the Tcl script "vars.tcl", the following result should appear on the screen:

```
>>tclsh vars.tcl
10+15
25
```

```
{}: expr $c
"": expr 10+15
[]: 25
```

When bracketing with "{}", Tcl interprets the string as it is; The result in this case is "expr $c". The string $c is replaced with its value when bracketing with """". The result in this case is "expr 10+15". Finally, "[]" identifies the sequence of execution. The string "expr $c" is executed first. The result in this case is "25".

Global Variables

In Example A.1, we briefly mentioned about local variables. But what was missing there is the notion of global variables. Global variables are common and used extensively throughout a program. These variables can be called upon by any procedure in the program. Example A.5 shows an example use of global variables.

Example A.5 (Global variables).

```
set PI 3.1415926536
proc perimeter {radius} {
    global PI
    expr 2*$PI*$radius
}
```

Since "PI" is defined outside of the procedure **perimeter**, the keyword **global** is used here to make "PI" global and available within the procedure. When called upon, this procedure simply calculates the perimeter of a circle based on the supplied input **radius**. Finally, we note here that no default values are automatically assigned to variables. Any attempt to call an uninitialized variable would lead to a runtime error.

Array

An array is a special variable which can be used to store a collection of items. An array stores both the indexes and the values as strings. For example, index "0" is not a number, but a numeric string. By default, an array in Tcl is an associative array. Example A.6 below shows various ways of string manipulation.

Example A.6 (Array assignment).

```
# Numeric indexing
set arr(0) 1
set arr(1) 3
set arr(1) 5
```

```
# String indexing
set wlan(datarate) 54000000
set wlan(protocol) "tcp"
```

Lists

A list is an ordered collection of elements such as numbers, strings or even lists themselves. The key list manipulations are shown below:

- **List creation**: A list can be created in various ways as shown in Example A.7 below.

  Example A.7. The following two statement are equivalent
  (i) set `mylist` `"1 2 3"`
  (ii) set `mylist` `{1 2 3}`

  From the above, a list can be created in three ways. First, it can be created by the reserved word `list` which takes list members as input arguments. Alternatively, it can be created by embracing the members within a pair of curly braces or a pair of quotation marks.
- **Member retrieval**: The following command returns the nth $(= \{0, 1, \cdots\})$ element in a list `mylist`:

  ```
  lindex $mylist $n
  ```

- **Member setting**: The following command sets the value of the nth element in a list `mylist` to be `<value>`:

  ```
  lset $mylist $n $value
  ```

- **Group retrieval**: The following command returns a list whose members are the nth member through the mth member of a list `mylist`:

  ```
  lrange $mylist $n $m
  ```

- **Appending the list**: The following command attaches a list `alist` to the end of a list `mylist`:

  ```
  lappend $mylist $alist
  ```

A.1.4 Input/Output

Tcl employs a so-called *Tcl channel* to receive an input using a command **gets** or to send an output using a command **puts**.

Tcl Channels

A Tcl channel refers to an interface used to interact to the outside world.
Two main types of Tcl channels include standard reading/writing channels
and file channels. The former are classified into stdin for reading, stdout for
writing, and stderr for error reporting. The latter needs to be attached to a
file before it is usable. The syntax for attaching a file to a Tcl file channel is
shown below:

```
open <filename> [<access>]
```

This command returns a Tcl channel attached to a file with the name
<filename>. The optional input argument <access> could be "r" for reading,
"w" for writing to a new file, or "a" for appending an existing file.

When a Tcl channel is no longer in use, it can be closed by using the
command close whose syntax is as follows:

```
close <channel>
```

where <chanel> is the Tcl channel which need to be closed.

The Commands gets and puts

The command puts and gets reads and writes, respectively, a message to a
specified Tcl channel. In particular, the command "gets" reads a line from a
Tcl channel, and passes every character in the line except the end-of-line character to the Tcl running environment. The Tcl channel could be a standard
channel or a file channel. The syntax of the command gets is as follows:

```
gets <channel> <var>
```

Here, all the characters in the current line from the channel channel will be
stored in the variable <var>.

The command "puts" writes a string <string> followed by an end-of-
line character to a Tcl channel <channel>. If <channel> is not specified, the
stdout will be used as a default channel. The syntax of the command puts
is as follows:

```
puts [-nonewline] ]<channel>[ <string>
```

where the nonewline option above specifies not to write an end-of-line character to the end of the string.

Normally, the command puts does not output immediately onto a Tcl
channel. Instead, it puts the input argument (i.e., string) in its buffer, and
releases the stored string either when the buffer is full or when the channel is
closed. To force the immediate outputting, flush is used. Note that while a
standard channel is opened and closed on the fly (i.e., upon an invocation of
"puts), a file channel needs to be closed explicitly using the command close.

Example A.8. Consider the following Tcl codes:

```
puts "Press any key to continue..."
gets stdin
set ch_in [open "input.txt" "r}]
set ch_out [open "output.txt" "a"]
set line_no 0
while {[gets $ch_in line] >= 0} {
      puts $ch_out "[incr line_no] $line"
}
close $ch_in
close $ch_out
```

In this example, the content of file input.txt is copied to file output.txt line by line. In addition, the line number is prefixed at the beginning of each new line.

A.1.5 Mathematical Expressions

Tcl implements mathematical expressions through mathematical operators and mathematical functions. A mathematical expression of either type must be preceded by a reserved word "**expr**". Otherwise, Tcl will recognize the operator as a character (see Lines 1 and 4 in Example A.2). A mathematical expression consists of an operator and operands. A list of most widely used operators is given in Table A.1. An operand can be either floating-point, octal or hexadecimal numbers. To be evaluated as octal and hexadecimal numbers, the numbers must be preceded by 0 and 0x, respectively.

As another means to implement mathematical operations, mathematics functions can be placed after the reserved word "**expr**". The built-in mathematical functions are shown below, where the input argument of a function is enclosed by parentheses.

Table A.1. Tcl mathematical operators.

Operators	Usage
$-+\sim!$	Unary minus, unary plus, bit-wise negation, logical negation
$* \, / \, \%$	Multiplication, division, remainder
$+-$	Addition, subtraction
$>$	Bit shift left, right
$<>=$	Less than, greater than, less than or equal, greater than or equal
$\&$	Bit-wise AND
$\wedge$	Bit-wise exclusive OR
$\|$	Bit-wise OR
$\&\&$	Logical AND
$\|\|$	Logical exclusive OR
x?y : z	If x is non-zero, then y. Otherwise, z.

abs(x)	cosh(x)	log(x)	sqrt(x)
acos(x)	double(x)	log10(x)	srand(x)
asin(x)	exp(x)	pow(x,y)	tan(x)
atan(x)	floor(x)	rand(x)	tanh(x)
atan2(x)	fmod(x)	round(x)	wide(x)
ceil(x)	hypot(x,y)	sin(x)	
cos(x)	int(x)	sinh(x)	

The detail of all the above functions is given in [27]

Example A.9. Examples of invocation of mathematical functions log10(x) and abs(x) are shown below.

```
>>tclsh
>>expr log10(10)
1.0
>>expr abs(-10)
10
>>expr 1+2
3
```

A.1.6 Control Structure

Tcl control structure defines how the program proceeds. This is carried out using the commands if/else/elseif, switch, for, while, foreach, and break/continue

if/else/elseif

An if/else/elseif command provides a program with a selective choice. A general form of this command is shown below:

```
if {<condition1>} {
    <actions_1>
} elseif {<condition2>} {
    <actions_2>
}
    .
    .
    .
else {
    <actions_n>
}
```

Here, the command first checks whether condition1 in the if statement is true. If so, it will take actions_1. Otherwise, it will check whether condition2 in the elseif statement is true. If so, it will take actions_2. If not, the process continues for every elseif statement. If nothing matches, actions_n defined under the else condition will be taken.

switch

The switch command is a good substitute of a long series of a if/else/elseif command. It checks the value of a variable against given patterns, and takes actions associated with a matched pattern. The structure of a switch command is shown below:

```
switch <value> {
    <pattern_1> {
        <actions_1>
    }
    <pattern_2> {
        <actions_2>
    }
    .
    .
    .
    default {
        <actions_n>
    }
}
```

In this case action_i will be taken if <value> matches with <pattern_i> where i= { 1,2,...,n-1}. If none of the predefined patterns matches with the value, the default actions (i.e., actions_n) will be taken.

while/for/foreach

The commands while, for, and foreach are used when actions need to be repeated for several times. The command while repeats actions until a predefined condition is no longer true. The command for repeats the actions for a given number of times. The command foreach repeats the actions for every item in a given list. The syntax of these three commands are as follows:

```
while {<condition>} {
    <actions>
}

for {<init>} {<condition>} {<mod>} {
    <actions>
}

foreach {<var>} {<list>} {
    <actions>
}
```

The `while` command repeats the `<actions>` as long as the `<condition>` is true. The `for` command begins with an initialization statement `init`. After taking `<actions>`, it executes the Tcl statement `<mod>` and checks whether the `<condition>` is `true`. If so, it will repeat the `<actions>`. Otherwise, the command will terminate. The command `foreach` repeats `<actions>` for every member in the `<list>`. In each repetition, the member is stored in variable `<var>` and can be used for various purposes.

break/continue

Commands `break` and `continue` are used in looping structures `while, for,` and `foreach`. They are used to prematurely stop the looping mechanism. Their key difference is that while the command `break` immediately exits the loop, the command `continue` simply restarts the loop.

Example A.10.

```
set var 0
while {$var < 100} {
    puts $var
    set var [expr $var+5]
    if {$var == 20}
      break
}
puts $var
```

In this example, the loop continues as long as `$var < 100`. However, the command `break` terminates the looping mechanism if `$var == 20`. Therefore, the above program will print out 20. If the reserved word `break` is replaced with a reserved word `continue`, the loop will restart after being stopped. In this case the program will print out 100.

A.1.7 Procedures

A procedure is usually used in place of a series of Tcl statements to tidy up the program. The syntax of a procedure is shown below:

```
proc <name> {<arg_1> <arg_2> ... <arg_n>} {
    <actions>
    [return <returned_value>]
}
```

The definition of a procedures begins with a reserved word `proc`. The procedure name is placed after the word `proc`. The input arguments are placed within a curly braces, located immediately after the procedure name. Embracing with a curly braces, the main body placed next to the input argument.

Here, the actions for the procedures are defined. Optionally, the procedure may return a `<returned_value>`, using a reserved word `return`.

After defining a procedure, one may invoke the procedure by executing the following statement:

```
set var [<name> <value_1> <value_2> <value_n>]
```

where `var` is set to the value returned from the procedure `<name>`, and the values `<value_1> <value_2> <value_n>` are fed as input arguments of the procedure.

A.2 Objected Oriented Tcl (OTcl) Programming

OTcl is an object-oriented version of Tcl, just like C++ is an object-oriented version of C [28]. The basic architecture and syntax in OTcl are much the same as those in Tcl. The difference, however, is the philosophy behind each of them. In OTcl, the concepts of classes and objects are of great importance. A class is a representation of a group of objects which share the same behavior(s) or trait(s). Such a behavior can be passed down to child classes. In this respect, the donor and the receiver of the behaviors are called a superclass (or a parent class) and a subclass (or a child class), respectively. Apart from inheriting behaviors from a parent class, a class defines its own functionalities to make itself more specific. This inheritance is the very main concept for any OOP including OTcl.

A.2.1 Class and Inheritance

In OTcl, a class can be declared using the following syntax:

```
Class <classname> [-superclass <superclassname>]
```

If the optional argument in the square bracket is present, OTcl will recognize class `<classname>` as a child class of class `<superclassname>`. Alternatively, if the option is absent, class `<classname>` can be also declared as a child class of class `<superclassname>` by executing

```
<classname> superclass <superclassname>
```

Note that, class `<classname>` inherits the functionalities (including procedures and variables) of class `<superclassname>`. In OTcl, the top-level class is class `Object`, which provides basic procedures and variables, from which every user-defined class inherits.

Example A.11. Consider a general network node. When equipped with mobility, this node becomes a mobile node. Declaration of a class `Node` and its child class `Mobile` is shown below. This declaration allows class `Mobile` to inherit capabilities of class `Node` (e.g., receiving packets) and to include more capabilities (e.g., moving) to itself.

```
1 Class Node
2 Class Mobile -superclass Node
```

A.2.2 Class Member Procedures and Variables

A class can be associated with procedures and variables. In OTcl, a procedure and a variable associated with a class are referred to as instance procedure (i.e., instproc) and an instance variable (i.e., instvar), respectively.

Instance Procedures

The syntax which declaver an instproc is shown below:

```
<classname> instproc <procname> [{args}] {
<body>
}
```

where instproc "instproc" is defined in the top-level class Object. Here, the name of the instproc is <procname>. The detail (i.e., <body>) of the instproc is embraced within curly braces. The input arguments of the instproc are given in <args>. Each input argument is separated by a white space. OTcl supports assignment of each input argument with a default value. That is, the input argument will be assigned with the default value if the value is not given at the invocation. Denote an input argument and its default value by <arg> and <def>, respectively. The argument declaration is as follows:

```
{<arg> def}
```

For example, let an instproc has two input arguments: <arg1> and <arg2>. The first input argument <arg1> is not given a default value. The default value for the second input argument <arg2> is given by <def>. To declare this instproc, we replace [args] above with "<arg1> {<arg2><def>}".

Once declared, an instproc is usually invoke through an object (whose class is <classname>) using the following syntax.

```
<object> <procname> [{args}]
```

Instance Variables

Unlike instprocs, instvars are not declared with the class name. Instead, they can be declared anywhere in the file. The syntax for the declaration is as follows:

```
$self instvar <varname1> [<varname2> ...]
```

where instproc "instvar" and an instvar "self" (which represents the object itself) are defined in the top-level class Object. More than one instvar can be declared within an OTcl statement. Syntactically, we simply put the names of all instvars (each seperated by a white space) after "$self instvar".

After the declaration, an instvar can be manipulated by using a command set with the following syntax

```
<object> set <varname> [<value>]
<classname> set <varname> [<value>]
```

When presented, the input argument <value> will be stored in the instvar <varname> associated with the object <object> or the class <classname>. In absence of the argument <value> the above statements return the value stored in the associated instvar <varname>.

Example A.12. Based on Example A.11, the followings define a packet reception instproc for class Node and a moving instproc for class Mobile.

```
3   Node instproc recv {pkt} {
4        $self instvar state
5        set state 1
6   #    $self process-pkt $pkt
7   }

8   Mobile instproc move {x y} {
9        $self instvar location
10       set location[0] $x
11       set location[1] $y
12  }
```

Upon receiving a packet pkt, a Node sets its state to be active (i.e., 1), and invokes instproc process-pkt to process the packet pkt. As a derived class of class Node, class Mobile inherits this instproc. It also defines an instproc move to move to a new coordinate (x,y). This instproc simply sets the new coordinate to be as specified in the input argument (Lines 10–11).

A.2.3 Object Construction and the Constructor

An object can be created (i.e., instantiated) from a declared class by using the following syntax:

```
<classname> <objectname>
```

In the object construction process, instprocs alloc and init of class Object is invoked to initialize the object. Instproc alloc allocates memory space to stored the initiated object. Usually, referred as a constructor, instproc init defines necessary object initialization. This instproc is usually overridden by the derived classes.

Example A.13. The constructors of classes `Node` and `Mobile` in Example A.11 are defined below.

```
13 Node instproc init {} {
14      $self instvar state
15      set state 0
16 }

17 Mobile instproc init {} {
18      $self next
19      $self instvar location
20      set location[0] 0
21      set location[1] 0
22 }
```

At the constuction, class `Node` sets its variable `state` to 0 (i.e., inactive). Class `Mobile` first invokes the constructor of class `Node` in Line 18 (see the details of function `next` in Section A.2.4). Then, Lines 20–21 set the location of the mobile node to be (0,0).

A.2.4 Related Instprocs

Class `Object` also defines the following instprocs.

Instproc `next`

Invoked from within an instproc, `next` searches up the hierarchy (in parent classes) for an instproc with the same name, and invokes the instproc belonging to the closest parent class.

Instproc `info`

This instproc returns related information based on the input argument. It can be invoked using one of the two following ways:

```
<object> info <arg>
<classname> info <arg>
```

The upper and lower invocations return the information about the object and the class, respectively. The choice of the input argument `<arg>` for these two invocations are shown in Tables A.2 and A.3, respectively.

Example A.14. Include the following code to the above definition of classes `Node` and `Mobile`, and save the code in a file "`node.tcl`".

Table A.2. Options of the info instproc for objects.

Options	Functions
class	Returns the class of the object.
procs	Return the list of all local methods.
commands	Return the list of both Tcl and C local methods defined on the object.
vars	Return the list of instance variables defined on the object.
args <proc>	Return the list of arguments of the instproc <proc> defined on the object.
body <proc>	Returns the body of the instproc <proc> defined on the object.
default <proc> ... <arg> <var>	Returns 1 if the default value of the argument <arg> of the instproc <proc> is <var>, and returns 0 otherwise.

Table A.3. Options of the info instproc for classes.

Options	Functions
superclass	Return the superclass of the current class.
subclass	Return the list of all subclasses down the heirachy.
heritage	Return the inheritance precedence list.
instances	Return the list of instances of the class.
instprocs	Return the list of instprocs defined on the class.
instcommands	Return the list of instprocs and OTcl commands defined on the class.
instargs <proc>	Return the list of arguments of the instproc <proc> defined on the class.
instbody <proc>	Return the body of the instproc <proc> defined on the class.
instdefault ... <proc> <arg> <var>	Return 1 if the default value of the argument <arg> of the instproc <proc> is <var>, and return 0 otherwise.

```
23 Node n
24 puts "The instance of class Node is [Node info instances]"
25 puts "The class of n is [n info class]"
```

By executing the file "node.tcl", the following result should appear on the screen.

```
>>ns node.tcl
n
The instance of class Node is n
The class of n is Node
```

Exercise A.15. Write OTcl codes which make use of the above options for instproc `info` in Tables A.2–A.3.

A.3 AWK Programming

AWK is a general-purpose programming language designed for processing of text files [29]. AWK refers to each line in a file as a *record*. Each record consists of *fields*, each of which is separated by one or more spaces or tabs. Generally, AWK reads data from a file consisting of fields of records, processes those fields with certain arithmetic or string operations, and outputs the results to a file as a formatted report.

To process an input file, AWK follows an instruction specified in an AWK script. An AWK script can be specified at the command prompt or in a file. While the strength of the former is the simplicity (in invocation), that of the latter is the functionality. In the latter, the programming functionalities such as variables, loop, and conditions can be included into an AWK script to perform desired actions. In what follows we give a brief introduction to the AWK language. The details of AWK programming can be found in [30].

A.3.1 Program Invocation

AWK can be invoked from a command prompt in two ways based on the following syntax:

```
>>awk [ -F<ch> ] {<pgm>} [ <vars> ] [ <data_file> ]
>>awk [ -F<ch> ] { -f <pgm_file> } [ <vars> ] [ <data_file> ]
```

where {} and [] contain mandatory and optional arguments, respectively. The bracket <> contains a variable which should be replaced with actual values at the invocation. These variables include

ch	Field separator
pgm	An AWK script
pgm_file	A file containing an AWK script (i.e., an AWK file)
vars	Variables used in an AWK file
data_file	An input text file

By default, AWK separates records by using a white space (i.e., one or more spaces or tabs). However, if the option "-F is present, AWK will use <ch> as a field separator.[1] The upper invocation takes an AWK script <pgm> as an input argument, while the lower one takes an AWK file <pgm_file> as an input argument. In both cases, variables <vars> and input text file <data_file> can be optionally provided. If an input text file is not provided, AWK will wait for input argument from the standard input (e.g., keyboard) line by line.

Example A.16. Defines an input text file "infile.txt" in the following. We shall use this input file for most of the examples in this section.

---

[1] For example, awk -F: uses a colon ":" as a field separator.

```
#infile.txt
Rcv 0.162 FromNode 2 ToNode 3 cbr PktSize= 500  UID= 3
EnQ 0.164 FromNode 1 ToNode 2 cbr PktSize= 1000 UID= 8
DeQ 0.164 FromNode 1 ToNode 2 cbr PktSize= 1000 UID= 8
Rcv 0.170 FromNode 1 ToNode 2 cbr PktSize= 1000 UID= 7
EnQ 0.170 FromNode 2 ToNode 3 cbr PktSize= 1000 UID= 7
DeQ 0.170 FromNode 2 ToNode 3 cbr PktSize= 1000 UID= 7
Rcv 0.171 FromNode 2 ToNode 3 cbr PktSize= 1000 UID= 4
EnQ 0.172 FromNode 1 ToNode 2 cbr PktSize= 1000 UID= 9
DeQ 0.172 FromNode 1 ToNode 2 cbr PktSize= 1000 UID= 9
Rcv 0.178 FromNode 1 ToNode 2 cbr PktSize= 1000 UID= 8
EnQ 0.178 FromNode 2 ToNode 3 cbr PktSize= 1000 UID= 8
DeQ 0.178 FromNode 2 ToNode 3 cbr PktSize= 1000 UID= 8
```

Note that in AWK, "#" marks the beginning of a comment line.

At the command prompt, we may run an AWK script to show the lines which contains "EnQ" as follows:

```
>>awk /EnQ/ infile.txt
EnQ 0.164 FromNode 1 ToNode 2 cbr PktSize= 1000 UID= 8
EnQ 0.170 FromNode 2 ToNode 3 cbr PktSize= 1000 UID= 7
EnQ 0.172 FromNode 1 ToNode 2 cbr PktSize= 1000 UID= 9
EnQ 0.178 FromNode 2 ToNode 3 cbr PktSize= 1000 UID= 8
```

Here, the <pgm> is specified as /EnQ/ and the <data_file> is specifies as infile.txt. An AWK script /EnQ/ looks for a line which contains a text EnQ and display the line on the screen.

A.3.2 An AWK Script

An AWK script contains an instruction for what AWK will perform. It asks AWK to look for a pattern in a record, and performs actions on a matched pattern. The syntax of an AWK script is as follows:

```
<pattern> {<actions>}
```

A <pattern> could be a logical expression or a regular expression.[2] An <actions> specifies actions for the matched pattern. Each actions in the curly braces is separated by a semi-colon (";"). As will be discussed later in this section, AWK provides a wide variety of <actions>.

---

[2] While a logical expression is usually implemented by an if statement, a regular expression returns true when finding a matched pattern. The formal definition of a regular expression can be found in [31].

A.3.3 AWK Programming Structure

The general form of an AWK program is shown below:

```
BEGIN {<initialization>}
<pattern1> {<actions>}
<pattern2> {<actions>}
.
.
.
END {<final actions>}
```

Prior to procession an input text file, AWK performs `<initialization>` specified in the curly braces located after the reserved word `BEGIN`. Then, for each record, it performs actions if the records match with the corresponding pattern. After processing the entire file, it performs `<final actions>` specified in the curly braces located after the reserved word `END`.

A.3.4 Pattern Matching

The first part of an AWK script is a pattern as specified in `<pattern>`. The pattern can be a logical or a regular expression. If this part evaluates to `true`, the corresponding action will be taken.

Logical Expressions

For a logical expression, the following operators could be necessary:

```
<  (less than)                  =  (equal)
<= (less than or equal)         != (Not Equal)
>  (greater than)               || (OR)
>= (greater than or equal)      && (AND)
```

Regular Expressions

A regular expression provides a concise and flexible means to represent a text of interest. It is used extensively in programming language such as AWK, Tcl, Perl, etc. Syntactically, a regular expression is enclosed within a pair of forward slashes ("/", e.g., `/EnQ/`). It supports much more functionalities in searching for a pattern as shown in Table A.4:

Exercise A.17. Write an input string which matches with *each* of the following regular expressions. The input string should not match with other regular expressions.

Table A.4. Special characters used in regular expressions.

Character	Description
//	Contain a regular expression (e.g., /text/)
^	Match the beginning of a record only (e.g., /^text/)
$	Match the end of a record only (e.g., /text$/)
[]	Match any character inside (e.g., [text])
[a-z]	Match any lower-case alphabet
[A-Z]	Match any upper-case alphabet
[0-9]	Match any number
[a-zA-Z0-9]	Match any alphabet or number
.	Match any character (e.g., /tex./)
*	Match zero or more character in front of it (e.g., /tex*/)
.*	Match any string of characters
?	Match zero or more regular expression in front of it (e.g., /[a-z]?/)
+	Match one or more regular expression in front of it (e.g., /[a-z]+/)

  (i) /^Node/
 (ii) /Node$/
(iii) /[Nn]ode/
 (iv) /Node./
  (v) /Node*/
 (vi) /Nod[Ee]?/
(vii) /Nod[Ee]+/

By default, a regular expression is matched against the entire record (i.e., line). To match a certain regular expression againt a given variable `var`, we use the following syntax:

```
$var ~ /<pattern>/
$var !~ /<pattern>/
```

While the upper command searches for a line which matches with `<pattern>`, the lower command searches for a line which does not match with `<pattern>`.

A.3.5 Basic Actions: Operators and Output

The key operators in AWK are shown below.

```
+ (addition)          ++ (increment)
- (subtraction)       == (decrement)
* (multiplication)    =  (assignment)
/ (division)          %  (modulo)
```

Like in C++, a combination of arithmatic operators and an assignment operator is also possible. For example, "a += b" is equivalent to "a = a+b". The combined operator in AWK include "+=", "-=", "*=", "/=", and "%=".

AWK outputs a variable or a string to a screen using either `print` or `printf`, whose syntax are as follows:

```
print <item1> <item2> ...
printf(<format>,<item1>,<item2>,...)
```

where `<item1>`, `<item2>`, and so on can be either variables or strings, `<format>` is the format of the output. Using `print`, a string needs to be enclosed within a quotation mark (`""`), while a variable could be indicated as it is.

Example A.18. Define an AWK file "`myscript.awk`" as shown below.

```
# myscript.awk
BEGIN{}
/EnQ/ {var = 10; print "No Quotation: " var;}
/DeQ/ {var = 10; print "In Quotation: " "var";}
END{}
```

Run this script for the input text file `infile.txt` defined in Example A.16. The following result should appear on the screen.

```
>>awk -f myscript.awk infile.txt
No Quotation: 10
In Quotation: var
No Quotation: 10
In Quotation: var
No Quotation: 10
In Quotation: var
No Quotation: 10
In Quotation: var
```

The above AWK script prints out two versions of variable `var`. The upper line prints out the value (i.e., 10) stored in variable `var`. In the lower line, variable `var` is enclosed within a quotation mark. Therefore, string `var` will be printed instead.

The command `printf` provides more printing functionality. It is very similar to function `printf` in C++. In particular, it specifies the printing format as the first input argument. The subsequent arguments simply provide the value for the place-holders in the first input argument. The readers are encouraged to find the detail of the printing format in any C++ book (e.g., [14]) or in [30].

AWK does not have a direct command for file printing. Rather, output redirection can be used in conjunction with `print` and `printf`. In a Unix-like system (e.g., Linux or Cygwin) a character ">" and ">>" can be used to redirect the output to a file. The syntax of the output redirection is shown below.

```
print <input_argument> > <filename>
print <input_argument> >> <filename>
```

Note that the command **print** can be replaced with the command **printf**. The difference between the above two lines is that while ">" redirects the output to a new file, ">>" appends the output to an existing file. If **<filename>** exists, the upper line will delete and recreate the file whose name is **<filename>**, while the lower line will append the output to the file **<filename>** without destroying the existing file.

Exercise A.19. Repeat Example A.18, but print the result in a file "**outfile .txt**". Show the difference when using ">" and ">>".

A.3.6 Variables

As an interpreter, AWK does not need to declare data type for variables. It can simply assign a value to a variable using an assignment operator ("="). To avoid ambiguity, AWK differentiates a variable from a string by quotation marks (" "" "). For example, **var** is a variable while **"var"** is a string (see Example A.18).[3]

AWK also support arrays. Arrays in AWK can have only one dimension. Identified by a square bracket (**[]**), indexes of an array can be both numeric (i.e., a regular array) or string (i.e., an associative array). Example of arrays are **node[1]**, **node[2]**, **link["1:2"]**, etc.

Apart from the above user-defined variables, AWK also provides several useful built-in variables as shown in Table A.5.

Table A.5. Built-in variables.

Variables	Descriptions
$0	The current record
$1,$2,...	The 1st, 2nd,... field of the record
FILENAME	Name of the input text file
FS	(Input) Field separator (a white space by default)
RS	(Input) Record separator (a newline by default)
NF	Number of fields in a current record
NR	Total number of records
OFMT	Format for numeric output (%6g be default)
OFS	Output field separator (a space by default)
ORS	Output record separator (a newline by default)

Exercise A.20. Based on the input file in Example A.16, develop an AWK script to show

---

[3] Unlike Tcl, AWK retrieves the value of a variable without a prefix (not like "$" in Tcl).

 (i) Total number of "EnQ" events,
 (ii) The number of packets that Node 3 receives, and
(iii) Total number of bytes that Node 3 receives.

A.3.7 Control Structure

In common with Tcl, AWK support three major types of control structures:
if/else, while, and for (see Section A.1.6). The syntaxes of these control
structures are as follows:

```
if(<condition>) <action 1> [else <action 2>]
while(<condition>) <action>
for(<initialization>;<condition>;<end-of-loop-action>)
  <action>
```

Again, when the actions contain more than one statement, these statements
must be embraced by a curly braces.

AWK also contains four unconditional control commands:

```
  break  Exit the loop
contine  Restart the loop
   next  Process the next record
   exit  Exit the program by executing the END operation
```

B

A Review of the Polymorphism Concept in OOP

B.1 Fundamentals of Polymorphism

As one of the main OOP concepts, polymorphism refers to the ability to invoke the same function with different implementation under different context. This concept should be simple to understand, since it occurs in our daily life.

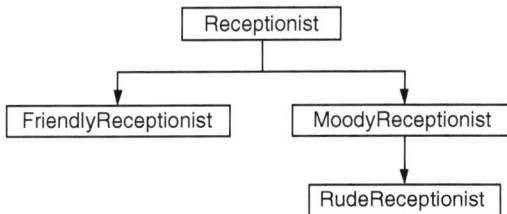

Fig. B.1. A polymorphism example: Receptionist class hierarchy.

Example B.1. Consider receptionists and how they greet customers. Friendly, moody, and rude receptionists greet customers by saying "Good morning. How can I help you today?", "What do you want?", and "What do you want? I'm busy. Come back later!!", respectively. We design a class hierarchy for receptionists as shown in Fig. B.1. The base class of the hierarchy is class `Receptionist`. Based on the personality, we derive classes `Friendly Receptionist` and `MoodyReceptionist` directly from class `Receptionist`. Also, we derive another class `RudeReceptionist` from class `Moody Receptionist`. The C++ code which represents these four classes is given below:

```
//receptionist.cc
1  #include "iostream.h"
```

```
2  class Receptionist {
3      public:
4      void greet() {cout<<"Say:\n";};
5  };

6  class FriendlyReceptionist : public Receptionist {
7      public:
8      void greet(){
9        cout<<"Say: Good morning. How can I help you today?\n"
10     }
11 };

12 class MoodyReceptionist : public Receptionist {
13     public:
14     void greet() { cout<<"Say: What do you want?\n"; };
15 };

16 class RudeReceptionist : public MoodyReceptionist {
17     public:
18     void greet(){
19         MoodyReceptionist::greet();
20         cout<<"Say: I'm busy. Come back later.\n";
21     };
22 };

23 main() {
24     FriendlyReceptionist f_obj;
25     MoodyReceptionist m_obj;
26     RudeReceptionist r_obj;
27     cout<<"\n------------ Friendly Receptionist ---\n";
28     f_obj.greet();
29    cout<<"\n------------ Moody Receptionist -------\n";
30     m_obj.greet();
31     cout<<"\n------------ Rude Receptionist -------\n";
32     r_obj.greet();
33     cout<<"-------------------------------------------\n";
34 }
```

Function main() instantiates three receptionist objects. Objects f_obj, m_obj, and r_obj are of classes FriendlyReceptionist, MoodyReceptionist, and RudeReceptionist, respectively (Lines 24–26). They greet a customer in Lines 28, 30, and 32 by invoking function greet() in Lines 8–10, 14, and

18–21, respectively.[1] By running `receptionist`, the following results should appear on the screen.

```
>>./receptionist
------------ Friendly Receptionist ---------
Say: Good morning. How can I help you today?

------------ Moody Receptionist ---------
Say: What do you want?

------------ Rude Receptionist ---------
Say: What do you want?
Say: I'm busy. Come back later!!
---------------------------------------------
```

Example B.2. Remove Line 14 in Example B.1 and run "`./receptionist`" again. The following result should appear on the screen:

```
>>./receptionist
------------ Friendly Receptionist ---------
Say: Good morning. How can I help you today?

------------ Moody Receptionist ---------
Say:

------------ Rude Receptionist ---------
Say:
Say: I'm busy. Come back later!!
---------------------------------------------
```

Since class `MoodyReceptionist` does not define function `greet` (Line 14 is removed), it uses the function `greet()` inherited from class `Receptionist` (i.e., printing "Say:" on the screen).

Examples B.1 and B.2 demonstrate the concepts of polymorphism through receptionists and how they greet customers. When invoking the same function (e.g., `greet()`), three objects of different classes act differently (e.g., by saying differently). Example B.1 shows a basic polymorphism mechanism, where each class has its own implementation. Examples B.2 shows that it is also possible not to override function `greet()`.[2]

---

[1] Note that in Line 19 function `greet()` of class `MoodyReceptionist` is invoked in the scope of class `RudeReceptionist` by using "`::`".

[2] For example, class `MoodyReceptionist` inherits function `greet()` from class `Receptionist`.

B.2 Type Casting and Function Ambiguity

In most cases, polymorphism is fairly straightforward. A derived class may inherit or override functions from the base class. When polymorphism involves type casting, the mechanism in Examples B.1 and B.2 may lead to different result. To see how, consider the following examples.

Example B.3. Replace function "main" in Example B.1 with the following:

```
1 main() {
2 FriendlyReceptionist *f_pt;
3 MoodyReceptionist *m_pt;
4 RudeReceptionist *r_pt;
5 f_pt = new FriendlyReceptionist();
6 m_pt = new MoodyReceptionist();
7 r_pt = new RudeReceptionist();

8 cout<<"\n------------ Friendly Receptionist ----\n";
9 f_pt->greet();
10 cout<<"\n------------ Moody Receptionist ----\n";
11 m_pt->greet();
12 cout<<"\n------------ Rude Receptionist ----\n";
13 r_pt->greet();
14 cout<<"-------------------------------------------\n";
15 }
```

With the above code, the result for running `./receptionist` would be the same as that in Example B.1. The major difference in the above `main()` function is the use of pointers (Lines 2–4), instead of regular objects (in Example B.1).

Example B.4. In Example B.3, replace Lines 3–4 with the following:

```
MoodyReceptionist *m_pt,*r_pt;
```

This is an example of ambiguity caused by type casting. The pointer `r_pt` is declared as a pointer to a `MoodyReceptionist` object; however, the statement "`new RudeReceptionist()`"' creates an object of type `RudeReceptionist`. When invoking a function (e.g., `greet()`), the key question is which class should function `greet()` be associated with: `MoodyReceptionist` (i.e., the declaration class) or `RudeReceptionist` (i.e., the construction class)? To answer this question, we can simply run "`./receptionist`", and obtain the following results:

```
>>./receptionist
------------ Friendly Receptionist ---------
Say: Good morning. How can I help you today?
```

```
------------ Moody Receptionist ---------
Say: What do you want?

------------ Rude Receptionist -------------
Say: What do you want?
---------------------------------------------
```

From the above result, the answer is the former one: `MoodyReceptionist`.

Consider the statement "`r_pt = new RudeReceptionist`". The latter part, "`new RudeReceptionist`", allocates memory space to an object of class `RudeReceptionist`, and returns a pointer to the created object. The former part "`r_pt = `" assigns the returned pointer to `r_pt`. Since `r_pt` is a pointer to a `MoodyReceptionist` object, this statement implicitly casts the created `RudeReceptionist` object to a `MoodyReceptionist`, before the pointer assignment process. It is now clear that the type of `r_pt` before and after the casting is `MoodyReceptionist*`. Therefore, function `r_pt->greet()` is associated with class `MoodyReceptionist`.

Unlike a regular object, a pointer needs two memory spaces: one for itself and another for the object that it points to. The former space is created at the pointer declaration, while the latter is created using "`new`". Function ambiguity occurs when the pointer is declared to point to an object of one type, but the pointed object is created to store an object of another type. By default, the pointer and the object will be associated with the declaration type, not the construction type.

B.3 Virtual Functions

The result in Example B.3 is different from that in Example B.1. When creating a pointer by executing "`new RudeReceptionist`", we expect the rude receptionist to say "What do you want? I'm busy. Come back later!!", not just 'What do you want?' as in Example B.4. To do so, a `RudeReceptionist` object needs to be associated with the construction type not the declaration type. In C++, such the association is carried out through *virtual functions*.

Unlike regular functions, virtual functions always belong to the construction type, regardless of type casting. C++ declares a virtual function by putting a keyword "`virtual`" in front of the function declaration. Note that, the virtuality property is inheritable. We only need to declare the virtual function once in the base class. The same function in the derived class automatically inherits the virtuality property.

Example B.5. In Example B.4, replace Line 4 in Example B.1 with the following line:

```
virtual void greet() {cout<<"Say:\n";};
```

which declares the function `greet()` of class `Receptionist` as virtual.

Since `r_pt` is created using "`new RudeReceptionist`", virtual function `r_pt->greet()` belongs to class `RudeReceptionist`. At the declaration "`Moo-dyReceptionist *r_pt`", the pointer `r_pt` is created by its default constructor. However, the space where `r_pt` points to (i.e., `*r_pt`) is created by the statement "`new RudeReceptionist`". Since a virtual function sticks to the construction type, the statement `r_pt->greet()` invokes function `greet ()` of class `RudeReceptionist`. After running `./receptionist`, we will obtain the same result as that in Examples B.1 and B.3.

B.4 Abstract Classes and Pure Virtual Functions

An *abstract class* provides a general concept from which more specific classes derive. Conforming to the polymorphism concept, it specifies "*what to do*" in special functions called *pure virtual functions*, and forces its derived classes to define their own "*how to do*" by overriding the pure virtual functions. Containing at least one pure virtual function, an abstract class is said to be *incomplete* since it does not have a "*how to do*" part. Consequently, no object can be initiated from an abstract class. By not implementing *all* virtual functions, the derived class would still be an abstract class (i.e., incomplete), and cannot initiate any object.

C++ declares a pure virtual function by putting "`virtual`" and "`=0`" at the beginning and the end of function declaration, respectively.

Example B.6. Consider again an example on receptionists and how they greet customers. We keep the class hierarchy in Fig. B.1 unchanged. To make class `Receptionist` an abstract class, we modify Example B.5 by removing Lines 4 in Example B.1 replacing the declaration of class `Receptionist` in Example B.1 with the following codes:

```
1 class Receptionist {
2     public:
3     virtual void greet()=0;
4 };
```

After running "`./receptionist`", we should obtain the same results as in Example B.1. In this example, three main components are related to the use of an abstract class.

- *A pure virtual function*: Function `greet()` is declared in class `Recepti onist` as a pure virtual function (Line 3 in Example B.6).
- *An abstract class*: Containing a pure virtual function `greet()`, class `Receptionist` is an abstract class. No object can be instantiated from class `Receptionist`. Class `Receptionist` therefore acts as a template class for classes `FriendlyReceptionist`, `MoodyReceptionist`, and `Rude Receptionist`.

Table B.1. Declaration with no implementation, declaration with no action, and invalid declaration.

Declaration	Example
Pure virtual declaration	`virtual void greet()=0;`
Declaration with no action	`virtual void greet() {};`
Invalid declaration	`virtual void greet();`

- *Implementation of pure virtual function*: Classes `FriendlyReceptionist`, `MoodyReceptionist`, and `RudeReceptionist` must provide implementation for function `greet()` (see Example B.1). Unlike Example B.2, removing the implementation (e.g., Line 16 in Example B.1) leaves the derived classes (e.g., `MoodyReceptionist`) an abstract class, and the instantiation (e.g., `m_pt = new MoodyReceptionist`) would cause a compilation error.

There are three similar declarations for a virtual function (see Table B.1). First, a pure virtual function is declared as explained above (e.g., `virtual void greet() = 0;`). Secondly, a (non-pure) virtual function of a derived instantiable class must contain implementation but may have no action. For example, "`virtual void greet() {};`" contains no action inside its curly braces. This function overrides the pure virtual function of its parent class, making the class non-abstract and instantiable. Finally, consider a class whose parent class is an abstract class. By opting out "`{}`" (i.e, "`virtual void greet();`"), the pure virtual function is left unimplemented and the class would still be an abstract class. Again, any object instantiation would lead to a compilation error.[3] *An important note for NS2 users: You cannot opt out both "=0" and "{}". If you do not want provide an implementation, leave the curly braces with no action after the declaration. Otherwise, NS2 will show an error at the compilation.*

B.5 Class Composition: An Application of Type Casting Polymorphism

Upto this point, the readers may raise few questions. That is, why do we need to cast an object to different type and use the keyword `virtual`? Wouldn't it be easier to declare and construct an object with same type? For example, can we not use Example B.3 instead of Example B.4? Doesn't it remove function ambiguity? The answer is "yes"; nevertheless, type casting makes the programming more scalable, elegant and interesting. For this reason, the programming with type casting is a common practice in NS2.

---

[3] Here, we assume that declaration and implementation are in one file. When declaration and implementation are separated in two files, you can opt out "{}" in a ".h" file and provide the implementation in another ".cc" file.

B.6 Programming Polymorphism with No Type Casting: An Example

Example B.7 below shows a scenario, which needs no virtual function. However, we will see later that Example B.7 leads to programming inconvenience as the program becomes larger.

Example B.7. Consider a company and how it serves a customer. The main functionality of the company is to serve customers. As a courtesy, the company greets every customer before serving. Assume that the company has one receptionist to greet the customer. The receptionist can be friendly, moody, or rude as specified in Example B.1. The following C++ code represents the company with the above description:

```
      //company.cc
1   class Company {
2       public:
3           void serve() {
4               greet();
5               cout<<"\nServing the customer ... \n";
6           };
7           void greet () {};
8   };

9   class MoodyCompany : public Company {
10      public:
11          MoodyCompany(){employee_ = new MoodyReceptionist;};
12          void greet(){employee_->greet();};
13      private:
14          MoodyReceptionist* employee_;
15  };

16  int main() {
17      MoodyCompany my_company;
18      my_company.serve();
19      return 0;
20  }
```

where class `MoodyReceptionist` is defined in Example B.1.

Class `Company` (Lines 1-8) has two functions. Function `serve()` in Lines 3–6 greets the customers by invoking function `greet()`. Then, it serves the customer by showing the message "`Serving the customer ...`" on the screen. The function `greet()` in Line 7 has no action in class `Company`, and is implemented by child classes of class `Company`.

Class `MoodyCompany` (Lines 9–15) derives from class `Company`. It has one moody receptionist stored in the variable `employee_` (Line 14). Class

MoodyCompany implements function `greet()` by having `employee_->greet()` in Line 12.

In the function `main()`, an object `my_company` of class `MoodyCompany` is instantiated in Line 17. Line 18 invokes function `serve()` associated with the object `my_company`. By running the executable file `company`, the following result will appear on the screen:

```
>>./company
Say: What do you want?
Serving the customer ...
```

which is quite expected from the code. Clearly, we do not need virtual functions in this example.

B.7 A Scalability Problem Caused by Non Type Casting Polymorphism

The main problem of polymorphism with non type casting is the scalability. As the inheritance tree becomes more complicated, we may need to develop a large number of classes. For example, suppose we would like to change the reception in the company to be a friendly receptionist. We will have to define another class as follows:

```
class FriendlyCompany : public Company {
    public:
        FriendlyCompany() { employee_ =
                            new FriendlyReceptionist}
        void greet() {employee_->greet();};
    private:
        FriendlyReceptionist* employee_;
};
```

Also, replace Line 17 in Example B.7 with

```
FriendlyCompany my_company;
```

By running "`./company`", the following result should appear on the screen:

```
>>./company
Say: Good morning. How can I help you today?
Serving the customer ...
```

The problem is that a new `Company` class (e.g., `FriendlyCompany`) is required for every new `Receptionist` class (e.g., `FriendlyReceptionist`). Furthermore, the company may have other types of employee such as technicians, managers, etc. If there are 10 classes for receptionists and 10 classes for technicians, we need to defines 100 classes for to cover all combination of employee types. In the next section, we will show how this scalability problem can be avoided by using class composition.

B.8 The Class Composition Programming Concept

Type casting acts as a tool which helps avoid the scalability problem. Instead of deriving all class combination (e.g., 100 classes of combinations of 10 receptionists and 10 technicians), we may declare an *abstract user class* object (e.g., `Receptionist`), and cast the abstract user class object to a more specific object (e.g., `FriendlyReceptionist`).

Example B.8. Consider a company and how it serves a customer in Example B.7. By allowing type casting, the code representing the company is given below:

```
      //company.cc
 1    class Company {
 2        public:
 3            void hire(Receptionist* r) {
 4                employee_ = (Receptionist*)r;
 5            };
 6            void serve() {
 7                employee_->greet();
 8                cout<<"\nServing the customer ... \n";
 9            };
10        private:
11            Receptionist* employee_;
12    };

13    int main() {
14        MoodyReceptionist *m_pt= new MoodyReceptionist();
15        Company my_company;
16        my_company.hire(m_pt);
17        my_company.greet();
18        return 0;
19    }
```

Also to bind function `greet()` to the construction type, we need to declare function `greet` of class `Receptionist` as virtual. Here, we replace Line 4 in Example B.1 with "`virtual void greet();`" or "`virtual void greet() = 0;`".

Class `Company` declares a variable `employee_` as a `Receptionist` pointer in Line 11. The company hires an employee by invoking function `hire(r)` in Lines 3–5. Taking a `Receptionist*` object, `r`, as an input argument, function `hire(r)` assigns an input `Recectionist` pointer to its private variable `employee_`. In Lines 6–9, the company `serves` the customers as it does in Example B.7.

In function "`main()`", an object of class `Company`, `my_company`, is created in Line 15. In Line 16, `my_company` hires an employee `m_pt` which is a pointer

to a `MoodyReceptionist` object. From Lines 3–5, function `hire(m_pt)` casts
the pointer `m_pt` to a `Receptionist` pointer. Since the function `greet()` of
class `MoodyReceptionist` is virtual, `employee_->greet()` is associated with
the construction type in Line 14 (i.e., class `MoodyReceptionist`). By running
"`./company`", we will obtain the following result:

```
>>./company
Say: What do you want?
Serving the customer ...
```

which is the same as that in Example B.7.

As shown in Fig. B.2, the class composition programming concept with
type-casting polymorphism consists of four main class types.

- An *abstract class* (e.g., `Receptionist`) is a template class.
- A *derived class* (e.g., classes `MoodyReceptionist`) derives from the above
  abstract class.
- An *abstract user class* (e.g., class `Company`) declares objects of the *abstract
  class* (e.g., `Receptionist`). It employs the functions of the abstract class
  without the need to know the detailed implementation of the abstract
  class. In Example B.8, class `Company` does not need to know what type
  of `Receptionist` the `employee_` is, nor how the `employee_` greets the
  customers.
- A *user class* (e.g., `main`) declares objects of the *derived class* (e.g.,
  `MoodyReceptionist`). It makes the abstract class more specific by binding (e.g., using function `hire(r)`) the abstract variable (e.g., `*employee_`)
  belonging to the abstract user object (e.g., `my_company`) to the derived object (e.g., `m_pt`).

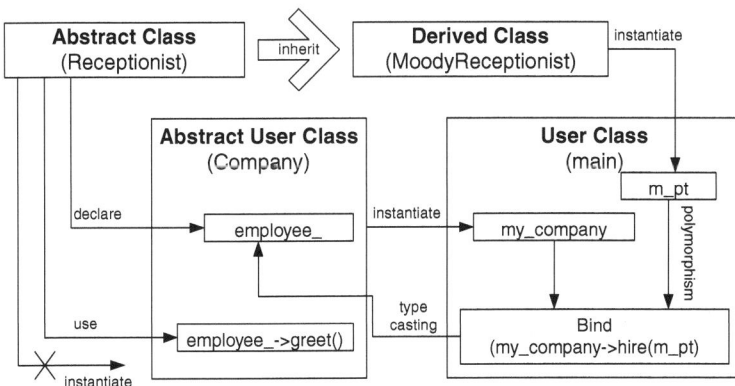

Fig. B.2. A diagram of the class composition concept with type casting polymorphism.

The concept of class composition is to have an abstract user class (e.g., Company) declare its variable from an abstract class (e.g., Receptionist) and later cast the declared object (e.g., employee_) to a more specific type (e.g., MoodyReceptionist). In particular, the mechanism consists of four following steps

(i) Declare an abstract class (e.g., Receptionist).
(ii) From within an abstract user class (e.g., Company), declare (e.g., Receptionist* employee_) and use (e.g., employee_->greet()) objects of the above abstract class.[4]
(iii) In a user class (e.g., main()),
    (a) Instantiate an object (e.g., my_company) of the abstract user class (e.g., Company).
    (b) Instantiate an object (e.g., *m_pt) of the derived class (e.g., Moody Receptionist).
(iv) Bind (e.g., using hire(r)) the abstract class object (e.g., *employee_) in the abstract user class (e.g., Company) to the object initiated from within the user class (e.g., *m_pt). Since the latter object class derives from the former one, the type casting is fairly straightforward.

To change the company's receptionist to be a friendly receptionist, we only need to change function main as follows, without having to modify other parts of the codes

```
int main() {
    FriendlyReceptionist *f_pt;
    f_pt = new FriendlyReceptionist();

    Company my_company();
    my_company.hire (f_pt);
    my_company.greet();
    return 0;
}
```

To see how type casting helps avoid scalability, consider the above example where a company may have one of 10 possible receptionist classes and one of 10 possible technician classes. Without type casting, we need to define 100 classes to cover all the combination of receptionists and technicians in addition to one based class Company. By allowing type casting, we can declare

---

[4] Again, declaration of a too specific (e.g. MoodyReceptionist as opposed to Receptionist) class in non-type-casting polymorphism leads to the scalability problem. As the entire program becomes larger, we need to redefine classes for every new class, hence substantially growing the total number of classes. To avoid the scalability problem, we need to declare classes to be as general as possible. This general class can later be cast as a more specific class.

two variables (of abstract classes `Receptionist` and `Technician`) in a company. In the main program, we can instantiate a receptionist and a technician from any of these `Receptionist` and `Technician` classes. After instantiating receptionist and technician objects from the derived class, we can cast the instantiated objects back to classes `Receptionist` and `Technician` and assign them to the company. Under the same scenario, the class composition concept requires only 20 classes for receptionists and technicians, and therefore, greatly alleviates the scalability problem.

References

1. A. S. Tanenbaum, *Computer Networks*, 3rd ed. Prentice Hall, 1996.
2. R. E. Shannon, "Introduction to the art and science of simulation," in *Proc. of the 30th conference on Winter simulation (WSC'98)*, 1989.
3. R. G. Ingalls, "Introduction to simulation: Introduction to simulation," in *WSC '02: Proceedings of the 34th conference on Winter simulation*. Winter Simulation Conference, 2002, pp. 7–16.
4. W. H. Tranter, et al., *Principles of Communication Systems Simulation*. Prentice Hall, 2004.
5. A. Papoulis and S. U. Pillai, *Probability, Random Variables and Stochastic Processes*, 2nd ed. McGrawHill, 2002.
6. W. H. Press, et al., *Numerical Recipes in C*, 2nd ed. Cambridge University Press, 1997.
7. R. M. Goldberg, *Parallel and Distributed Simulation Systems*. John Wiley & Sons, Inc., 2000.
8. J. Banks and I. J. S. Carson, *Discrete-Event Systems Simulation*. Prentice-Hall, Inc., 1984.
9. The Network Simulator Wiki. [Online]. Available: http://nsnam.isi.edu/nsnam/index.php/
10. The Network Simulator – ns-2. [Online]. Available: http://www.isi.edu/nsnam/ns/
11. M. Greis. Tutorial for the Network Simulator NS2. [Online]. Available: http://www.isi.edu/nsnam/ns/tutorial/
12. J. Chung and M. Claypool. Ns by example. [Online]. Available: http://nile.wpi.edu/NS/
13. The Network Simulator Wiki–Contributed Code. [Online]. Available: http://nsnam.isi.edu/nsnam/index.php/Contributed_Code
14. H. Schildt, *C++: The Complete Reference, 4th Edition (Kindle Edition)*, 4th ed. McGraw-Hill/Osborne Media, 2002.
15. K. Fall and K. Varadhan. (2007, Aug.) The ns manual (formerly known as ns notes and documentation). [Online]. Available: http://www.isi.edu/nsnam/ns/ns-documentation.html
16. Réseaux et Performances. NS 2.26 source original: Hierarchical index. [Online]. Available: http://www-rp.lip6.fr/ns-doc/ns226-doc/html/hierarchy.htm

17. T. H. Cormen, et al., *Introduction to Algorithms*, 2nd ed.   MIT Press and McGraw-Hill, 2001.
18. M. Mathis, et al., *TCP selective acknowledgement options*, RFC 768 Std., 1996.
19. J. Kurose. The TCP/IP course website. [Online]. Available: http://www.networksorcery.com/enp/protocol/udp.htm
20. J. F. Kurose and K. W. Ross, *Computer Networking: A Top-Down Approach.* Pearson Addison-Wesley, 2008.
21. V. Paxson and M. Allman, *Computing TCP's Retransmission Timer*, RFC 2988 Std., November 2000.
22. P. L'Ecuyer, "Good parameters and implementations for combined multiple recursive random number generators," *Operations Research*, vol. 47, no. 1, pp. 159–164, 1999.
23. D. Libes, "A debugger for Tcl applications," in *Tcl/Tk Workshop*, June 1993. [Online]. Available: http://expect.nist.gov/tcl-debug/tcl-debug.ps.Z
24. The GDB Developers. GDB: The GNU project debugger. [Online]. Available: http://www.gnu.org/software/gdb/
25. T. Nandagopal, S. Lu, and V. Bhargharvan "A unified architecture for the design and evaluation of wireless fair queuing algorithms," *MOBICOM99*, pp. 132–142, 1999.
26. S. Sanfilippo. An introduction to the Tcl programming language. [Online]. Available: http://www.invece.org/tclwise
27. How can I do math in Tcl, Welcome to the Tclers Wiki!, http://wiki.tcl.tk/528
28. Berkeley Continuous Media Toolkit. OTcl tutorial. [Online]. Available: http://bmrc.berkeley.edu/research/cmt/cmtdoc/otcl/
29. A. Robbins and D. Gilly, *Unix in a Nutshell: System V Edition.*   O'Reilly & Associates, Inc., 1999.
30. An AWK primer. [Online]. Available: http://www.vectorsite.net/tsawk.html
31. Wikipedia. Regular expression. [Online]. Available: http://en.wikipedia.org/wiki/Regular_expression

General Index

Code Index

Made in the USA
Lexington, KY
01 April 2011